BIOCHEMISTRY RESEARCH TRENDS

# GLYCOSYLPHOSPHATIDYLINOSITOL-ANCHORED PROTEINS AND THEIR RELEASE FROM CELLS

## FROM PHENOMENON TO MEANING

# BIOCHEMISTRY RESEARCH TRENDS

Additional books in this series can be found on Nova's website under the Series tab.

Additional e-books in this series can be found on Nova's website under the eBooks tab.

BIOCHEMISTRY RESEARCH TRENDS

# GLYCOSYLPHOSPHATIDYLINOSITOL-ANCHORED PROTEINS AND THEIR RELEASE FROM CELLS

## FROM PHENOMENON TO MEANING

GÜNTER A. MÜLLER

nova science publishers
New York

**Library of Congress Cataloging-in-Publication Data**

ISBN: 978-1-53613-966-2

*Published by Nova Science Publishers, Inc. † New York*

# CONTENTS

Nothing tends so much to the advancement of knowledge as the application of a new instrument. The native intellectual powers of men in different times are not so much the causes of the different success of their labours, as the peculiar nature of the means and artificial resources of their possession.

Sir Humphrey Davy 1812

# Preface

Starting with the first description of anchorage of a subset of cell surface proteins in eukaryotic cells from yeast to mammals by a glycosylphosphatidylinositol moiety covalently attached to the carboxy-terminus of the protein moiety, a considerable body of experimental evidence for the potential of glycosylphosphatidylinositol-anchored proteins (GPI-AP) of being released into the extracellular environment has been accumulating over the past three decades. Detailed studies revealed that the release involves a multitude of distinct configurations ranging from monomeric and oligomeric versions to macromolecular assemblies of the GPI-AP which have either lost or retained (part of) the glycosylphosphatidylinositol anchor. The configurations reported so far encompass soluble protein moieties of the GPI-AP, carrier protein-bound monomers, phospholipid-free multimeric aggregates and non-vesicular/particulate phospholipid- and cholesterol-harboring micelle-like complexes as well as membrane vesicles (exosomes, microvesicles) and particles (surfactant-like particles, lipoprotein-like particles, nodal vesicular particles, milk fat globules). The structural and morphological diversity of these configurations of released GPI-AP is reflected in the multitude of different molecular mechanisms underlying the release. These involve (i) cleavage of the protein moiety or GPI anchor by proteases or phospholipases, respectively, (ii) endocytosis of GPI-AP expressed at the plasma membranes, subsequent budding into multivesicular bodies and final exocytosis of exosomes with GPI-AP exposed at their surface, (iii) bulbing and shedding of microvesicles from plasma membrane domains expressing the GPI-AP, (iv) synthesis of lipid droplets at the luminal or cytoplasmic leaflet of endoplasmic reticulum membranes with inserted GPI-AP, followed by vesicular transport and exocytosis of lipoprotein-like particles or bulbing and shedding of the plasma membranes yielding milk fat globules, respectively, both harboring the GPI-AP and less well understood (v) assembly of GPI-AP with other GPI-AP, exclusively, or with carrier proteins or with phospholipids and cholesterol and finally (vi) direct transfer of GPI-AP from donor to acceptor cells through intimate contact of their plasma membranes without the generation of soluble intermediates. Release of GPI-AP may occur in a spontaneous and constitutive fashion or be regulated and controlled by certain endogenous or exogenous signals or environmental stimuli. The putative functional and (patho)physiological implications of the diverse modes of GPI-AP release are presented with emphasis on the structures and configurations involved as well as the underlying molecular mechanisms. In addition, the potential relevance of GPI-AP release as innovative biomarkers for disease prediction, diagnosis and stratification as well as novel vehicles for drug delivery is discussed. Moreover, the mere description of the release of GPI-AP from cells using modern instrumentation in the absence of any knowledge about the underlying causal (molecular) mechanisms may be regarded as an example for "hermeneutic phenomenology" within molecular life sciences. The three-step study procedure should be based on a minimal, but carefully considered and selected bias rather than on a strong working hypothesis (i), which will lead to a mere database rather than to a narrative argumentation of the order of the experiments (ii), and on a subsequent minimal interpretation of the resulting database rather than on causal mechanistic explanations and speculative working models as typically presented in scientific publications (iii). The final goal of "hermeneutic phenomenology" of life processes represents the creation of a careful, respectful and cautious path from the pure description of the phenomenon to the realization of its meaning.

# ABBREVIATIONS

| | |
|---|---|
| AChE | acetylcholinesterase |
| AP | alkaline phosphatase |
| CEA | carcinoembryonic antigen |
| DAF | decay accelerating factor |
| ER | endoplasmic reticulum |
| ERAD | ER-associated destruction |
| ERES | ER-exit sites |
| EV | exosomes and microvesicles |
| ECM | extracellular matrix |
| GC-MS | gas chromatographic and mass spectroscopic analysis |
| GLEC | GPI-AP- and lipid-harboring extracellular complexes |
| GPCR | G protein-coupled receptor |
| GPI | glycosylphosphatidylinositol |
| GPI-AP | GPI-anchored protein |
| GPI-GFP | GPI-anchored green fluorescent protein |
| GPI-HBP1 | GPI-anchored HDL-binding protein-1 |
| GPI-PLC/D | GPI-specific phospholipase C/D |
| LPP | lipoprotein-like particles |
| MFG | milk fat globules |
| NAFLD | non-alcoholic fatty liver disease |
| NASH | non-alcoholic steatohepatitis |
| NVP | nodal vesicular particle |
| PC | phosphatidylcholine |
| PC/PI-PLC/D | PC/PI-specific PLC/D |
| PI | phosphatidylinositol |
| PIG | phosphoinositolglycans |
| PLAP | placental AP |

| | |
|---|---|
| PLC/PLD | phospholipase C/D |
| PNH | paroxysmale nocturnal hemoglobinuria |
| SAW | surface acoustic waves |
| SLP | surfactant-like particles |
| SRP | signal recognition particle |
| T2D | type 2 diabetes |
| tER | transitional ER |
| TGN | *trans*-Golgi network |
| THP | Tamm-Horsfall glycoprotein |
| uPAR | urokinase plasminogen activator receptor |
| VSG | variant surface glycoprotein |

*Chapter 1*

# Glycosylphosphatidylinositol-Anchored Proteins at the Boundary of Cell and Environment, Interior and Exterior Worlds – A "Historical" Introduction

It is now clear that cell surface proteins play important roles in mediating effects and exchange between cells and their environment in both directions, the most critical ones relying on the transfer of information and material. Cell surface proteins are associated with plasma membranes by means of a variety of structures and mechanisms. One of the more recently discovered modes of attachment involves the covalent linkage of a glycolipid to the protein. In this case, no part of the polypeptide moiety comes into contact with the plasma membrane. The glycolipid moiety is constituted by a glycosylated variant of phosphatidylinositol (PI), a typical constituent of biological membranes, of particular relevance for signal transduction processes between environment and cells. So far, more than 200 glycosylphosphatidylinositol-anchored proteins (GPI-AP) have been identified in eukaryotic organisms. It is remarkable, that this protein class constitutes a functionally and evolutionarily diverse group of biological macromolecules with the only common feature apparently relying on the expression and structure of the glycosylphosphatidylinositol (GPI) anchor as well as the mechanism of its biogenesis. Whereas meanwhile this has been accepted in general, the possibility of release from the cell surface and biological membranes, which has been attracting increasing interest during the last two decades, as a functional and (patho)physiological characteristics common to all GPI-AP remains to be analyzed in future biochemical, cell biological, pharmacological and bioinformatical studies.

# 1. Structure of Biological Membranes

Although the early models of membrane structure have long been superseded by various versions of the fluid mosaic model, it is interesting to look back at some of the important milestones in the development of biological membrane theory to see how concepts were created and since then have evolved. When surface tension measurements suggested that proteins, in additions to lipids, were present in the bilayer, Stein and Danielli (1956) introduced their version of a protein-containing biological membrane. The thoughts which motivated this model were not elaborated on, but viewed six decades later, the model proposed by Stein and Danielli contains several novel ideas which have now received full experimental support. Most strikingly, they postulated that the lipid bilayer was not continuous but was interrupted at intervals by protein complexes which could provide communication routes across the hydrophobic barrier. Although Stein and Danielli probably did not envisage the type of membrane architecture, now thought to be true, they did create an important and novel concept.

It is surprising, therefore, that Stein's and Danielli's view of membrane organization was so precipitously ignored in favor of the second major development in membrane structure – the Robertson's "unit membrane theory," a version of which had earlier been proposed by Millington and Finean (1962). Here, the view of a bilayer surface covered with a layer of protein was retained, but added to it was the plerophory that this protein was asymmetrically distributed between the two surfaces of a continuous bilayer (Robertson 1964). In the articulation of this important – and correct – new concept of asymmetry, which decades later received its most refined manifestation in the identification of GPI-AP, the idea of a bilayer punctuated by protein was omitted.

Interestingly, both views of the membrane appeared to envisage that the surfaces of the bilayer were completely coated with protein. However, it is not clear why this was thought to be obligatory but it is now known that it is not true. In general, about 40% of the surface area of the membrane probably consists of exposed lipid. Thus we arrive *via* a number of other hypothetical models and intermediaries (Finean et al. 1971) at the modern version of the biological unit membrane – essentially a derivative of the fluid mosaic model of Singer and Nicolson (1972). Incorporated into this model are the earlier ideas of transbilayer disposition and asymmetric distribution of the protein components together with exposure of the lipids and, most dramatic of all, its very dynamic behavior. The lipid itself may be asymmetric depending on the type of membrane. The carbohydrate components of membranes are never free but covalently bound to either protein or lipid thereby giving rise to a great deal of potentially useful variability. This carbohydrate residues are thought always to be found on the surface of the membrane which is external to the cell, corresponding to the extracellular space, rather than on the intracellular membrane surface, equivalent to the cytoplasm.

## 1.1. Peripheral Association or Integral Insertion of Membrane Proteins

On the basis of relatively simple experiments in which membranes were subjected to various salt and/or pH conditions, it was suggested that they possess essentially only two major classes of protein. Those which can be removed from the bilayer under the above conditions were regarded as peripheral or extrinsic because solubilization did not require disruption of bilayer integrity. The presumption is that these proteins interact either with protein, lipid or carbohydrate residues exposed at the membrane surface and that no significant part of their covalent structure penetrates into the intramembranous hydrophobic domain. Although there may be a variety of different modes of peripheral association with a membrane surface, this definition appears to hold. A good test is brief exposure of the membrane to ice-cold NaOH at pH 11.0, the so-called carbonate extraction, when, even though the interactions between proteins may have a major hydrophobic component, denaturation will destroy any protein-protein interaction and thereby cause liberation of any non-transmembrane polypeptide. This behavior is thought to be specific for peripheral proteins rather than to occur with bilayer fatty acid-protein interactions as operative for transmembrane proteins. The vast majority of peripheral proteins appear to have a cytosolic orientation and are of monomeric nature. At variance, those that are observed on the outer surface of the membrane are usually subunits of larger complexes, e.g., $\beta_2$-microglobulin.

The earliest Danielli and Davson (1935) view of the unit membrane represented the protein component as a globular entity, subsequently superseded by the Stein and Danielli and Robertson preferences for extended structures. There is now no reason to be dogmatic. Since peripheral proteins are in essence no different from classical water-soluble ones, which are typically located in the cytoplasm or extracellular space, they can exhibit the full range of structural variations. When removal of the protein from the lipid constituent of the membrane cannot be achieved by high salt or alkaline pH treatment but requires proteolysis or – better – disintegration of the bilayer with detergents or chaotropic agents, such polypeptides are usually referred to as integral, transmembrane or intrinsic species. The presumption is that some part of their covalent proteinaceous structure penetrates into the hydrophobic interior of the phospholipid bilayer, providing an almost permanent anchor in the membrane. Such transmembrane proteins can usually – but not always – be identified by the degree to which they are accessible for labeling with photosensitive hydrophobic probes (Wells and Findlay 1980, Findlay 1987) or by the efficacy with which they partition into a detergent-enriched phase, such as Triton X-114 (Bordier 1981). Importantly, on the basis of these criteria, i.e., carbonate or salt extraction, hydrophobic labeling, TX-114 partitioning, GPI-AP behave as integral membrane proteins. The mode of their membrane insertion will be described in the following.

## 1.2. Modes of Integration of Membrane Proteins into the Lipid Bilayer

Various types of association of membrane proteins with the phospholipid bilayer have been identified so far. They can be divided into four major classes, which may also reflect their biogenetic routes.

### *1.2.1. Monotopic Proteins*

The first class of protein is integrated into the bilayer through part of its polypeptide chain but the intramembranous domain amounts to only a single transmembrane segment or in a few cases perhaps two. As such it represents only a small proportion (10-20%) of the total mass of the protein. The major portion is located in the aqueous compartments of either the cytoplasm or the extracellular space and there performs the biological function of the transmembrane protein, which may be either a catalytic, structural, binding or signaling activity. The hydrophilic domain can usually be removed from the membrane by proteolysis and, as soluble protein fragment, will usually still retain its physiological activity. The intramembranous domain anchoring the protein into the hydrophobic milieu of the phospholipid fatty acid layer is likely always to be in the form of an α-helix due to the unavailability of water molecules with hydrogen-bonding capabilities, although there could be some β-structure in the phospholipid head group region. The extremely short supply of water molecules in the hydrophobic domain enables fulfillment of the H-bonding potential of the carbonyl and amide groups within the peptide bonds best by internal H-bonding involving the formation of an $\alpha$-helix, only. This type of structure would orient all the amino acid side chains towards the fatty acid environment, a position consistent with their observed non-polar character. It is possible to conceive of other structural forms, such as realized in the ionophore gramicidin with a β-helix. But such structures require contributions by other transmembrane segments or the involvement of non-standard amino acids or isomers. A prominent example of a monotopic membrane protein represents the caveolin proteins (see below).

Although it is not essential for all H-bonding potential to be fulfilled all the time, it is generally believed that this is, nevertheless, the usual state of affairs. For this reason, researchers in the field of integral membrane protein structure have, for meanwhile several decades, winced at the plethora of diagrams filling the literature which show proteins occupying only one monolayer of the membrane or attached to the membrane by short polypeptide loops dipping down into the hydrophobic phase. However, those structures would require the polypeptide chain to form regular β-bends or non-regular coil or turn conformations. In soluble proteins, fulfillment of H-bonding potential in such situations is largely achieved using water and to a lesser extent polar amino acid side chains. H-bonding of this kind is not really available at the heart of the bilayer, leaving only the very improbable involvement of other proteins. However, such protein-protein associations would have to be permanent, not temporary as was envisaged in so-called "mobile receptor" concepts. For instance, the basic protein myelin was always shown as possessing small intramembranous proteinaceous loops. However, studies with hydrophobic probes convinced the scientific community that this mechanism of membrane integration was most unlikely (Harris and Findlay 1983). In summary, it is currently much more sound to assume that proteinaceous anchors in the bilayer exist as α-helices and that all turns or loops of structure occur in the aqueous polar phase.

### *1.2.2. Bitopic Integral Proteins*

Although this second class of membrane proteins is very similar to that of polytopic proteins, it differs in that the single transmembrane segment is more towards the middle of the sequence rather than very near to one end. It follows that there is a significant proportion of the mass of the protein at both surfaces of the membrane. Both aqueous domains can be expected to have some kind of activity and this activity is essentially independent of events within the bilayer. However, a very important difference from the monotopic proteins is that the intramembranous segment is critical for whatever mechanisms exist for communication and signal transmission between the two extramembranous domains. The most common examples of this type of structure are tyrosine protein kinase receptors of the growth factor receptor type, such as the insulin receptor, and receptors involved in internalization and endocytic processes, such as the transferrin receptor.

The critical feature is that the activity of the intracellular domain is effectively controlled by events at the extracellular surface. The fact of binding of a growth factor or transferrin to the external domain is relayed to the internal domain which as a result is "activated." The mechanism by which this information is transferred across the phospholipid bilayer has been deciphered during the past three decades. It now seems clear that the conformation of the transmembrane helix does not become altered although it is possible to conceive that structural changes at the external surface may pull/push amino acid side chains into new environments thereby precipitating conformational alterations in the cytosolic domain. Since both classes of receptors always appear to form dimers, multimers or larger aggregates in course of binding their particular ligand, it is thought that this behavior may be the key to information transfer and involves longer-lived protein association. Loss of the external domain, as can be provoked experimentally by limited proteolysis or ectopic expression of recombinant amino-terminally truncated version, leads to transformation of a wild-type growth factor receptor into an unregulated, constitutively and fully activated oncogene.

### *1.2.3. Polytopic Integral Proteins*

The third class of membrane protein traverses the bilayer a significant number of times and thus a larger proportion of its mass is located between the boundaries of the two membrane leaflets. As a very general rule, the single feature which usually distinguishes these intramembranous segments from those seen in water-soluble cytoplasmic or secretory proteins is that intrinsic hydrophobic stretches in the latter polypeptides are much shorter. In most cases one should not expect, however, much more than a small increase in the overall proportion of hydrophobic residues or overall hydrophobicity in this class of integral membrane protein as compared to water-soluble polypeptides. For the reasons discussed above, the regions linking the intramembranous sections are only located in the aqueous phase. Thus far, examples ranging from two transmembrane segments, e.g., the C subunit of the $F_0$ ATPase complex (Senior 1988), to up to 24 for the sodium gate in nerve axons (Noda et al. 1984) have been identified. Current data suggest that there is no real constraint on how few or how many of these intramembranous traverses can be realized for polytopic membrane proteins. What is singularly lacking in the vast majority of cases, however, is any real experimental evidence to substantiate the topographic models constructed from application of the various hydropathy plots. The present set of algorithms are at best only indicative and should be treated with caution. Several mistakes have already been made by relying entirely on this approach. Thus more discriminating methods are urgently required still.

With the larger number of protein sequences being produced, the diagnostic features which reveal transmembrane segments have become more apparent and no doubt this positive development has not come to its end so far. For instance, it will be possible in near future to loosely define discriminators which identify residues at the non-polar/polar membrane-membrane interface. Such indicators could well depend on factors such as the particular bilayer surface and the direction or orientation of the polypeptide chain. Thus the characteristics of the sequences emerging from the bilayer may be different from those entering the bilayer at the same surface. It is interesting, for example, that in opsins, positively charged side chains are often seen at the amino-terminal ends of transmembrane segments and negative charges at the carboxy-terminal ends (Findlay and Pappin 1986, Findlay et al. 1990). Another important factor to introduce is the sidedness of the intramembranous segments. A long transmembrane helix may well contain a regularly spaced series of polar residues, indicative of an intraprotein orientation which may be associated with the formation of a hydrophilic pore or ligand binding site. It would be useful to have all those features incorporated into a single program for detection of the intramembranous segments. In the case of the G-protein-linked receptors, such an approach is entirely feasible despite the fact that the degree of conservation between the proteins may be as low as 15% (Attwood et al. 1989). Discriminators can be constructed for each of the seven transmembrane segments of the serpentine-class of receptors, which recognize only the corresponding segment in other members of this huge protein family. The implication of this observation is that this group, despite their great variability, nevertheless retains their general structural organization and segment orientation. A similar situation may be found in other families, such as the mitochondrial inner membrane transporters.

## 1.3. Structure of Transmembrane Segments

Because of the long-lasting lack of crystallographic data on integral membrane proteins, the general features which govern the packing of the transmembranous segments remained unclear until recent times. Bacteriorhodopsin (Henderson and Unwin 1975) has provided a good picture of a seven-helix structure. The photosynthetic reaction center (Deisenhofer et al. 1985, Allen et al. 1987) gave further insights, but the organization of this system needed further careful analysis before general rules could be extrapolated to simpler systems. What the latter in particular has confirmed, however, is that the transmembrane segments are largely α-helical. Biophysical data and the general features of protein structure and bilayer properties have long made this element of secondary structure the favored candidate, but confirmation was vital. That is not to say that these transmembranous helices are always regular. As has been observed for bacteriorhodopsin (Leifer and Henderson 1983) and predicted for rhodopsin (Pappin and Findlay 1984) some time ago, distortions from regularity can be expected, particularly in conjunction with sites of considerable functional importance. A factor which may be relevant when considering the packing restraints is the degree of flexibility which the protein can be expected to entertain. Thus, receptors, such as those linked to G-proteins and transport proteins, both of which may need to undergo substantial conformational change, may exhibit more "flexible" packing and less regularity, perhaps associated with the presence of proline residues. Proteins which provide a highly structured environment for a permanently associated prosthetic group, e.g., the photosynthetic reaction center, may display a more rigid organization.

The cohesion of the transmembranous domain has been a matter of speculation for some time. In water-soluble proteins, hydrophobicity plays a major part in protein folding and stability, but with integral membrane proteins this entropic driving force may not be so profound. It has been speculated instead that disulphide bonds and polar interactions, such as salt bridges, may provide the principal stabilizing influences. The sequences of numerous putative transmembranous domains, protein modification studies (Davison and Findlay 1986a and b) and protein engineering (Karnik et al. 1988) all tend to suggest that such bonds are not sufficiently ubiquitous to constitute a general feature of integral membrane proteins. One is left with the general, somewhat unsatisfactory, assumption that all the usual bonding influences in proteins still play the dominant role. Inspection of the photosynthetic reaction center and model building studies with the opsin receptor family and a proton transport system (Findlay et al. 1990), all suggest that there is no particularly ubiquitous or unique feature of the faces involved in helix-helix interaction. Both polar and non-polar residues appear to be involved. There is a suggestion that aromatic side chains are found more often than one would expect, leading to the possibility of some stabilization through ring interactions *via* stacking. It is also possible that highly structured turns at the bilayer surface and the increased strength of the helix dipole in the transmembrane helices are two factors which have added importance for the cohesion of transmembranous domains. Finally, there are two features which may perhaps turn out to be of general importance for integral membrane proteins. These are an interhelical angle of about 15° and the absence of any bridging structure, i.e., none of the linking hydrophilic loops will straddle or cross-over any other.

## 1.4. Multimeric Organization

In contrast to earlier suggestions that integral membrane proteins, particularly those involved in transport phenomena, will take up a dimeric organization, it now appears that a number of different states are possible ranging from monomers such as rhodopsin (Brett and Findlay 1979) up to hexamers and dodecamers such as proton channels (Senior 1988). Very large oligomers, characteristic of structural proteins, have not so far been observed. The functional unit appears in most cases to be the monomer, but such a monomer may consist of more than one type of polypeptide chain (e.g., $Na^+/K^+$ ATPase). However, one group of molecules which is likely to be a major exception to this generalization is constituted by membrane channels, particularly those involved in facilitating ion movements. It is possible to roughly classify transport proteins into two broad categories. One type provides a route for the substrate across the bilayer which directly involves the interaction of side-chains belonging to a number of different transmembrane helices in the polypeptide chain. The process of transport can be linked to an enzyme reaction, in that there is a binding stage, which precipitates conformational changes resulting in product release. In the case of membrane transport, the conformational changes effect physical translocation of the substrate rather than its chemical transformation.

The second type of transport protein appears to have a quite different organization. Here the transport route (or channel) is lined by identical or equivalent transmembrane segments. As such, there are major issues of symmetry which constrain the packing of the channel-lining helices. This is exemplified by the apparent tetrameric arrangement that may occur with the four-fold internally duplicated $Na^+$-channel (Noda et al. 1984) and result in a very tight packing motif with no apparent water-filled pore. In a five-membered complex, a situation postulated for the nicotinic acetylcholine receptor class (Stephenson 1988), no symmetrical packing arrangement of regular helices will prevent the creation of a restricted interhelical gap. It is just possible to conceive of such a channel containing highly structured water and, in such a scenario, the complex could retain high specificity and be subject to conformationally dependent regulation. In contrast, no such selectivity could be imposed on a six-membered aggregate where the channel will have a diameter of 10-12Å and should allow bulk flow of both water and solutes. The gap junction appears to be an example of this kind of six-membered complex

(Unwin and Zampighi 1980). The regulation of such an entity raises interesting problems, especially since it is clear that such junctions can be shut down. The only evidence to date suggests that the relative orientation of the subunits can be altered to compress at least one end of the central pore (Unwin and Ennis 1984). Model building studies indicate that it is possible to rotate or twist the various helices about a pivot within the bilayer, such that at least one entry point to the channel can be quite severely restricted (Findlay et al. 1990). However, it is not easy to envisage large enough movements to completely close access to the channel unless regular helical structures are substantially disturbed in that region.

One other major channel type remains to be discussed. These are the relatively non-selective pores found in the outer membrane of bacteria and mitochondria. Generally referred to as porins, these proteins do not have the long stretches of hydrophobic residues characteristic of more usual integral membrane proteins (Jap 1989). Instead, a polar/non-polar alternating sequence is more common, leading to the suggestion, which has received considerable support from structural analysis, that this group of transporters consists not of α-helices but of β-strands, which traverse the membrane numerous times. The alternating side chains would then be in either the hydrophobic phase of the lipid bilayer or in the aqueous phase of the central pore. To fully satisfy the H-bonding potential of amino acid residues in the peptide bond, the β-sheets should be arranged in roughly parallel fashion describing a closed system similar to the circumference of a circle. This type of structure would thus be energetically stable and account for the thin rim of protein encompassing a large pore, thought to be characteristic of the porin-type complex. Finally, a new and exciting twist to the self-assembly of proteins to produce ion transport channels has been provided by the observation that peptides, representing or modeled on putative channel-lining helices, can catalyze specific ion movements in reconstituted systems (Lear et al. 1988, Oiki et al. 1988). This opened up the prospect of detailed analysis of relatively simple model systems to elucidate the factors involved in assembly and the mechanisms responsible for specific ion transport.

## 1.5. Biogenesis

Whereas a generally accepted mechanism exists for the extrusion of secreted proteins through the bilayer of the endoplasmic reticulum (ER), and this mechanism can be reasonably adapted to cope with proteins which are anchored in the bilayer by a single transmembrane domain or a carboxy-terminal (glyco)lipid tail (see chapter 2), no such agreement exists on the process by which polytopic polypeptides are integrated and assembled in the lipid milieu. A number of variants of the signal-hypothesis concept have been proposed but the mechanisms advanced call for either a very complex assembly process and/or have many unsubstantiated aspects.

It is by no means ruled out, therefore, that the signal(s) present in the sequence of the polytopic protein serve only to direct the nascent synthetic complex to an appropriate bilayer. Once there, insertion into the phospholipid milieu may be largely entropically driven. It may occur by sequential steps ranging from one helix at a time to the whole protein. A double helix hairpin is the most popular option at the moment and transient and polypeptide-induced opening of the secYEG translocon at the interface oriented towards the hydrophobic membrane interior. Alternatively, there may be quite a different and new (energy-dependent) process. This brief general discussion illustrates the general theme contained throughout this historical introduction – namely that there are a wide variety of ways in which proteins can be associated with the membrane phospholipid bilayer, in general, and anchored at the cell surface, in particular, among them transmembrane domains and glycolipids. Thus, there are likely to be a number of mechanisms for achieving this organization as well as functions underlying a specific mode of cell surface association, including release from the cell surface.

## 2. Lipid-Linked Proteins

In this fourth class of membrane proteins, no part of the polypeptide chain makes meaningful contact with the hydrophobic intramembranous environment. Anchoring is achieved by fatty acid moieties which are linked directly to the protein or constitute part of a phospholipid or glycolipid appendage. In addition to GPI-AP, several other modes of covalent attachment of fatty acids to proteins have been reported so far. It was much easier to demonstrate that fatty acids were associated with isolated membrane proteins than it was to establish that the association is of covalent nature, or than it was and still is today to elucidate the position and type of the covalent linkage. The first clear demonstration came from Handtke and Braun (1973) who determined that the amino-terminus of the Gram-negative lipoprotein constituent of the peptidoglycan cell wall was modified mainly with palmitic acid. Moreover, the side-chain and the amino-terminal residue cysteine was coupled to diacylglycerol *via* a thioether linkage. Integration into the bilayer was thus doubly or even triply indemnified. Subsequently, membrane-bound penicillinases from Gram-positive bacteria were found to undergo the same kind of modifications (Lampen and Nielson 1984).

Hydroxylamine-stable incorporation of radiolabeled myristic acid has been demonstrated in a number of viral and eukaryotic proteins (Grand 1989). This stability to base/hydroxylamine treatment is taken to imply an amide link, almost certainly involving the amino-terminus, but the side-chain of lysine is also a theoretical possibility. Furthermore, fatty acid attachment to the amino-terminus is a highly specific process involving an invariant glycine residue. Mutagenesis of the amino-terminal glycine of $p60^{src}$ to alanine or glutamic acid residues abolished this modification (Kamps et al.

1985). The attachment appears to occur during protein biosynthesis since it fails to take place in cells treated with the translation inhibitors, emetine or cycloheximide, but still occurs in those exposed to the glycosylation inhibitor tunicamycin (Magee and Courtneidge 1985). It looks as though only one residue of myristic acid is attached *per* amino group, which makes it a little surprising that a single fatty acid chain of rather short length (14:0) can have such a profound localizing effect. Certainly, it does determine the membrane location of p60$^{src}$ (Pellman et al. 1985), but equally there have been reports of the myristoylation of soluble proteins such as calcineurin (Aitken and Cohen 1984). It may be that membrane-associated myristoylated proteins have additional sites which stabilize binding to the bilayer. Interestingly, all myristoylated proteins are located at the cytosolic surface of the plasma membranes or in the cytosol.

While most myristoylation is insensitive to hydroxylamine or base (e.g., 0.1 M KOH), attachment of most palmitic acid is, in contrast, sensitive to both these agents, leading to the suggestion that an ester or thioester linkage is involved. Thus serine, threonine or cysteine residues are capable of being modified and in almost all cases, these residues are not at the amino-terminus of the protein. Another important difference is that palmitoylation is not totally dependent on protein synthesis (O'Brien et al. 1987), implying that it is a late or turnover process. Examples of proteins which appear to undergo this modification include the *ras* oncogene family (Sefton et al. 1982), viral membrane glycoproteins (Grand 1989), ankyrin (Staufenbiel and Lazarides 1986), the transferrin receptor (Omary and Trowbridge 1981) and rhodopsin (O'Brien and Zatz 1984, Ovchinnikov et al. 1988). For p21$^{ras}$ palmitoylation is thought to be vital for membrane association, but for most other proteins the reasons for the formation of this modification are not easy to surmise, since many of these proteins are integral by virtue of their primary structure, and their activities appear not to be compromised by failure to carry out this palmitoylation. Indeed, it has been suggested that in the case of rhodopsin, palmitoylation may be incidental (Karnik et al. 1988). Therefore, the attachment of either myristic or palmitic acid should not be regarded as inevitably indicating a membrane location.

Finally, it seems to be true that proteins which utilize fatty acid moieties as their only means of association with the membrane do not appear to directly involve the bilayer in any aspect of their activities other than simply localization. They do not seem so far to facilitate, for instance, the movement of either information or material across the bilayer. They usually possess enzymic or binding activity of some kind and this activity is not significantly regulated or influenced by the lipid bilayer or events occurring within it.

### 2.1. Discovery of GPI-AP

Phosphatidylinositol-specific phospholipase C (PI-PLC) has played a crucial role in the discovery and analysis of GPI anchors. However, the existence of a bacterial PLC with specificity for inositol phospholipids could not be demonstrated until some time after it had been noted that alkaline phosphatase (AP) might be anchored at cellular membranes in an unusual fashion. Thus, the earliest investigations about a GPI-AP led to the discovery of the PI-PLC rather than the other way around. Historically, investigations about the pathogenesis of anthrax led to the observation that injection of culture filtrates from *Bacillus anthracis* into laboratory animals causes a marked elevation in levels of serum AP (Slein and Logan 1960). Further study of this phenomenon revealed that the factor responsible for the elevation of serum AP was produced by *Bacillus cereus* and *Bacillus thuringiensis*, suggesting that the AP-releasing factor is distinct from the lethal toxin/protective antigen components of the *Bacillus anthracis* cultures (Slein and Logan 1962). Several other bacterial species were examined and found to produce little or none of the AP-releasing activity (Slein and Logan 1962).

Further investigation indicated that the factor produced by the *Bacillus* species can also release AP *in vitro* from slices of rabbit bone and kidney and from suspensions of leukocytes (Slein and Logan 1962). The development of this *in vitro*-assay for the AP-releasing factor enabled partial purification from *Bacillus cereus* culture filtrates (Slein and Logan 1963 and 1965). The releasing factor was thereby distinguished from the "lecithinase" or phosphatidylcholine (PC)-degrading PLC activity produced by some *Bacillus* and *Clostridium* species. Furthermore, it co-eluted from a DEAE-cellulose column with a novel PLC activity which appeared to have specificity for PI and this enzyme was therefore proposed to be responsible for AP release (Slein and Logan 1965). Although these findings were intriguing and interesting, this set of observations was not pursued and the results were not confirmed until the PI-PLC was purified from *Bacillus cereus* in 1976 (Ikezawa et al. 1976). Two novel PLC entities, one of which with specificity for sphingomyelin, the other for PI, were found in the culture filtrates of *Staphylococcus aureus* around the same time period (Doery et al. 1965). Although these activities had been partially purified, the effect of the staphylococcal PI-PLC on AP release was not determined, presumably because the investigators were unaware of the work on *Bacillus cereus* PI-PLC described above. Furthermore and perhaps surprisingly, elevation of serum AP is not a hallmark of staphylococcal infection. In 1977, the staphylococcal PI-PLC was purified to homogeneity (Low and Finean 1977a) and its ability to release AP from membranes demonstrated (Low and Finean 1977b). The PI-PLC turned out to be produced by coagulase-positive (i.e., pathogenic) strains of *Staphylococcus aureus*, only, suggesting that it might be a toxin. It was claimed that the PI-PLC and the δ-haemolytic activities were the same protein (Caird and Wiseman 1970). However, this idea was challenged when the PI-PLC was purified and found to be devoid of haemolytic activity. At present, there is no evidence that any of the other staphylococcal exotoxins, e.g., α- or β-haemolysins, enterotoxins, exfoliatin or toxin-1, are related to the PI-PLC (Marques et al. 1989). In fact, the potential role of this enzyme in staphylococcal pathogenesis remains an intriguing question for future investigation.

The discovery by chance of bacterial PI-PLC and its AP-releasing effects *in vitro* were confirmed about a decade later with purified PI-PLC from *Bacillus cereus* (Ikezawa et al. 1976) and *Staphylococcus aureus* (Low and Finean 1977c). It was subsequently shown with the *Bacillus cereus* PI-PLC that the purified enzyme manages also to cause the expected elevation in serum AP when injected into rabbits (Ohyabu et al. 1978). However, there was still no obvious mechanistic explanation for those results. AP was believed to be an integral, i.e., transmembrane protein since it was well-known of being most effectively solubilized by the organic solvent butanol. The results did not fit any simple prediction according to the "fluid mosaic model" of membrane structure, which was widely accepted during those times (see above). How could a PLC with specificity for the polar head group of a quantitatively very minor membrane phospholipid, i.e., PI, release a protein that was supposedly anchored at the lipid bilayer by hydrophobic interactions? Two seemingly obvious and somewhat trivial explanations for these results were offered and investigated first, (i) the PI-PLC preparations commonly used are contaminated by a protease or (ii) PI-PLC treatment causes the membrane to fragment due to cleavage of PI and possibly to a very minor degree highly abundant PC as consequence of its limited specificity.

Ad (i): Protease activity could not be detected in the purified PI-PLC (Ikezawa et al. 1976, Low and Finean 1977c). However, since at those times it was already known that a number of membrane enzymes, including AP, can be released from membranes in an active form, the possibility of protease contamination was considered seriously. The subsequent demonstration of AP release by PI-PLC preparations purified from two other species of bacteria, *Clostridium novyi* (Taguchi and Ikezawa 1978) and *Bacillus thuringiensis* (Taguchi et al. 1980) weakened the protease hypothesis even further since it seemed to be highly unlikely – albeit not entirely impossible – that a specific protease (undetectable in "broad range" assays for non-specific endoproteases) is produced by four different bacterial species and subsequently becomes enriched during four distinct purification procedures. Furthermore, the AP-releasing activity and the PLC activity of the *Staphylococcus aureus* preparations co-eluted during analytical cation exchange chromatography and were each inhibited by isotonic salt (Low and Finean 1977c).

Ad (ii): Gross disruption of the membrane also seemed to be an unlikely explanation since the released AP was not sedimented by ultracentrifugation, which would have removed membrane fragments (Ikezawa et al. 1976, Low and Finean 1977c). Furthermore, phospholipids and several other (integral) plasma membrane enzymes were found to resist PI-PLC-induced release (Low and Finean 1978). However, more direct evidence was obtained when it was shown by gel filtration under non-denaturing conditions that the released AP had a molecular weight in the range of 100-200 kDa, meaning that its size is small in comparison to membrane fragments (Low and Finean 1978, Ohyabu et al. 1978). The most obvious explanation for these collective observations was that the AP utilized the PI as a means of attachment to the membrane and was released as a direct consequence of PI hydrolysis by the PI-PLC. During the course of these studies, it was discovered that 5'-nucleotidase (Taguchi and Ikezawa 1978, Low and Finean 1978) and acetylcholinesterase (AChE)(Low and Finean 1977a) were also released by PI-PLC in a fashion similar to that of AP. The fact that these three enzymes were released from a wide variety of mammalian tissues, membrane fractions and intact cells showed that the release phenomenon reflects some fundamental feature of membrane structure as well as a novel mechanism for anchoring of enzymes and other proteins at the cell surface. Conversely it appeared unlikely that the phenomenon of protein release represents a specific response of a particular cell type to stimulation by the PI-PLC. Although this prediction of attachment *via* a PI moiety as a more general mode of membrane protein anchorage has turned out to be correct, it was several years before additional PI-PLC-releasable proteins were identified.

Strikingly, treatment of membranes with phospholipases exhibiting various substrate specificities, such as towards the major membrane phospholipid PC, or with a variety of agents, such as high salt, chelators, high and low pH, designed to disrupt the (non-covalent) interactions between membranes and peripheral (but not transmembrane) proteins failed to release detectable amounts of AP (Ikezawa et al. 1976, Low and Finean 1977b, Low and Zilversmit 1980). In addition, it was observed that the butanol-solubilized AP lost the ability to bind phospholipid vesicles after treatment with PI-PLC. These results suggested that the sensitivity of protein(s) to PI-PLC is not a property of the lipid bilayer, such as loss of structural integrity or of (non-covalent) interactions between its constituents, but relies on a structure intimately associated with the protein and capable of surviving extraction with carbonate, butanol etc. Collectively these data may be interpreted as indirect evidence for attachment of AP, 5'-nucleotidase and erythrocyte AChE to cellular membranes by covalent linkage between sites on the proteins and polar head groups of PI molecules located at the (outer leaflet of the) phospholipid bilayer (Low and Zilversmit 1980). Stronger evidence for this data-driven proposal was obtained several years later. AChE from *Torpedo* electric organ purified to homogeneity by affinity chromatography in the presence of detergent retained its sensitivity to hydrolysis by PI-PLC using reconstitution into liposomes and TX-114 partitioning as criteria (Futerman et al. 1985a).

Initial attempts were made to provide direct proof for covalent attachment of the PI-PLC releasable proteins to PI (Shukla et al. 1980) by isolating hepatocytes from [$^3$H]*myo*-inositol-labeled rats and purifying the proteins after their membrane release by PI-PLC action through chromatography on Sephadex G-25. Disappointingly, no significant increase in [$^3$H]-labeled protein was observed in comparison with untreated controls (Shukla et al. 1980).

Retrospectively, the lack of a positive result obtained in this experiment may be explained by subsequent studies which indicated that inositol is a relatively poor biosynthetic label for the GPI anchor. For instance, cultured cell lines expressing high levels of AP required metabolic labeling with [$^{3}$H]*myo*-inositol at 200 μCi/ml for 5-15 h before significant incorporation of radiolabel could be demonstrated (Howard et al. 1987, Takami et al. 1988). Direct proof for covalent linkage of the PI-PLC-releasable proteins to PI was obtained only when GC-MS was utilized to identify *myo*-inositol in the carboxy-terminal anchoring domain of AChE purified from *Torpedo* (Futerman et al. 1985b) or AP purified from human placenta (Low et al. 1987). This demonstration not only indicated that PI was a critical component of the structure anchoring diverse proteins to the surface of the cells, but also opened the door for preliminary identification of many other GPI-AP by their accessibility to PI-PLC cleavage as well as initial description of the release of GPI-AP with intact GPI anchor from plasma membranes, which rely on molecular mechanisms distinct from phospholipase action (see below).

## 2.2. Structure of GPI-AP

Several modes of anchorage of proteins at the surface of eukaryotic cells are known (see above and Figure 1), such as the penetration of the plasma membrane lipid bilayer by one to several proteinaceous transmembrane domains (monotopic, bitopic and polytopic transmembrane proteins), the mere peripheral association with the extracellular domain of other plasma membrane proteins (peripheral membrane proteins) or the insertion of fatty acyl chains covalently linked to certain (amino- or carboxy-terminal) amino acids of the polypeptide chain into the outer leaflet of the plasma membrane (e.g., palmitoylated or farnesylated proteins). In addition, certain proteins in eukaryotic cells are modified with a specific glycolipid species, the GPI anchor (Figure 1), which becomes embedded in the outer phospholipid layer of plasma membranes thereby mediating attachment of the protein moiety to the cell surface as so-called GPI-AP (Orlean and Menon 2007). Variant surface glycoprotein (VSG), merozoite surface antigen and promastigote surface protease from the protozoan parasites *Trypanosoma brucei* (Ferguson et al. 1985 and 1988), *Plasmodium falciparum* (Haldar et al. 1985) and *Leishmania major* (Schneider et al. 1990), respectively, represent the first GPI-AP for which the structure of their anchor was elucidated by a combination of (bio)chemical cleavage and subsequent two-dimensional NMR. During the following decades GPI-AP have been detected in eukaryotic organisms ranging from yeast (*Saccharomyces cerevisiae*: Nuoffer et al. 1991, Müller et al. 1992), plants (*Pyrus communis*: Oxley and Bacic 1999) and mammals (Low and Zilversmit 1980, Ikezawa and Taguchi 1981, Medof et al. 1986, Webb and Todd 1988). Meanwhile, databases relying on algorithms for the *in silico* prediction of GPI anchorage argue for 1-2% of the translated proteins in mammals representing GPI-AP (Eisenhaber et al. 2001, Poisson et al. 2007). So far, the state-of-the-art (bio)chemical generation of GPI anchor fragments and their mass spectroscopic MS/MS analysis have revealed the GPI modification of more than 150 proteins in mammalian cells including receptors, enzymes and adhesion molecules (Paulik and Bertozzi 2008).

The glycolipidic structure of the GPI anchor is highly conserved from yeast to man and constituted by amphiphilic PI as the phospholipid component and a hydrophilic glycan core. This highly conserved glycan core consists of a non-acetylated glucosamine and three mannose residues connected *via* specific glycosidic linkages, one end of which glycosidically linked to the 6-hydroxyl group of PI and the other non-reducing end invariably amide-linked to the carboxy-terminus of the protein moiety *via* an ethanolamine-phosphate-diester bridge, resulting in the following "overall"-structure: Polypeptide-CO-HN-Et-(P)6Manα(1→2)Manα(1→6)Manα(1→4)GlcNH$_2$α(1→6)-*myo*-Inositol-1-(P)-diacyl(alkyl)glycerol (Ferguson et al., 1985 and 1988, Englund 1993, Paulik and Bertozzi 2008) (Figure 1). Thus, the amino group of the ethanolamine creates the point of attachment of the GPI anchor to the protein moiety (Ferguson et al. 1988). In mammalian cells the GPI anchors are modified by additional ethanolamine-phosphate side branches coupled to the first glycan core mannose residue in all GPI-AP analyzed so far (Englund 1993, Orlean and Menon 2007) and, in addition, to the second and fourth glycan core mannose residues in certain GPI-AP (Taron et al. 2004). Additional heterogeneity is introduced by the presence of *N*-acetylgalactosamine, which can be linked to galactose and/or sialic acid residues at the fourth glycan core mannose in certain GPI-AP, such as rat brain Thy-1 and CD59 (Homans et al. 1988, Rudd et al. 1997) or bovine liver CD73 (Taguchi et al. 1994). Furthermore, some mammalian GPI-AP harbor an *N*-acetylated glucosamine-phosphate residue at the second glycan core mannose (Fukushima et al. 2003).

The phospholipid constituent is built from a PI moiety with two long-chain fatty chains (McConville and Ferguson 1993, Redman et al. 1994), either as diacyl-PI, exclusively, or as mixture of diacyl and 1-alkyl-2-acyl-PI, the latter being usually the major type. The majority of GPI anchors is equipped with saturated fatty acids. However, certain GPI-AP, such as CD55, CD59 and AChE in human erythrocytes, are known to harbor an unsaturated fatty acyl chain at the *sn*2 position and a palmitoyl chain esterified to the second hydroxyl group of the inositol moiety (Roberts et al. 1988, Rudd et al. 1997). As consequence, GPI-AP with three (i.e., two saturated and one unsaturated) fatty acids attached, are thought to be anchored at the plasma membranes rather tightly, e.g., at the membranes of erythrocytes during the life-time in the blood. At variance, in the majority of (non-erythrocyte) GPI-AP the inositol-bound fatty acid residue typically becomes cleaved off in course of esterase action immediately after attachment of the GPI anchor to the protein moiety.

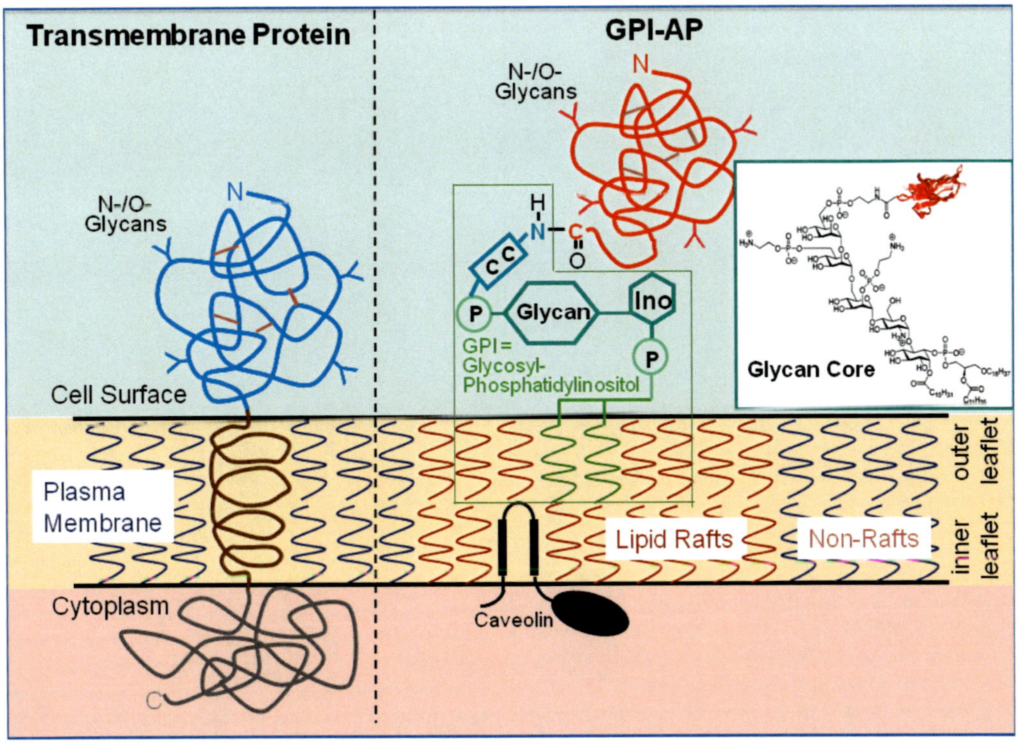


Figure 1. Principle structures and topology of cell surface proteins. Transmembrane proteins predominantly reside in non-raft areas of the plasma membranes (phospholipids in dark blue). In the bitopic topological configuration, a single extracellular (amino-terminal) domain harboring *N*- and *O*-glycosidically linked carbohydrate chains (in light blue) and a single cytoplasmic (carboxy-terminal) domain (in grey) is connected *via* a single transmembranous domain (in brown). In contrast, GPI-AP are typically enriched with lipid rafts (phospholipids in brown) with the extracellular protein moiety harboring *N*- and *O*-glycosidically linked carbohydrate chains (in red) and are inserted *via* their GPI anchor constituted by PI and the glycan core (in green) in the outer leaflet of the plasma membrane bilayer, only. The structure of the glycan core of human erythrocyte AChE is shown in the inset. The monotopic membrane protein, caveolin-1, which penetrates only the inner leaflet of the plasma membrane bilayer *via* its single intramembranous domain with two cytoplasmic domains, a small and a large one exposed at the cytoplasmic surface of the plasma membranes (in black), is often found associated with lipid rafts together with GPI-AP.

*Chapter 2*

# Biogenesis and Cell Surface Expression of GPI-AP

The biogenesis of GPI-AP can be divided into three sequential steps involving (i) the translocation of the protein moiety across the ER membrane with subsequent coupling of the GPI anchor, (ii) the transport of the mature GPI-AP from the ER to the Golgi apparatus with concomitant fatty acid remodeling of the GPI anchor and finally (iii) cell surface expression of the GPI-AP.

## 1. Fabrication and Assembly of GPI-AP

The biogenesis of GPI-AP involves the pre-fabrication of the total GPI anchor at the luminal face of the endoplasmic reticulum (ER) membrane by the stepwise transfer of the corresponding carbohydrate and ethanolamine-phosphate residues from activated precursors to PI and the glycan core, respectively, and subsequent coupling of the completed GPI anchor to the polypeptide moiety upon its complete translation and transport into the ER (Fujita and Kinoshita 2012). Production and coupling of the GPI anchor requires ten reactions in total which are catalyzed by more than 20 gene products, at least (Leidlich et al. 1994). The last step of transfer and amide coupling of the pre-fabricated GPI anchor to the polypeptide moiety is carried out by a membrane-bound multisubunit enzyme complex, the GPI transamidase, which is constituted by the PIG-K (GPI8), GAA1, PIG-S, PIG-T and PIG-U subunits.

## 1.1. Signal Sequences

GPI transamidase recognizes precursor sequences of GPI-AP harboring a GPI attachment signal sequence at their carboxy-terminus for post-translational processing to the mature GPI-AP protein moiety and simultaneous coupling to the anchor. The attachment signal sequence consists of a carboxy-terminal stretch of hydrophobic amino acids that is clipped off by proteolytic cleavage. The newly generated carboxy-terminal amino acid, the GPI attachment site or ω-site, is used for coupling of the pre-fabricated GPI anchor. Both proteolytic cleavage of the GPI-AP polypeptide precursor moiety and formation of the amide bond between the newly created carboxy-terminus and the amino group of the GPI anchor are intrinsically coupled and simultaneously catalyzed by dual action of the GPI transamidase complex at the luminal face of the ER (McConville et al. 1993, Ferguson et al. 1999, Ikezawa 2002). The PIG-K subunit of the GPI transamidase operates as catalytic subunit in the recognition and cleavage of the GPI attachment signal peptides and exhibits sequence homology to cysteine protease family members. The other four subunits are also critical or even essential for the cleavage (of the GPI-AP polypeptide precursor) and coupling (of the GPI anchor precursor) reactions (Orlean and Menon 2007).

Similar to the amino-terminal protein export signal sequences guiding precursors of secretory and membrane proteins into the lumen of the ER for subsequent transport along the typical secretory (Palade) pathway, the GPI attachment signal sequences from various GPI-AP precursors do not share a consensus sequence. Thus a unique primary amino acid sequence is apparently not deciphered by the GPI transamidase. Nevertheless, GPI attachment signal sequences harbor several critical structural features, among them (i) a total length of 20-30 residues between the ω-site and the carboxy-terminus of the GPI-AP polypeptide precursor, (ii) an amino acid with small side chain (i.e., Gly, Ala, Ser, Asn, Asp and Cys) operating as ω-site, (iii) the amino acid positioned two residues carboxy-terminal to the ω-site as one with a small side chain (i.e., Gly, Ala, Ser), (iv) a spacer sequence constituted by six or more hydrophilic residues starting three residues carboxy-terminal to the ω-site and (v) a carboxy-terminal stretch of hydrophobic amino acids, which spans the ER phospholipid bilayer in completion to enable insertion of the GPI-AP polypeptide precursor into the ER membrane prior to coupling of the GPI anchor (Moran and Caras 1991, Gerber et al. 1992).

Meanwhile, algorithms such as Big-PI (Eisenhaber et al. 1999), GPI SOM (Fankhauser and Mäser 2005) and FragAnchor (Poisson et al. 2007) as well as web-based prediction tools (Buscaglia et al. 2004, Macrae et al. 2005, Omaetxebarria et al. 2006, Nakayasu et al. 2009, Fujita et al. 2009) are available and widely used for the prediction of GPI anchorage of cell surface proteins and the determination of their ω-sites based on the sequences of their polypeptide precursors. Moreover, the experimental demonstration of GPI anchorage and identification of the ω-site have made considerable progress by mass chromatographic and MS/MS spectroscopic comparative analyses of peptides derived from the putative GPI-AP and displaying or lacking the glycan core. After proteolytic digestion of the polypeptide moiety, the peptide-glycans are prepared from the GPI-peptides by cleavage with (G)PI-specific phospholipases C or D ((G)PI-PLC/D) for removal of the GPI anchor diacyl- or acyl-alkyl-glycerol moieties, whereas peptide-ethanolamine is generated by dephosphorylation with aqueous hydrogen fluoride for removal of the complete GPI anchor through diester cleavage (Masuishi et al. 2013). Importantly, the GPI-attachment signal sequences seem to harbor the entire information required and adequate for post-translational modification with GPI anchors, since soluble proteins can be transformed into GPI-AP upon recombinant addition of the GPI-attachment signal sequence to their carboxy-terminus.

Interestingly, the processes of translocation of the GPI-AP polypeptide precursor across the ER membrane into the ER lumen and coupling of the GPI anchor seem to be segregated, but tightly coordinated. Upon insertion of the hydrophobic stretch of the GPI attachment signal sequence into the ER membrane, completion of the translocation of the GPI-AP polypeptide precursor into the aqueous ER lumen with the concomitant danger of misfolding and aggregation and the resulting exclusion from further transport to the cell surface has to be prevented. This may be achieved either by an intrinsic stop-transfer function of the hydrophobic stretch of the GPI attachment signal sequence or - more likely - by its recognition through the transamidase. In the latter case, it seems reasonable to assume a close association of the transamidase with the (signal recognition particle-independent) translocation machinery (translocon) for efficient recognition and processing of GPI attachment signal sequences of the newly translocated GPI-AP polypeptide precursors (Dalley and Bulleid 2003). Interestingly, the hydrophobic stretch within the GPI attachment signal sequence, predicted to act as transmembrane domain according to its overall hydrophobicity, apparently does not act as the sole recognition criterion since it failed to substitute for a typical stop-transfer sequence of a bitopic transmembrane protein. Interestingly, the positions of single charged amino acids within GPI attachment signal sequences appear to be responsible for either ignorance or recognition by the stop-transfer decoding mechanism involving the ribosome and/or translocon (Delahunty et al. 1993) and thereby to decide about either complete translocation across or insertion into the ER membrane with subsequent GPI anchor coupling. In conclusion, the GPI attachment signal sequence harbors the information for GPI anchor coupling encoded by its hydrophobic domain as well as additional features, such as the positioning of flanking charged residues, that prevent recognition as a typical transmembrane domain (Dalley and Bulleid 2003).

## 1.2. Translocation Components and Structural Requirements

Recently, a haploid genetic screening approach was used to analyze the requirements for ER targeting of GPI-AP, which was performed with the cellular prion protein (PrPc) and CD59 (Davis et al. 2015). The screens revealed the differential involvement of a subset of GPI anchor-modifying factors for CD59 (PIGN, PGAP2, PIGF) as well as of ER transport components for PrPc (SEC62, SEC63) in the biogenesis and trafficking of mammalian GPI-AP dependent on the structure of their protein and anchor moieties. The authors concluded that multiple and remarkably differing biosynthetic and trafficking pathways rather than a single universal route operate during the biogenesis of structurally similar and topologically related GPI-AP in specific fashion (Davis et al. 2015). Specific requirements of GPI-AP for transport into the ER in comparison to typical secretory and transmembrane proteins were elucidated in course of a combination of microscopic screening for cell surface expression and hydropathy plot analysis of the yeast secretome (Ast et al. 2013). Remarkably, GPI-AP are translocated across the ER membrane independent of the well-characterized signal recognition particle- (SRP-) dependent pathway and instead use the less studied SRP-independent protein targeting and translocation system (Müller and Zimmermann 1987) involving a network of cytosolic proteins and factors (Ast et al. 2013).

In addition to the contribution of a multitude of components of the cellular secretory pathway machinery, the correct folding of the protein moiety of GPI-AP seems to be essential for their proper cell surface expression. It has been demonstrated that in course of exposure to acute cellular stress, subsequent binding of the resulting misfolded GPI-AP to resident ER chaperones may lead to saturation of the ER folding capacity (Satpute-Krishnan et al. 2014). This is counteracted by the forced dissociation of the misfolded GPI-AP from the chaperones and subsequent interaction with ER export receptors triggering their quantitative vesicular transport from the ER *via* the Golgi apparatus to the cell surface. However, expression of misfolded GPI-AP at the cell surface is of transient nature, only, due to their rapid internalization and trafficking to lysosomes for degradation. Thus, misfolded GPI-AP are efficiently removed from the ER without causing deleterious effects by permanent residence at the cell surface prior to the onset of consequences of the GPI-AP-induced ER stress and overloading of the export pathway for other secretory and (trans)membrane proteins, such as their accumulation and destruction in the ER (Satpute-Krishnan et al. 2014).

These findings are in agreement with misfolded GPI-AP representing poor substrates for the ER quality control mechanisms which typically target terminally misfolded proteins for ER-associated destruction (ERAD). Consequently, it has been proposed that a GPI anchor sterically obstructs ERAD and instead destines the GPI-AP protein moiety for elimination in the vacuole/lysosome in course of its detour *via* the cell surface. Importantly, the apparent bypassing of the ERAD does not rely on the presence of the

GPI anchor *per se*, but is due to its canonical remodeling, which occurs in all GPI-AP (Kinoshita et al. 2008, Sikorska et al. 2016). This remodeling seems to provide a protein-independent ER export signal which takes place independent of folding of the GPI-AP protein moiety and to lead to efficient ER export of even misfolded species. Consequently, it was concluded that the ER quality control for the entire class of GPI-AP is less efficient in comparison to secreted and transmembrane proteins (Sikorska et al. 2016). In contrast, the operation of a specific ER-associated proofreading mechanism which tests GPI-AP for intactness of their polypeptide structure and conformation and efficiently releases the properly processed ones, only, has been proposed (Castillion et al. 2011, Fujita et al. 2011).

## 2. Transport to the Golgi Apparatus and Fatty Acid Remodeling

After coupling of the GPI anchor to the protein moiety in the ER and prior to departure from the ER, the inositol-bound fatty acyl chain is clipped off by the deacylase PGAP1 (Tanaka et al. 2004). Interestingly, in most cell types with the exception of human erythrocyte lineage cells PGAP1 is equipped with a catalytic serine-containing lipase motif. Thereafter, the correctly folded GPI-AP entities move along the typical secretory pathway from the ER to the Golgi apparatus, which is mediated by the p24 family proteins, Emp24p and Erv25p in yeast (Muniz et al. 2000) and p23 in mammalian cells (Takida et al. 2008). For this passage, they become packaged into coat protein II (COPII) vesicles, that bud from specialized transitional ER (tER) or ER-exit sites (ERES)(Jensen and Schekman 2011). The coupling of the GPI anchor to the protein moiety represents a prerequisite for departure of the GPI-AP from the ER. Protein moieties lacking a GPI anchor and thereby behaving as soluble entities are typically destined for proteasomal degradation. Apparently, in yeast, protozoan parasites and mammalian cells GPI anchors can be regarded as signal for export from the ER (Doering and Schekman 1996, McDowell et al. 1998, Mayor and Riezman 2004). The deacylation of GPI-AP by PGAP1 apparently ensures the efficacy of this vesicle-mediated transport step (Tanaka et al. 2004), albeit the steady-state levels of GPI-AP at the cell surface in PGAP1-deficient CHO cells and blood cells from PGAP1-knock-out mice were not affected (Ueda et al. 2007). In addition to inositol deacylation, another step of structural remodeling of GPI anchors, the esterase-like removal of the side chain ethanolamine-phosphate moiety by PGAP5, is necessary for the productive sorting to and incorporation into ERES/tER (Fujita et al. 2009). The catalytic activity of PGAP5, a dimetal-containing phosphoesterase which remodels the glycan core of GPI-AP, has been demonstrated to be required for the efficient transport of GPI-AP from the ER into the early secretory pathway. Thus, correct remodeling of the GPI anchor seems to be prerequisite for its function as signal for the controlled departure of the corresponding GPI-AP from the ER.

GPI-AP do not protrude into the cytoplasm as is the case for (monotopic, bitopic and polytopic) transmembrane proteins with their cytoplasmic domains. These are often engaged in the direct interaction with cytosolic COPII coat proteins, that is required for active sorting and productive packaging of transmembrane proteins into ER-to-Golgi transport vesicles. For GPI-AP, this typical function of cytoplasmic domains is therefore taken over by the type-I membrane proteins p24, which act as adaptor proteins or cargo receptors by forming a bridge between the correctly remodeled GPI anchor on the luminal leaflet of the ER membrane (*via* their glycan- and lipid-binding globular GOLD domain and a coiled-coil domain) and the COPII/Sec24 proteins on the cytoplasmic leaflet of the ER membrane (*via* their cytoplasmic domains)(Strating and Martens 2009, Fujita et al. 2011). Thus, the initiation of the anterograde transport from the ER to the Golgi apparatus by the GPI-AP-carrying COPII/Sec24 vesicles seemingly depends on the formation of heterooligomeric complexes of p24, which is controlled by the luminal pH of the ER and Golgi organelles. In fact, formation of the complexes between GPI-AP and p24 occurs at the neutral or mildly alkaline conditions prevalent in the ER lumen. Subsequent retrograde transport of p24 from the Golgi apparatus back to the ER for the next cycle of GPI-AP shuffling necessitates its dissociation from the GPI-AP, that is provoked by the mildly acidic pH in the lumen of the *cis*-Golgi stacks.

Interestingly, the multimeric p24 proteins as single-spanning type I (bitopic) membrane proteins with about 85% of their total mass residing in the ER lumen and only 5% protruding into the cytosol as short cytoplasmic tails and the GPI-AP with their protein moiety located in the ER lumen, exclusively, are both completely asymmetrically distributed across the ER membrane. According to the classical model, this asymmetric distribution of membrane proteins provokes considerable resistance to membrane bending into bulbing vesicles with their membrane enclosing the major portion of the protein mass. The formation of those vesicles would suffer from over-crowding of their luminal membrane leaflet that becomes squeezed during the budding event (Copic et al. 2012). Therefore, it has been postulated that in case of abundance of proteins at the donor luminal membrane leaflet, an extensive COPII coat at the cytoplasmic membrane leaflet may facilitate budding through the generation of mechanical force sufficient to overcome the rigidity of the lipid bilayer toward deformation into the direction of the protein-loaded leaflet (Derganc et al. 2013). In analogy, COPII proteins may facilitate the packaging of GPI-AP at the ER into vesicles for anterograde transport to the Golgi apparatus. Recent findings about the biogenesis at and transport from the ER of GPI-AP including the enzymes and auxiliary factors involved have been reviewed previously in great detail (Takeda and Kinoshita 1995, Eisenhaber et al. 2003, Kinoshita et al. 2013, Paladino et al. 2015).

# 3. Apical/Basolateral Sorting and Cell Surface Expression

Upon arrival of the GPI-AP at the Golgi apparatus, the unsaturated fatty acyl chains in 1-alkyl-2-acylglycerols at the sn2 position as well as in diacylglycerols of the GPI anchor become exchanged for stearic acid (Maeda et al. 2007) by the sequential action of the membrane proteins Per1p in yeast and PGAP3 in mammals (Fujita et al. 2006). During this remodeling lyso-GPI-AP intermediates are generated, which subsequently become re-acylated with stearic acid by the Golgi membrane protein PGAP2 (Tashima et al. 2006). Similar fatty acid remodeling processes occur in *Saccharomyces cerevisiae* (Bosson et al. 2006) and *Trypanosoma brucei* (Jaquenoud et al. 2008), however involving saturated long-chain (C26:0) and medium-chain (C14:0) fatty acids, respectively. Moreover, in *Saccharomyces cerevisiae* the diacylglycerol moiety becomes replaced by ceramide in course of an additional remodeling reaction at the Golgi apparatus, which is catalyzed by the PGAP2 yeast homolog, Cwh43p (Ghugtyal et al. 2007, Umemura et al. 2007).

## 3.1. Targeting to Distinct Plasma Membrane Domains

After having reached the Golgi apparatus, the GPI-AP are transported from the *trans*-Golgi network (TGN) to plasma membranes. Recently it was demonstrated using fibroblasts from knockout mice and inhibition of anterograde transport by treatment with Brefeldin A that expression of the SNARE TI-VAMP7 at membrane lipid rafts of the TGN is prerequisite for homeostasis of both (glyco)sphingolipids and the Golgi apparatus as well as effective transport of GPI-AP to the cell surface (Molino et al. 2015). In polarized cells, such as epithelial or endothelial cells, at least two distinct plasma membrane domains, the apical and basolateral ones, have to be discriminated and targeted by vesicles. These sort and segregate membrane proteins during their transport from the TGN to the cell surface along either the apical or basolateral cognate routes in course of deciphering corresponding sorting signals. One of the first signals for apical sorting postulated was the GPI anchor on basis of the fact that the majority of GPI-AP are expressed at the extracellular leaflet of the apical plasma membrane domains in most epithelial cells studied as well as in the axonal region in neurons (Lisanti et al. 1989, Brown et al. 1992, de Hoop and Dotti 1993, Paladino et al. 2006). Interestingly, apical sorting of GPI-AP seems to depend on their accumulation at lipid rafts (see below for structural details) formed at the TGN membrane. In agreement, it is well known that GPI-AP acquire detergent resistance, i.e., the property to resist extraction from membranes by cold non-ionic detergent for 1 h at 4°C, in course of arrival and subsequent remodeling of the GPI anchor fatty acids at the Golgi apparatus (Maeda et al. 2007, Fujita and Kinoshita

2010). Furthermore, blockade of sphingolipid biosynthesis or dissociation of the lipid raft structure by cholesterol depletion was demonstrated to interfere with apical targeting of GPI-AP (Lipardi et al. 2000, Paladino et al. 2004).

However, currently the association of GPI-AP with lipid rafts is assumed to be required but not to be sufficient for apical sorting since a number of GPI-AP can be recovered from the basolateral plasma membrane domain either exclusively or in addition to the apical domain (Benting et al. 1999, McGwire et al. 1999, Sarnataro et al. 2002). The oligomerization of GPI-AP *per se* or in combination with transmembrane proteins, such as the p24 family members, may provide the second signal required for apical sorting (Delacour et al. 2006 and 2007, Mishra et al. 2010). The oligomerization, which seems to involve the GPI-AP protein moiety as well as the fatty acids of the GPI anchor (Seong et al. 2013), becomes stabilized by incorporation into lipid rafts and interaction with lectins, such as galectins. Furthermore, it is conceivable that coat proteins and cargo receptors support the budding of secretory vesicles equipped with the typical lipid raft components, such as GPI-AP, from the TGN, similar to the formation of ER-to-Golgi transport vesicles (as described above).

### 3.2. Defective Targeting to the Plasma Membranes

Functional inactivation of the biogenesis of GPI-AP in course of gene disruption (in yeast and CHO cells) and knock-out (in mice) for certain components required for GPI anchor biosynthesis or coupling of the GPI anchor and protein moiety, such as PIGM (Almeida et al. 2006) or PIG-A (Takeda et al. 1993, Sato et al. 2007), revealed that the cell surface expression of GPI-AP is of critical importance for the vitality of yeast (Leidich et al. 1994), for the survival and virulence of parasitic protozoa in the host (Nagamune et al. 2000), for cell wall synthesis, morphogenesis, development of root meristems, germination, shoots and pollen tube growth in plants (Lallanne et al. 2004, Gillmor et al. 2005) and for embryogenesis (Nozaki et al. 1999), neurogenesis, immune responses (Hazenbos et al. 2004 and 2011), fertilization (Ueda et al. 2007) and skin development (Tarutani et al. 1997) in mammals. Moreover, naturally occurring somatic mutations in hematopoietic stem cells (PIG-A) and autosomal recessive mutations (PIG-M, PIG-V, PIG-N) in genes encoding enzymes of GPI biosynthesis, which consequently lead to GPI-deficient cells, have been correlated to several human diseases, such as the acquired hemolytic disease, paroxysmal nocturnal hemoglobinuria (Kawagoe et al. 1996, Bessler et al. 2001), dysregulated homeostasis of blood coagulation and neurological function (Almeida et al. 2006), hyperphosphatasia mental retardation syndrome (Krawitz et al. 2010) and multiple congenital anomalies-hypotonia-seizures syndrome (Maydan et al. 2011). Defective deacylation of the inositol ring in PGAP1-knock-out mice was reported to lead to neonatal death associated with otocephaly and malformation of the jaw in most cases, whereas the few survivors displayed a considerably reduced body weight and male sterility (Ueda et al. 2007). These findings hint to a critical role of the GPI anchor lipid structure in the physiological function of GPI-AP during development. Taken together, the important structural and functional aspects of GPI anchor remodeling are reflected in the known striking consequences of mutations in genes involved in remodeling of the GPI lipid moiety. In humans those mutations are known to cause a variety of diseases encompassing neurological abnormalities and were summarized previously in several reviews (Kinoshita et al. 2008, Fujita and Kinoshita 2012, Kinoshita and Fujita 2016).

In conclusion, the trafficking of GPI-AP from the ER to the cell surface relies on the structural and compositional remodeling of the GPI anchor for the generation of transport signals that actively induce and control the translocation of GPI-AP *via* the Golgi apparatus to the plasma membranes. These signals seem to confer specific interactions of the GPI-AP, the protein moiety of which facing the lumen of the secretory organelles, with the corresponding organellar membranes. These interactions manage to trigger the differential trafficking of GPI-AP away from the path taken by typical transmembrane proteins. In particular, in polarized epithelial cells, GPI-AP are predominantly sorted to the apical surface at the TGN under the control of specific factors by clustering into lipid rafts, which may operate as apical sorting platforms. Importantly, the experimental evidence hints to differences in the plasma membrane organization of GPI-AP between polarized and non-polarized cells. These findings support the operation of distinct mechanisms that regulate GPI-AP organization in different cell types. The mechanisms engaged by GPI-AP for their selective transport along the secretory pathway with emphasis on the control of their export from the ER and the TGN have been reviewed recently (Muniz and Riezman 2016). The potential relevance of disturbances in the mechanisms regulating the biosynthesis, intracellular trafficking and plasma membrane organization of GPI-AP for the development of diseases has been discussed elsewhere (Paladino et al. 2015). The latest findings about the molecular mechanisms responsible for the polarized trafficking and transport of GPI-AP to distinct plasma membrane domains in polarized epithelial cells as well as its putative relationship to GPI-AP sorting at the TGN have been summarized recently (Zurzolo and Simons 2016).

*Chapter 3*

# Intrinsic Biophysical, Biochemical and Physiological Features of GPI-AP

Since the GPI anchor represents a rather complex and evolutionarily well conserved structure from yeast to man, it may seem surprising that a unifying understanding of the function of GPI anchoring is still lacking at present. The high evolutionary conservation of GPI anchors strongly argues for functional characteristics and physiological roles which are common for all GPI-AP. According to one argumentation, the GPI anchor was required for the fulfillment of specific tasks by and optimal functioning of certain proteins in response to specific environmental cues in the past, but is dispensable under the altered conditions of present life. This assumption would also fit to the experimental evidence available which suggests that membrane association of proteins through GPI anchors has been introduced at an earlier time point during evolution than that *via* transmembrane domains. With regard to an alternative opinion, the considerable energy input required for the complex biogenesis of GPI-AP (see above) must inevitably be coupled to an intrinsic advantage for survival *vs.* the energetically less demanding production of transmembrane or peripheral membrane proteins *via* the selection pressure continuously operating during evolution from the past to the future. Accordingly, GPI anchorage of proteins cannot be replaced by other modes of membrane attachment and cell surface expression, such as fatty acylation, prenylation or transmembrane domains. No doubt, a synthesis of these seemingly opposing alternatives is conceivable with certain proteins being functional only with a GPI (rather than with a transmembrane) anchor and others fulfilling their physiological role as both GPI-AP and transmembrane protein (Nosjean et al. 1997).

# 1. Positioning of the Protein Moiety Towards the Cell Surface

Only limited information is available at present about the position and orientation of the protein moiety of GPI-AP relative to the cell surface. A rather short distance between them has been suggested on the basis of biochemical and modeling studies using the protein moiety of the GPI-AP 5'-nucleotidase or CD73 from lymphocytes (Sharom et al. 1985). In fact, the data obtained by measurement of the discontinuity in the Arrhenius plot upon transformation of the bilayer from the solid gel to the fluid liquid-crystalline phase are compatible with a very short distance or even argue for direct contact of the protein moiety with (one leaflet of) the membrane bilayer (Brasitus and Schachter 1980). Furthermore, the glycan core moiety of the GPI anchor of Thy-1 was visualized and found to be positioned between the polar phospholipid head groups and the surface of its protein moiety in a compact and tightly folded configuration (Barboni et al. 1995) or in a glycan-binding pocket within the protein moiety itself (Rademacher et al. 1991). Both studies strongly suggest positioning of the Thy-1 protein moiety very close to or even in contact with the bilayer.

Moreover, the distance between the phospholipid bilayer surface and the protein moiety of the GPI-AP human placental alkaline phosphatase (PLAP) was measured using the biophysical technique of fluorescence resonance energy transfer and a purified and defined reconstituted liposomal system (Lehto and Sharom 2002). Upon the incorporation of fluorescently labeled PLAP into the liposomes equipped with intercalated fluorescent lipid probes, the fluorescence transfer from the PLAP donors to the lipid acceptors was found to be quenched in concentration-dependent fashion. From measurement of the energy transfer, the shortest distance between the PLAP protein moiety and the water-lipid interface of the membrane bilayer was calculated to range from 10 to 14 Å (Lehto and Sharom 2002). This very close positioning of the protein moiety of PLAP above the lipid bilayer may lead to its laying on the cell surface with direct physical contact and thereby support the transmission of alterations in the composition and structure of the plasma membrane outer leaflet to the GPI-AP protein moiety with consequences for its structural features and functional roles.

# 2. Clustering

An early experimental finding typical for GPI-AP is their intrinsic association with specific biochemically defined domains of the plasma membrane, the so-called (detergent-insoluble and glycolipid-enriched) microdomains or lipid rafts (Brown and London 1998, Varma and Mayor 1998). These are characterized by (i) resistance towards solubilization (by cold non-ionic detergents, such as TX-100, for 1 h at 4°C) on the basis of sedimentation (in course of centrifugation at 100.000x*g* for 1 h) as the experimental criterion (Mayor and Riezman 2004, Hanzal-Bayer and Hancock 2007, Danielsen and Hansen 2008), (ii) enrichment of certain phospholipids, glycolipids and (glyco)sphingolipids with preferably long-chain saturated fatty acids (at the outer membrane leaflet) as well as cholesterol (at the inner membrane leaflet) *vs.* the neighboring membrane, i.e., non-raft, areas or other microdomains (Figure 2), (iii) small size (10-200 nm), (iv) low buoyant density during sucrose gradient centrifugation as the experimental criterion and (v) intermediary biophysical state between the solid gel phase and the fluid liquid-crystalline phase. Moreover, other lipid-modified proteins are enriched in plasma membrane lipid rafts too, in particular, dually acylated and palmitoylated proteins, such as tyrosine kinases of the src family, however residing at the inner leaflet and thus opposite to the GPI-AP at the outer leaflet.

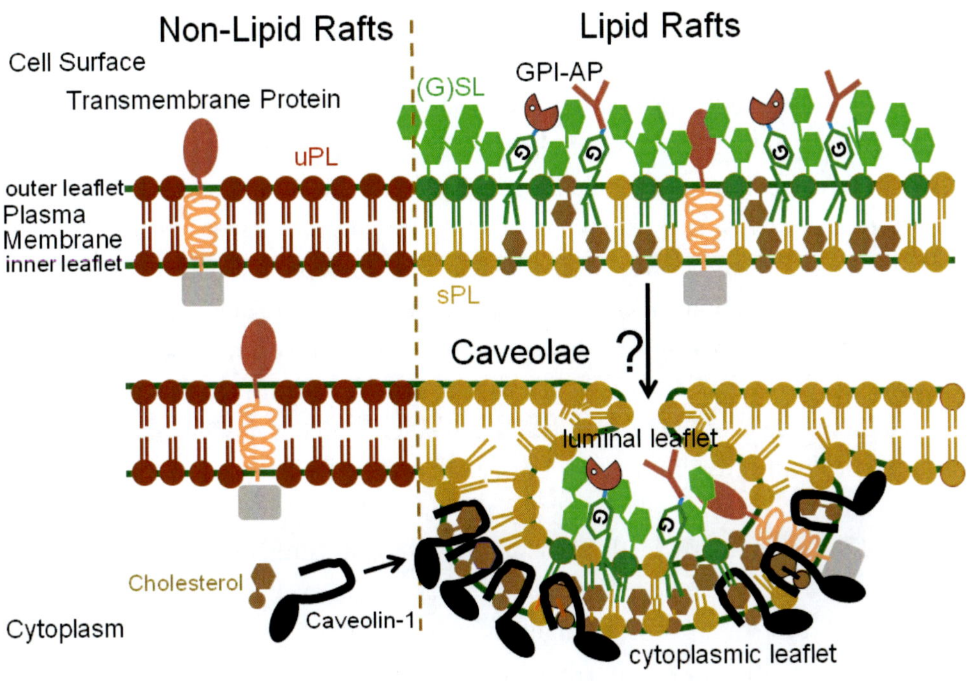


Figure 2. Lipid raft localization of GPI-AP. Lipid rafts are characterized by the relative enrichment of GPI-AP and (glyco)sphingolipids (GSL) at the outer leaflet of the plasma membrane bilayer as well as of caveolin-1, phospholipids with saturated phospholipids (sPL) and cholesterol at the inner leaflet compared to non-lipid rafts harboring predominantly typical transmembrane proteins and phospholipids with unsaturated fatty acids (uPL). It has been proposed that lipids rafts may operate as site for the formation of caveolae in course of bulbing of the plasma membranes into the cytoplasm and subsequent incision and fusion (and shedding in case of liberation of closed caveolar vesicles into the cytoplasm). Thereby the resulting caveolae would be constituted by the typical lipid raft components with the anchor of the GPI-AP being inserted in the luminal leaflet of the caveolar membrane bilayer, their protein moieties facing the caveolar lumen and caveolin-1 as well as cholesterol being enriched at the cytoplasmic leaflet of the caveolar membrane bilayer (Busija et al. 2017, Lamaze et al. 2017). This model for the biogenesis of caveolae still awaits confirmation by experimental data.

The presence of at least one long-chain saturated fatty acid in the glycerophospholipid, (glyco)sphingolipid or ceramide moieties of GPI-AP, introduced in course of fatty acid remodeling (see above) with the aid of the two Golgi-membrane proteins PGAP2 (Tashima et al. 2006) and PGAP3 (Maeda et al. 2007) in conjunction with the presence of high concentrations of cholesterol seems to be critical for the accumulation of GPI-AP in lipid rafts. In agreement, unremodeled GPI-AP, expressed in PGAP2- and PGAP3-double mutant CHO cells and consequently harboring unsaturated fatty acid chains, cannot be efficiently extracted from the cell surface by cold non-ionic detergents in contrast to the remodeled GPI-AP of wild-type CHO cells (Maeda et al. 2007).

The altered extraction behavior of unremodeled GPI-AP coincided with significantly reduced capability for the formation of oligomers, which indicates that remodeling of GPI anchors (i) is a prerequisite for spontaneous oligomerization even in physiological membrane environment, (ii) determines the degree of oligomerization through the fatty acid composition, (iii) stabilizes and regulates the lateral expansion of lipid rafts, (iv) contributes to the control of signal transduction pathways and (v) triggers trafficking of GPI-AP to the apical plasma membrane domain of polarized cells. The organization of GPI-AP into assemblies with diameters of less than 120 nm and constituted by two molecules of GPI-AP, at least, with relatively high local density in the steady or basal state, i.e., in the absence of (multivalent) ligands (in case of receptors or binding proteins) and cross-linking agents, such as antibodies, (in case of enzymes or structural proteins) has been demonstrated previously using a combination of pair-correlation and photoactivated localization microscopy (Sengupta et al. 2011). The data can be reconciled with the formation of transient GPI-AP homodimers provoked by GPI-GPI interactions. In contrast, GPI-AP residing outside of lipid rafts are thought to freely diffuse within the plane of the plasma membrane outer leaflet as monomers.

At present, lipid rafts are thought to operate as highly heterogeneous and dynamic membrane microdomains or platforms for the initiation and coordination of signaling processes which operate efficiently through the reversible attraction of certain signaling molecules, such as dually acylated Src family kinases, and exclusion of others, such as G protein-coupled receptors (GPCR)(Ilangumaran et al. 2000, Rajendran and Simons 2005, Simons and Gerl 2010). The signaling molecules located at the cytoplasmic leaflet become activated in response to communication *via* (non-covalent) interaction with GPI-AP residing at the extracellular leaflet leading to the transduction of signals across the plasma membrane into the cytoplasm (Stefanova et al. 1991, Malek et al. 1994). It is assumed that adaptor proteins provide the physical link and thereby the mechanistic coupling between the GPI-AP and the intracellular signaling molecules located at opposite leaflets of the plasma membrane (Chen et al. 2009). Effective coupling merely through direct (hydrophobic) contacts between the termini of the opposing fatty acyl chains at the midst of the membrane interior seems to be unlikely.

The migration behavior of GPI-AP within the plane of the plasma membrane was visualized by novel single-molecule detection technologies. The data suggest that initially GPI-AP, such as the GPI-anchored receptor CD59, form either monomers or nanoclusters of 5-10 nm size in conjunction with (glyco)sphingolipids, cholesterol (Figure 2) and elements of the actin cytoskeleton (Goswami et al. 2008). Importantly, upon specific stimulation of the cells, GPI-AP residing at those nanoclusters tend to aggregate into larger clusters or microdomains of 10-300 nm size, which are surrounded by components of the cytoskeleton (Kusumi et al. 2011). Concomitantly with this transformation of nanoclusters into microdomains, the recruited GPI-AP and certain signaling proteins, such as $G_{\alpha}$-proteins, Src family tyrosine kinases or phospholipase C$\gamma$, seem to constitute a signaling cascade which becomes prepared for transmembrane signal transduction (Suzuki et al. 2007 and 2012). It remains to be clarified whether GPI-AP in conjunction with (glyco)sphingolipids and cholesterol are entrapped in lipid raft microdomains transiently, i.e., in response to exogenous stimuli (e.g., receptor ligands, antibodies, crosslinking reagents) only, or are located therein permanently (with a certain probability and at a certain portion), i.e., at steady or basal state (Ahmed et al. 1997, Eggeling et al. 2009, Pinaud et al. 2009). In any case, GPI-AP equipped with two saturated fatty acyl chains are capable of efficient association with (glyco)sphingolipids and cholesterol, oligomerization into nanoclusters and formation of microdomains of liquid-ordered state, which may be dependent on or independent of the activation state of the cells.

However, since the first proposal of lipid rafts operating as membrane organizing principle with consequences for cellular functionality and physiology, serious questions and deep doubts have been raised about their existence *in vivo*, precise size, exact composition, real lifetime and physiological relevance. In general, it is accepted (more or less) that lipid rafts act as short-lived molecular assemblies for the induction of interactions between different lipidic and proteinaceous molecules, among them GPI-AP, within a rather restricted plasma membrane area. Thereby they are thought to trigger the transient local accumulation and regional compartmentalization of those molecules at higher molecular order with consequences on intracellular (signaling) processes (Lingwood and Simons 2010). In fact, the creation of membrane assemblies harboring GPI-AP has recently been re-investigated using the incorporation of synthetic fluorescent GPI-AP analogs into the outer leaflet of CHO cell plasma membranes (Raghupathy et al. 2015). The observation of a decrease in fluorescence anisotropy according to Förster resonance energy transfer effects argues for the formation of nanoclusters at a scale of $<$ 100 nm, which is dependent on the length of the saturated acyl chains ($\geq$ 18 carbon atoms) of the GPI anchor as well as on the presence of cholesterol and phosphatidylserine at the inner plasma membrane leaflet. Furthermore, nanoclustering with a remarkably high degree of molecular order was found to be considerably upregulated in the presence of high cholesterol at the inner leaflet as well as the ectopic expression of synthetic protein domains linking inner leaflet phosphatidylserine and cytoplasmic actin filaments.

Apparently, the resulting immobilization of phosphatidylserine induces interleaflet coupling of phosphatidylserine and GPI-AP, which could be facilitated by the presence of saturated long fatty acyl chains in both of them.

In conclusion, these data strongly argue for the formation of nanoclusters in course of interleaflet or transbilayer coupling in the presence of cholesterol and saturated long fatty acyl chains as a consequence of immobilization of either the outer or the inner leaflet interaction partner, i.e., of the GPI-AP through crosslinking by exogenous ligands and antibodies or of phosphatidylserine through capturing by endogenous cortical actin, respectively. At physiological level, the actin dynamics may determine and control the temporal and spatial stabilization of the nanoclusters. It remains to be clarified as to whether the actin-induced organization of GPI-AP into nanoclusters (Goswami et al. 2008) and the underlying molecular mechanism (Raghupathy et al. 2015) as studied with cell membranes *in vivo* (Sharma et al. 2004, Honigmann et al. 2014a) is related to the incorporation of GPI-AP into detergent-resistant lipid rafts or into phase-separated model membranes of ternary lipid mixtures as defined *in vitro* (Simons and Ikonen 1997).

However as outlined by Sezgin and coworkers (Sezgin et al. 2015), the molecular mechanism underlying the separation of plasma membranes in lipid raft or microdomain *vs.* non-lipid raft or non-microdomain phases cannot be fully explained by the successful and missing, respectively, interaction and immobilization of inner leaflet phospholipids at certain membrane areas with cytoplasmic components, such as actin filaments. From plasma membranes prepared as so-called giant vesicles from living cells, which harbor the authentic membrane proteins and lipids but lack any cytoskeletal elements, structures or domains can be isolated which contain the raft-characteristic equipment of lipidic and proteinaceous components using the criterion of detergent solubilization at low temperature (Sezgin et al. 2012).

Apparently, plasma membrane heterogeneity is due to several distinct mechanisms. At least one of them seems to rely on assemblies of specific membrane (glyco)phospholipids, membrane proteins and cortical cytoskeletal elements and another one is apparently independent of the cytoskeleton. In addition, different experimental set-ups and conditions, including the cell type studied and the expression level of the lipid raft components, may considerably contribute to the heterogeneity of plasma membrane areas. This may also explain the failure to observe the formation of an ordered membrane phase or the enrichment of nanoscopic-ordered domains in course of micropatterning experiments using single-molecule tracking of GPI-anchored green fluorescent protein (GPI-GFP) upon their direct insertion into plasma membranes of living cells (Sevcsik et al. 2015). Importantly, detailed analysis of the effect of the GPI-GFP micropatterning on the local membrane environment revealed that immobilization of the GPI-GFP prevents the lateral diffusion of other membrane components, but does not affect their lateral mobility over distances exceeding their physical size. Based on measurements using this experimental set-up for variation of the expression ratio between specific cytoplasmic and plasma membrane components as well as GPI-AP, it was concluded that the separation of plasma membrane lipid rafts *vs.* other membrane phases does not represent a general determinant for the organization of plasma membrane proteins, including GPI-AP (Sevcsik et al. 2015).

In conclusion, generalizations of characteristics found for a given cell type and experimental configuration with regard to plasma membrane organization and assembly as seemingly typical lipid rafts solely on the basis of the criterion of detergent insolubility should be strictly avoided. Nevertheless, lipid rafts expressing GPI-AP with long fatty acyl side chains may well be considered as nanoclusters triggered by (i) fixation of inner leaflet lipids to the cortical actin cytoskeleton (Kusumi et al. 2010, Honigmann et al. 2014b, Raghupathy et al. 2015), (ii) interleaflet coupling involving specific lipids and cholesterol (Spillane et al. 2014) and (iii) membrane curvature (Jarsch et al. 2016). In any case a putative physiological or pathological role of these nanoclusters critically depends on their (maximal) lifetime within the plasma membrane of a given cell type as well as on the effect of the clustering (for a sufficient period of time) of relevant GPI-AP on their structural and functional characteristics.

## 3. Lateral Mobility

The diffusion coefficient of a variety of cell surface molecules over a (physiological) temperature range of 24-37°C was measured with fluorescence correlation spectroscopy (Saha et al. 2015). Exogenously incorporated fluorescent lipid probes with short acyl chains displayed an elevation of the diffusion coefficient over this temperature range as expected. In sharp contrast, the diffusion of fluorescently labeled and incorporated GPI-AP was independent of temperature, but regained temperature-dependence in course of (i) appearance at plasma membrane blebs, (ii) expression at total plasma membranes of cholesterol-depleted cells and (iii) acute interference with actin dynamics or myosin activity. Moreover, a recombinant model transmembrane protein with a cytosolic actin-binding domain was found to diffuse laterally independent of temperature, too. Strikingly, a similar promotion of temperature-independent diffusion by cortical actin was revealed for GPI-AP during analysis of the effect of disruption of the filamentous dynamic actin nucleator formin. In contrast, alteration of the cortical actin mesh size or inactivation of the branched actin nucleator Arp2/3 did not led to temperature-dependent diffusion of GPI-AP (Saha et al. 2015). These findings argue for a critical role of active fluctuations of the dynamic cortical actin filaments in limiting the lateral mobility of GPI-AP within the plane of the plasma membrane, which is driven by mere thermal diffusion, through the indirect interaction of the GPI-AP with actin.

The integration of the data about (i) the organization (nm- to μm-scale), (ii) the dynamics (μsec- to sec-scale) and (iii) the short-range interactions of GPI-AP obtained with a variety of biophysical techniques including fluorescence correlation spectroscopy, fluorescence recovery after photobleaching and single particle tracking as well as a number of high resolution methods including fluorescence (or Förster) resonance energy transfer microscopy strongly argues for a model of the membrane bilayer where certain membrane constituents are intrinsically connected with the underlying dynamic cytoskeleton. This type of plasma membrane organization is termed "active actin-membrane composite" cell surface to emphasize the critical impact of the cytoskeleton on the lateral mobility of plasma membrane proteins including the GPI-AP (Saha et al. 2016). Nevertheless, this model of the dynamic coupling of the inner leaflet to the underlying cortical actin network fails to explain the nature of the apparent intimate interactions between (the peripheral proteins and phospholipids of) the inner leaflet and (the phospholipids and GPI-AP of) the outer leaflet of the plasma membrane bilayer.

The most relevant and recently published experimental data on the diffusion, lateral mobility, clustering, lipid raft association, actin binding and transmembrane signal transduction *via* GPI-AP, which are based predominantly on advanced single-molecule fluorescence imaging, may be reconciled as follows: (i) Under basal conditions, i.e., in the absence of specific ligands, crosslinking reagents or antibodies as receptor agonists or clustering tools, plasma membrane lipid rafts harbor GPI-AP as monomers or transient homodimers (triggered by interactions between their protein moieties), which are stabilized by expression of the GPI anchor and the presence of cholesterol (Suzuki 2015). (ii) Under basal conditions, GPI-AP do not form heterodimers which emphasizes the critical role of the GPI-AP protein moiety in homodimerization. (iii) In response to ligation or crosslinking, i.e., in the presence of ligands or antibodies, GPI-AP monomers and transient homodimers act as precursors for stable homodimers and homooligomers, which are apparently of small size and short half-life time and termed homodimer rafts (Suzuki et al. 2012). (iv) In course of high ectopic cell surface expression or ligation/crosslinking of the GPI-AP, the GPI-AP homodimers or homooligomers become recruited into stable homo- and hetero-GPI-AP oligomer rafts held together through the hydrophobic interactions which are typical for lipid rafts. For instance, ligation of CD59 causes the formation of stable oligomer rafts which apparently consist of up to four copies of CD59 and manage to trigger intracellular $Ca^{2+}$ responses, provided the GPI anchor had not been cleaved lipolytically and cholesterol had not been depleted (Suzuki et al. 2012).

Together these findings are compatible with small transient GPI-AP homodimer rafts operating as basic building blocks for the assembly and physiological function of more extended and more stable lipid rafts harboring GPI-AP homooligomers at the outer leaflet. Subsequently they undergo clustering and immobilization into larger multi-component assemblies, which are enriched in and clustered together "in correlation" with cytoskeletal actin filaments as well as phosphatidylserine at the inner leaflet, and are possibly identical with typical lipid rafts. These may serve as signaling platforms for the rapid and short-lived transduction of ligand messages across the plasma membranes to the intracellular signaling machinery, constituted by various effector molecules, and subsequent downstream processing (Suzuki 2015). Thus, the combination of theoretical simulations and experimental findings strongly argues for the operation of a straightforward molecular mechanism for nanocluster formation, which is capable of inter-leaflet or transbilayer coupling in both directions, i.e., for signal transmission from the extracellular milieu *via* the plasma membrane inner leaflet to the cell interior and *vice versa* from the cytoplasm *via* the cell surface to the environment (Sezgin et al. 2015).

## 4. Signal Transduction

The homo- and hetero-oligomerization of GPI-AP, which apparently determine and control their incorporation into lipid rafts and thereby restrict their lateral mobility, have been associated with their specific physiological function since the past two decades (Simons and Toomre 2000, Mayor and Riezman 2004, Paldino et al. 2004). In addition, a multitude of observations suggest that the oligomerization process, the physical confinement within lipid rafts and the dynamics of lateral movement within the plasma membrane outer leaflet of the GPI-AP as well as the mutual coupling of these events can be modified by factors distinct from the interaction of cytoskeletal elements with lipid raft components (see above). The factors encompass (natural or artificial) clustering agents (e.g., ligands for GPI-anchored receptors, chemical cross-linkers, polyvalent antibodies) present in the extracellular environment as well as the composition of the extracellular matrix (ECM)(Robinson 1991a). For instance, the urokinase plasminogen activator receptor (uPAR), a GPI-AP engaged in the control of cell adhesion, migration and proliferation, has been studied as a functionally active fluorescent protein conjugate in living cells for the potential effect of ECM proteins and uPAR ligands on the dynamics and dimerization of uPAR at the cell surface (Caiolfa et al. 2007). It turned out that vitronectin induces the formation of dimers and restricts the diffusion rate of uPAR at the basolateral plasma membrane domain of polarized epithelial cells. Moreover, the endocytosis and recycling of uPAR upon ligation by the urokinase plasminogen activator seemed to control the lateral movement of uPAR within and its confinement to lipid rafts as well as to determine the distribution between uPAR monomers and dimers (Caiolfa et al. 2007). It is tempting to speculate that cell surface protein assemblies constituted by ECM components are critical for the regulation of the dynamics and localization of GPI-AP monomers *vs.* dimers and oligomers. Moreover, differentiation of cells expressing a corresponding chimeric uPAR escaped negative control by uPA, accompanied by a concomitant size shift of the lipid rafts containing the chimeric GPI-AP.

At least for GPI-AP with non-interactive external protein moieties, such as the carcinoembryonic antigen (CEA), the type of GPI anchor seems to be critical for their cell surface expression, oligomerization and localization at lipid rafts as well as transcellular signaling. Accordingly, exchange of the CEA-specific GPI anchor for the unrelated GPI anchor of the neural cell adhesion molecule did not support the corresponding features specifically displayed by CEA (Nicholson and Stanners 2006). Importantly, co-expression of defective and functional CEA protein moieties equipped with the GPI anchor of CEA yielded cells which completely failed to transduce external signals *via* the interaction of the ECM with integrin arguing for interference of transmembrane signaling *via* authentic CEA by the simultaneous presence of inactive CEA in intimate neighborhood (mediated by GPI anchorage at common lipid rafts) in dominant fashion. Consequently, the authors argued that the GPI anchors of GPI-AP play a critical role in the function of their protein moieties by guiding them to a specific subset of plasma membrane lipid rafts, which controls the interaction of the GPI-AP with appropriate (extra- and intracellular) signaling components (Nicholson and Stanners 2006).

## 5. Additional Functional Properties

During the past three decades a number of reasons for employing GPI anchors for the attachment of proteins at the cell surface beyond those already presented above have been proposed: The cell surface of protozoa is known to be covered by a dense coat constituted by many copies of a few very specialized GPI-AP, such as the trypanosomal VSG, whereas those of mammalian cells are equipped with GPI-AP of high structural variety expressed at significantly lower density. Genetic interference with GPI anchor biogenesis is lethal for protozoa and yeast and blocks the development of mammalian embryos (see above), but is compatible with the life of already fully developed mammalian cells. Thus, apparently GPI anchors and GPI-AP mediate differential functions in different cells and organisms, among them (physical) protection against environmental cues (e.g., viruses, bacteria, immune cells), sensitivity towards external stimuli (e.g., cytokines, mechanical constraints) and (enzymic) activities (e.g., concentration, transport, degradation) directed against extracellular components (e.g., of plasma, urine, lymph). This pronounced heterogeneity is in curious conflict with the exquisite conservation of the structure of the GPI anchor during evolution from yeast to man as reflected in the presence of an invariable glycan core and, in most cases, of saturated long fatty acyl chains.

*Chapter 4*

# RELEASE OF GPI-AP

The origin of the hypothesis that one of the (major) physiological roles of GPI anchorage of cell surface proteins relies on the possibility of constitutive and/or controlled release of the (protein moiety of) GPI-AP into the extracellular space can be dated back to the period immediately following the first biochemical identification and structural characterization of GPI anchors and gained further credit by the broad acceptance of their existence in most eukaryotic cells (Low 1989, Lauc and Heffer-Lauc 2006). Initially, the molecular mechanism envisaged for the release was thought to be based solely on cleavage of the protein moiety or GPI anchor by specific cell-associated or extracellular proteases or phospholipases, respectively (Figure 3). Later, the release of GPI-AP equipped with the complete GPI anchor, which does not rely on proteolytic or lipolytic processing, was found to occur in a wide variety of cells, in addition. The physiological function of the released GPI-AP may critically depend on the mode of its release, i.e., whether it operates as hydrophilic anchor-less or amphipatic anchor-containing version.

## 1. CLEAVAGE OF THE GPI ANCHOR

The engagement of a glycophospholipid moiety as the principle for membrane attachment of GPI-AP intuitively suggests that the GPI anchor may provide a pre-determined cleavage site that permits the liberation of the protein moiety into the extracellular milieu or in the case of unicellular eukaryotic organisms, such as the yeast *Saccharomyces cerevisiae*, the covalent linkage of the liberated protein moiety to carbohydrate structures of the "extracellular" cell wall (Müller et al. 1996) upon appropriate action of (endogenous or exogenous) enzymes (Low and Saltiel 1988, Low 1989 and 1990, Paulik and Bertozzi 2008). Importantly, for some GPI-anchored enzymes and receptors the removal of the GPI anchor *in vitro* was found to elicit considerable changes between the soluble anchor-less and the amphipatic GPI-anchored version with regard to substrate specificity and kinetics of catalysis (Braun-Breton et al. 1988, Walter et al. 1992, Gmachl et al. 1993, Brewis et al. 1994, Lehto and Sharom 1998) and in the affinity, specificity and kinetics of ligand binding, respectively (Tozeren et al. 1992, Müller and Bandlow 1994, Wang et al. 1996) in cell-free test systems as well as intact cells. Thus it is conceivable that the GPI-anchored versions of certain GPI-AP form a pool of functionally inactive/active precursors/products which undergo activation/inactivation due to acquiring the maximal/minimal enzymic activity or optimal/adverse binding characteristics of the GPI-AP upon removal of their GPI anchor. In any case, it remains to be clarified whether the modification of the enzyme/receptor moiety with the GPI anchor *per se* or the anchorage at plasma membranes or both in concerted action is responsible for the specific functional characteristics of the complete GPI-AP compared to the anchor-less and soluble protein moiety of the GPI-AP.

A recent example for the functionality of an anchor-less soluble version of a GPI-AP is represented by CD109, which acts as a novel transforming growth factor-β (TGF-β) co-receptor and in the membrane-anchored state negatively regulates TGF-β signaling through TGF-β receptor degradation *via* a SMAD7/Smurf2-mediated mechanism (Li et al. 2016). Interestingly, an interaction between a recombinant soluble anchor-less version of CD109 and the TGF-β receptor of high affinity and low dissociation rate constant was detected. In addition, soluble CD109 was found to antagonize TGF-β-induced Smad2/3 phosphorylation, transcription and cell migration. These results suggest that the soluble version of CD109 competes with TGF-β for binding to the TGF-β receptor and thereby impairs TGF-β signaling and TGF-β-induced cellular responses (Li et al. 2016). However, the generation of a soluble version of CD109 *in vivo* remains to be demonstrated and the underlying molecular mechanism, such as proteolytic or lipolytic cleavage, to be elucidated.

### 1.1. Proteolytic Cleavage

Since for decades proteolytic cleavage has been widely observed as mode for the release of the extracellular (and often large) domains of a multitude of (bitopic and polytopic) transmembrane proteins from the cell surface, it cannot be regarded as a releasing mechanism specific and typical for GPI-AP (Figure 3). Moreover, so far proteolytic processing has been observed for a few GPI-AP, only. For instance, the GPI-anchored NKG2D ligand, ULBP3, is released from tumor cells by a metalloprotease of the ADAM family, which cleaves the protein moiety within its carboxy-terminal stalk region (Waldhauer et al. 2008, Fernandez-Messina et al. 2010). The human Tamm-Horsfall glycoprotein (THP) is anchored at the luminal cell surface of the thick ascending limb of both Henle and convoluted distal kidney tubules *via* a GPI anchor, from which it becomes released into the urine. Compatible with the lipid raft residence of a typical GPI-AP, THP was recovered from kidney lysates in a Triton X-100 insoluble form of low buoyant density (Cavallone et al. 2001). The apparent lower electrophoretic mobility of kidney THP compared to urinary THP was eliminated in course of treatment of the former with a phospholipase. The authors proposed a proteolytic cleavage event at a site within the juxtamembrane region or near the carboxy-terminus of the extracellular domain of kidney THP as the molecular mechanism for the release of THP into urine (Cavallone et al. 2001). Taken together, it seems likely that proteolytic cleavage of GPI-AP at sites amino-terminal to the GPI anchor represents a (minor) physiological reason for the expression of a subset of cell surface proteins as GPI-AP. However, since this requirement for release is also met by typical transmembrane proteins, it is tempting to speculate that anchorage of cell surface proteins by GPI fulfills physiological roles, which are not covered (adequately) by other types of membrane proteins.

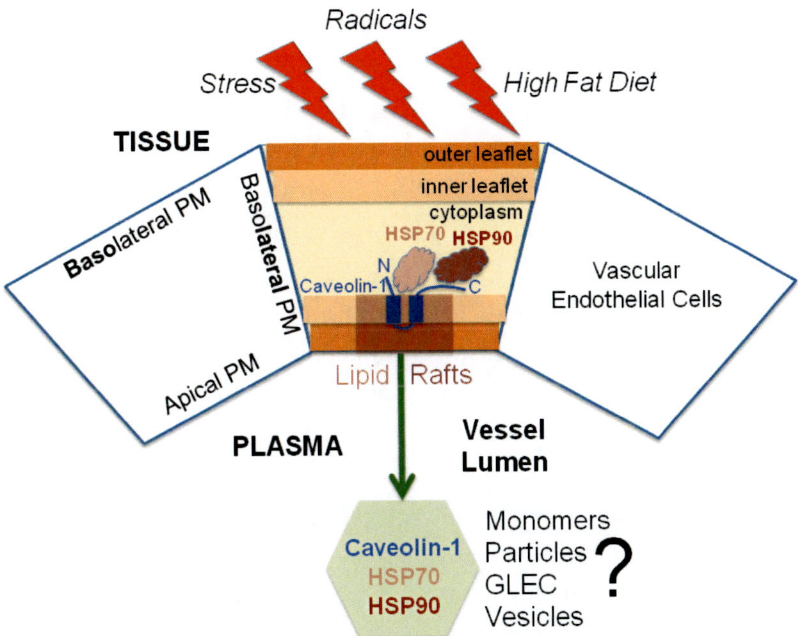


Figure 3. Release of caveolin into plasma. A number of studies provide evidence that the monotopic membrane proteins of the caveolin family appear in the circulation in complex with heat shock proteins, such as HSP70 and HSP90, in particular in response to certain exogenous stimuli, such as (metabolic) stress, (oxygen) radicals and high-fat diet (Radons and Multhoff 2005, Zhang et al. 2006, Evdokimovskaya et al. 2010 and 2012, Yuan et al. 2011, Gumulec et al. 2012, Uyy et al. 2013). It is conceivable that polar vascular endothelial cells forming the walls of blood vessels operate as the source for HSP-caveolin complexes. The route of their release *via* the apical plasma membranes (PM) rather than basolateral PM into the vessel lumen and the underlying molecular mechanism, which may involve lipid rafts, as well as their detailed structure, which may be of multimeric, particle-like, GLEC-like or vesicle-like nature, remains to be determined.

## 1.2. Lipolytic Cleavage

During the past three decades the number of studies about soluble versions of GPI-AP has been increasing steadily. The majority of them are apparently generated by lipolytic cleavage of their GPI anchor, among them trypanosomal VSG, mammalian AP and human ecto-5'-nucleotidase (CD73). The release of soluble VSG was observed during differentiation of the bloodstream to the procyclic forms of *Trypanosoma brucei* (Bülow et al. 1989). The measured enzymic activity of and immunofluorescence staining for CD73 associated with human umbilical vein endothelial cells were found to be reduced by about 50% upon incubation with tumor necrosis factor α (Kalsi et al. 2002). The abrogation of this effect by the PLC inhibitor neomycin strongly argues for a tumor necrosis factor α-induced activation of an endogenous PLC, which causes cleavage of the GPI anchor of CD73. The resulting loss of CD73 from the endothelial cell surface with the accompanying diminished formation of adenosine (from AMP) could contribute to endothelial activation in course of (tumor necrosis factor α-mediated) inflammation. Furthermore, CD73 was shown to be released from the surface of human endothelial cells upon challenge with anti-CD73 monoclonal antibodies, mimicking ligand binding, which was correlated to the initiation of intracellular tyrosine phosphorylation (Airas et al. 1997). The finding that the major portion of CD73 in lymphocytes is susceptible to cleavage by exogenously added bacterial PI-PLC, provided additional evidence that the ligand-induced release of lymphocyte CD73 involves the lipolytic cleavage of its GPI anchor.

Recently, it was reported that the plasma of newborns is characterized by significantly elevated conversion of adenosine monophosphate to adenosine, catalyzed by both 5'-nucleotidase and AP activity, compared to that of adults (Pettengill et al. 2013). AP is abundantly expressed at the surface of circulating neonatal blood cells and represents the predominant source for the dephosphorylation of adenosine monophosphate in these cells. Consequently, it was proposed that the enhanced lipolytic release from the surface of blood cells into blood and thereby the generation of soluble versions of CD73 and AP in newborns compared to adults are responsible for the elevated adenosine levels and upregulated extracellular purine metabolism in blood, which may promote an anti-inflammatory immunological status.

Since the identification of GPI-AP a variety of phospholipases with cleavage specificity C or D has been detected which manage to separate the protein moiety and GPI anchor of GPI-AP in cell-free or cellular test systems under various experimental requirements (e.g., absence or presence of detergent, use of sonication or liposomal reconstitution, basal or stimulated state of the cell). (G)PI-specific phospholipase C ([G]PI-PLC) and GPI-specific phospholipase D (GPI-PLD) cleave the GPI anchor at different sides of the phosphodiester bond within PI. Thereby diacylglycerol or phosphatidic acid, respectively, is liberated and a terminal glycan core-inositol structure left at the protein moiety, which harbors or lacks a (cyclic) phosphate residue, respectively (Chan et al. 1988, Ferguson and Williams 1988, Cross 1990). Thus briefly, the bond between the phosphate and glycerol residues is cleaved by a (G)PI-PLC, that between the inositol and phosphate residues by a GPI-PLD (Figure 3).

### *1.2.1. (G)PI-PLC*

A number of bacteria, among them *Bacillus thuringiensis*, *Bacillus cereus*, *Staphylococcus aureus*, *Clostridium novii* and *Listeria monocytogenes*, secrete a PI-PLC, which manages to hydrolyze mammalian GPI anchors in addition to the natural and endogenous substrate PI. A few parasitic protozoans, among them *Trypanosoma brucei* and *Leishmania*, express a GPI-PLC in their cytoplasm, which upon treatment of intact cells or plasma membrane vesicles is able to transform membrane-associated amphiphilic GPI-AP into their soluble hydrophilic versions (Sharom and Radeva 2004). Both bacterial and trypanosomal (G)PI-PLC have been and are still used for the detection and characterization of GPI-AP on a routine basis.

Importantly, the action of bacterial PI-PLC is affected by covalent modification of the GPI anchor, with acylation (e.g., palmitoylation) of the *myo*-inositol ring interfering with the hydrolytic cleavage reaction. This can be explained by the requirement of bacterial PI-PLC of the formation of a 1,2-cyclic phosphate bridge on the inositol ring during catalysis, which becomes prevented by a fatty acid moiety at position 2 (Stieger et al. 1986, Roberts et al. 1988). In fact, the formation of inositol-1,2-cyclic phosphate, which can be monitored by antibodies directed against this epitope of high antigenicity, the so-called cross-reacting determinant (Zamze et al. 1988, Broomfield and Hooper 1990, Jäger et al. 1990), has been used successfully to follow the lipolytic cleavage and membrane release of GPI-AP in course of (G)PI-PLC action (Lee et al. 1992, Müller et al. 1993, Fukuoka and Kobayashi 2001). Furthermore, glycosylations in neighborhood to the carboxy-terminus of the protein moiety of GPI-AP have been proposed to impair their accessibility towards cleavage by bacterial PI-PLC (Guther et al. 1994). The nature of the membrane phospholipids also seems to affect the efficacy of PI-PLC action since the lack of (glyco)sphingolipids in membranes of CHO cells was reported to result in hypersensivity of the GPI-AP CD14 for exogenous lipolytic cleavage (Hanada et al. 1993). This may be explained by clustering of the GPI-AP together with cholesterol and (glyco)sphingolipids in lipid rafts, leading to steric hindrance and impaired accessibility of the GPI anchor cleavage site for bacterial PI-PLC. Importantly, release of a candidate protein from the cell surface upon treatment of the cells with exogenously added (G)PI-PLC does not demonstrate unambiguously the direct membrane anchorage of this particular protein by a GPI anchor. In fact, it may lack a GPI anchor and be of soluble hydrophilic nature but interact with a GPI-AP and thereby acquire the status of a GPI-AP. This type of "indirect" GPI anchorage was demonstrated for the non-GPI-anchored basic fibroblast growth factor and lipoprotein lipase, which both interacting strongly with the

GPI-AP heparan sulfate proteoglycan (Brunner et al. 1991 and 1994, Müller et al. 1994b).

The expression of endogenous GPI-PLC activity in mammalian cells was postulated before its actual detection on basis of the apparent physiological need of downregulation of the cell surface expression of GPI-AP which was assumed to be accompanied necessarily by upregulation of the level of their soluble protein moieties in the extracellular space (Low and Zilversmit 1980, Low and Saltiel 1988). Surprisingly and seemingly at variance with this proposed function, GPI-PLC activity was detected at intracellular locations, associated with both the soluble cytoplasmic and the particulate membrane fractions, e.g., prepared from human liver (Rhee et al. 1989, Stinson and Hamilton 1994). Interestingly, the particulate versions was shown to behave as integral plasma membrane protein (Mensa-Wilmot et al. 1995), apparently with the catalytic domain facing the cytoplasmic leaflet of the plasma membrane bilayer. Very recently, the identification of the glycerophosphodiesterase GDE3 displaying six transmembrane domains and a large catalytic ectodomain as a GPI-PLC was reported which manages to cleave the GPI anchor of uPAR with accompanying release from the cell surface and loss of function (van Veen et al. 2017). uPAR is engaged in tissue remodeling as well as tumor cell migration, invasion and adhesion *via* mediating the proteolytic degradation of the ECM and binding to vitronectin/integrin, respectively. Importantly, it can be released from the cell surface of polarized cells at basolateral lipid rafts in course of protease and/or GPI-PLC action. The levels of (over)expression of GDE3 in cultured breast cancer cells, in a xenograft model and in patients suffering from cancer were found to be positively correlated with increases in the levels of soluble uPAR and reductions in neoplastic transformation, tumor growth and mortality (van Veen et al. 2017). The data provided sound evidence that the GPI-PLC GDE3 operates as a negative regulator of the uPAR (tumor-promoting) signaling network and that cell-autonomous lipolytic cleavage of GPI-AP can modify cell behavior.

Interestingly, another six-transmembrane ecto-phosphodiesterase, termed GDE2, was found to cleave and release a GPI-AP, too. The GPI-AP RECK normally blocks the protease-mediated shedding of a Notch ligand and thereby inactivates the ligand-Notch signaling axis, which is responsible for stimulation of differentiation in contacting neuronal cells. Recently, GDE2 (encoded by GDPD5 gene) was reported to promote neuroblastoma differentiation and embryonic neurogenesis in a cell-autonomous manner as well as to suppress cell motility and RhoA-driven neurite retraction (Matas-Rico et al. 2016a). Furthermore, a strong association between GDPD5 expression and favorable outcome in neuroblastoma patients could be delineated. This may rely on the lipolytic cleavage of the GPI-AP glypican-6, which as a typical heparan sulfate proteoglycan may operate as coreceptor for signaling pathways of proliferation rather than differentiation. Strikingly, GDE2 fails to attack uPAR (van Veen et al. 2017), which may be related to its cleavage specificity D which differs from that of GDE3. Consequently, GPI anchor cleavage has been attributed a role in neuronal differentiation (Matas-Rico et al. 2016b).

Glypicans are meanwhile recognized to act as coreceptors and modulators for a number of intracellular signal transduction pathways, in particular of those which are initiated by cell surface receptor tyrosine kinases or phosphatases (Filmus et al. 2008). Their mode of action relies on the facilitation of the recruitment of growth factor receptors (tyrosine kinases) at the cell surface. Consequently, lipolytic release of glypicans may cause redistribution of growth factor receptor (tyrosine kinases) leading to inactivation of the corresponding signaling pathway downstream of the receptor. Alternatively, glypicans could be engaged as membrane-tethered ligands through the direct interaction with receptor tyrosine kinases or transmembrane type-II receptor tyrosine phosphatases. With regard to the latter case, it has been proposed that the GDE2-catalyzed release of glypican-6 may modify phosphotyrosine-mediated signaling processes with consequences on neuronal differentiation (Matas-Rico et al. 2016b).

The expression of an endogenous (isoform of) GPI-PLC was described for the yeast *Saccharomyces cerevisiae* and a number of mammalian insulin target cells in culture, such as 3T3-L1 adipocytes, primary rat adipocytes, rat L6 myocytes and human endothelial cells on the basis of release of a number of GPI-AP, among them AP, lipoprotein lipase, the glycolipid-anchored cAMP-binding and -phosphodiesterase Gce1, the 5'-nucleotidase CD73 and a membrane dipeptidase from the cell surface into the culture medium, in particular in response to stimulation by certain metabolites such as glucose, hormones such as insulin, and growth factors (Chan et al. 1988, Romero et al. 1988, Bandlow et al. 1995, Movahedi and Hooper 1997, Müller et al. 1993 and 2005a), upregulation of glucose transport (Müller et al. 1994a) or exposure to certain drugs, such as the anti-diabetic sulfonylurea glimepiride (Geisen 1988, Müller et al. 1994c and 2005c, Ingham et al. 2014). Glimepiride lowers blood glucose predominantly by stimulation of insulin release from pancreatic $\beta$-cells (Geisen 1988) and, in addition to a minor degree, by mimicking metabolic insulin action in peripheral tissues (Müller and Geisen 1996, Müller 2000), such as activation of glucose transport in rat adipocytes (Müller and Wied 1993) and muscle cells (Bähr et al. 1995) and inhibition of lipolysis in adipocytes (Müller et al. 1993 and 1994a). Upon treatment of primary or cultured rodent adipocytes with pharmacological concentrations of glimepiride, a potent amphiphilic-to-hydrophilic conversion and release from the cell surface as soluble entities was observed for a subset of GPI-AP, presumably caused by activation of an endogenous GPI-PLC with catalytic ectodomain (Müller et al. 1993 and 1994a).

The functional and physiological significance of the hormone- and drug-induced lipolytic cleavage of GPI-AP in mammalian cells is not well understood at present, albeit initially a role of the cleavage fragments derived from the GPI anchor and consisting of the glycan core with or without the terminal phosphate residue and a few amino acids from the carboxy-terminus of the protein moiety, the so-called (phospho)inositolglycan

(-peptides), as soluble mediators (second messengers) for metabolic insulin action has been proposed (Lawrence et al. 1986, Saltiel et al. 1986 and 1988, Low and Saltiel 1988, Lisanti et al. 1989, Mato 1989, Farese 1990, Romero and Larner 1993, Gaulton and Pratt 1994, Lazar et al. 1994, Fonteles et al. 1996, Shaskin et al. 1997, Frick et al. 1998, Müller and Frick 1999, Müller et al. 2001a and b, Müller 2002, Larner et al. 2010). In fact, a multitude of the metabolic actions of insulin *in vitro* can be mimicked by direct incubation of insulin target cells, such as adipocytes, myocytes and hepatocytes, with phosphoinositolglycan(-peptides), which had been prepared from human acetylcholinesterase (Deeg et al. 1993), trypanosomal VSG (Misek and Saltiel 1992 and 1994) or yeast Gce1p (Müller et al. 1992 and 1997) or synthesized by chemical means (Frick et al. 1998), arguing for their involvement in insulin signal transduction. However, using a synthetic inositol derivative, GPI-2350, which inhibits the bacterial, trypanosomal and serum (G)PI-PLC/D with high potency and selectivity (towards other lipases and hydrolases)(Müller et al. 2005) and almost completely blocked the insulin- and glimepiride-inducible release of Gce1 and LPL from the surface of intact rat adipocytes by an endogenous GPI-PLC, a major role for this enzyme in metabolic insulin signaling and action in insulin target cells could be excluded. In contrast, activation of the GPI-PLC turned out to be indispensable for the insulin-mimetic effects of glimepiride *via* the previously described insulin receptor-independent cross-talk from lipid rafts to the insulin receptor substrate-1 protein (Müller and Frick 1999, Müller 2000, Müller et al. 2005).

Based on the criteria for lipid rafts mentioned above, certain GPI-anchored, (dually) acylated and transmembrane signaling proteins were found to be enriched in lipid rafts *vs.* non-raft areas of the plasma membranes. Furthermore, lipid rafts seem to be heterogeneous in composition and structure since subspecies of higher (hcRafts) and lower cholesterol (lcRafts) content can be distinguished from one another on the experimental basis of their lower and higher buoyant density, respectively (Müller 2002, Müller and Welte 2002, Müller et al. 2005). Strikingly, the insulin/glimepiride-induced redistribution of a subset of GPI-anchored as well as (dually) acylated signaling proteins from hcRafts to lcRafts (Müller et al. 2001a, b and 2002) was found to be compromised by GPI-2350. These data hint to a role of the mammalian GPI-PLC in the control of the localization and activation of signaling proteins within lipid rafts during signal transduction across the plasma membrane, in general. In particular in insulin target cells such as adipocytes, the analysis of the role of the endogenous GPI-PLC in insulin-like (rather than insulin-induced) signaling and metabolic action as well as of the effects of (small molecule) compounds and drug candidates on the activity of the GPI-PLC could reveal promising targets for future anti-diabetic drugs (Müller and Frick 1999, Müller 2002, Müller and Welte 2002).

Recently, four PI-PLC isoforms were identified in the unicellular eukaryotic organism *Paramecium caudatum*, which apparently are associated with different $Ca^{2+}$-channels at distinct subcellular locations where they exert varying and specific functions

(Staudt et al. 2016). Most importantly, all isoforms were found expressed at the cell surface, too and could be recovered from salt/ethanol washes of cells together with the lipolytically cleaved versions of a subset of GPI-AP. Furthermore, the PI-PLC isoforms were shown to be secreted into medium supernatants of living cells where they have access to plasma membrane-embedded GPI-AP of cells floating in the surrounding medium in short distance (Staudt et al. 2016). Thus, these PI-PLC isoforms may also contribute to the release of certain GPI-AP from the surface of the PLC-secreting or neighboring eukaryotic cells.

During the last three decades, a multitude of GPI-AP, among them CEA (Sack et al. 1988), renal dipeptidase from kidney proximal tubules (Park et al. 2001 and 2002), tissue non-specific alkaline phosphatase (TNAP)(Ermonval et al. 2015), growth arrest specific 1 (GAS1)(Ayala-Sarmiento et al. 2016) and CD14 (Ingham et al. 2014), have been demonstrated to be lipolytically released from normal and cancer cells and tissues *in vitro* and *in vivo* in the basal state or under certain (patho)physiological conditions or in response to certain hormones and stimuli, among them epidermal growth factor (Clemente et al. 1995), interleukin-2 (Eardley and Koshland 1991), adrenocorticotropic hormone (Vila et al. 1995), thyroid stimulating hormone (Jacquemin 1991) and nerve growth factor (Ohmichi et al. 1992). The identity of the mammalian GPI-PLC cleaving those GPI-AP remains to be elucidated in most cases as is true for the molecular mechanism for their activation, which may involve tyrosine kinase signaling pathways (Clemente et al. 1995).

In contrast, in *Drosophila melanogaster* the GPI-PLC responsible for the release of the glypican Dally-like (DLP) through cleavage of its GPI anchor and regulated by a number of morphogenic factors, such as the Wnt and Hedgehog proteins, was identified as Notum (Kreuger et al. 2004). Remarkably, Notum belongs to the hydrolase protein superfamily with a catalytic serine and apparently accepts the GPI anchor of other glypicans, such as glypican-3, and additional GPI-AP as substrates, thereby causing the release of their protein moieties from the cell surface (Traister et al. 2007). Remarkably, Notum manages to cleave and release from the intact cell surface the glypicans at the interface between the native membrane and the authentic extracellular environment in the absence of any detergent or solvent.

Interestingly, a systematic whole-organism screen for GPI anchor-cleaving activities was performed with transgenic mice expressing GPI-GFP overall in the body (Kondoh et al. 1999). In fact, fluorescence imaging revealed the presence of soluble circulating GFP in the extracellular compartments and interstitial fluids of several organs, such as testis and pancreas. This finding hints to the expression and action of several hydrolases with the potential for separating the fluorescent protein moiety from the GPI anchor in mice. However, on the basis of this approach, a differentiation between lipolytic and proteolytic cleavage specificity of the enzymes involved was not possible. Nevertheless, this screen enabled the purification of an enzyme from testis germ cells which specifically cleaves the GPI anchor of AP between the second and third mannose residue of the glycan core. Subsequently, this apparent mannosidase activity was attributed to the testicular isoform of the angiotensin converting enzyme (tACE). The typical substrates of this zinc protease are the bioactive peptides angiotensin I and bradykinin (Kondoh et al. 2005), which control blood pressure in concerted fashion (Turner and Hooper 2002). Strikingly, upon inactivation of its protease function by site-directed mutagenesis or pharmacological inhibition, tACE still managed to cleave the GPI anchor and to release a variety of GPI-AP from intact membranes, whereas overexpression of tACE was capable to restore the fertility of ACE-deficient sperm (Kondoh et al. 2005). Interestingly, the release of GPI-AP through the mannosidase activity of tACE *in vitro* was only observed in course of depletion of cholesterol from the cell membranes. This requirement can be explained by the need for a special presentation of the substrate GPI-AP within the phospholipid bilayer with regard to its fluidity and packing as well as lateral mobility of the GPI-AP as is the case for other phospholipases, such as GPI-PLD (see below). Cholesterol lowering would increase the fluidity and decrease the packing density and the expression of lipid rafts which in combination may favor tACE action towards GPI-AP. In this case, the typical raft localization of GPI-AP would cause their physical segregation from (non-raft) tACE and thereby prevent access of tACE to the GPI-AP. Unfortunately, so far two research groups have failed to demonstrate cleavage of GPI-AP by tACE (Leisle et al. 2005, Fuchs et al. 2005). Differences in the experimental conditions, such as membrane cholesterol, could explain this discrepancy.

### *1.2.2. GPI-PLD*

GPI-PLD was initially detected as abundant protein in human serum (Davitz et al. 1987, Low and Prasad 1988) and then intensively characterized with regard to its biochemistry (Stieger et al. 1991, Deeg and Davitz 1995) and molecular biology (Flores-Borja et al. 2004). In serum GPI-PLD, which is of amphipatic nature, is bound to high density lipoproteins and thereby apparently shielded from the aqueous milieu and concomitantly kept in inactive state (Hoener and Brodbeck 1992). Activation requires binding of apolipoprotein A-1 (Hoener et al. 1993). Albeit GPI-PLD is found mainly in plasma, a minor portion seems to be associated with membranes, which is presumably due to a number of hydrophobic stretches (Heller et al. 1992). *In vitro*, GPI-PLD is able to cleave the GPI anchor of a number of GPI-AP as test substrates, thereby separating the phosphatidate residue from the inositolglycan-protein moiety (Low and Huang 1991). Strikingly, GPI-PLD activity *in vitro* in the test tube strictly depends on the presentation of the GPI-AP in the membrane-solubilized state, rather than in the native membrane. In contrast, in intact cells GPI-PLD is active in the ER membrane during synthesis and assembly of the GPI anchors as well as in lipid rafts upon its ectopic expression (Mann et al. 2004). Thus, in contrast to *Drosophila* Notum and mammalian GPI-PLC, serum GPI-PLD cannot release native GPI-AP from the intact cell surface into the circulation, but apparently accepts the loosely inserted GPI anchor during its biosynthesis or depends on destabilization of the firmly embedded GPI anchor within the plasma membrane outer leaflet, in particular in lipid rafts due to downregulation of cholesterol.

In fact, as critical parameters for substrate availability for PI-PLC lipid fluidity and lipid packing were identified as is true for bacterial phospholipases (Lehto and Sharom 2002) as well as the type of the membrane phospholipids (Moon et al. 1999). This may explain the limited experimental evidence available for GPI-PLD activity towards GPI-AP in intact membranes. In greater detail, endogenous GPI-PLD is apparently capable of specifically releasing neuronal cell adhesion molecule (NCAM) from differentiating myoblasts (Mukasa et al. 1995), uPAR from ovarian cancer cells (Wilhelm et al. 1999) and CEA from human colon cancer cells (Naghibalhossaini and Ebadi 2006). Moreover, GPI-PLD activity was measured in the lysosomal fraction of liver cells (Hari et al. 1996), which may contribute to the intracellular degradation of GPI-AP under the specific (solubilizing and disintegrating) conditions of lysosomes leading to elevated lipid fluidity and reduced lipid packing. It remains to be clarified whether serum GPI-PLD is engaged in the release of GPI-AP from intact plasma membranes or intracellular membranes in mammalian tissues, in general or in exceptional cases, only. A general release could even cause problems if occurring in spontaneous, uncontrolled and unspecific fashion for all GPI-AP. Therefore, it seems to be more attractive to assume a role of serum GPI-PLD (in complex with HDL) in the removal of GPI-AP with complete GPI anchor which had escaped from tissue cells into the circulation by some releasing mechanisms (see below). This would contribute to the prevention of deleterious effects exerted by the amphipatic structure of GPI-AP, such as aggregation, unspecific interaction with the surface of vascular endothelial cells and blood cells, detergent-like effects. Complexes of GPI-PLD, HDL and GPI-AP may be taken up by liver cells for degradation in lysosomes which provide optimal conditions for the GPI-PLD activity by facilitating exposure and adequate presentation of the GPI anchor. Interestingly, a role of GPI-PLD in the control of blood glucose homeostasis was proposed (Riakwar et al. 2010)(see below).

## 1.3. Putative Physiological Roles of GPI Anchor Cleavage

Taken together, at present some issues about the expression and specificity of enzymic activities accepting GPI anchors as substrates remain to be clarified. Those encompass the intriguing question regarding the physiological relevance of the lipolytic release of GPI-AP *per se* and the function of the lipolytically released protein moieties. From a theoretical point of view the following consequences of the lipolytic release of GPI-AP can be considered: (i) its removal from the cell surface for functional inactivation and degradation/turnover; (ii) the alteration of the structural, conformational and functional properties of its protein moiety (e.g., with regard to ligand binding or catalytic activity) in course of its separation from the GPI anchor *per se* and/or loss of the intimate neighborhood to the cell surface; (iii) the acquisition of novel properties of the GPI-AP-less cell surface, (iv) the generation of cleavage products from the GPI anchor with signaling function, such as second messengers (e.g., phosphoinositolglycans) and/or (v) the generation of the protein moieties as soluble anchor-less version which exert a specific physiological function, such as transport protein and hormone (e.g., adipokines, myokines, hepatokines), in certain body fluids (e.g., blood, urine, tears, lymph) and interstitial spaces or in certain tissue cells in course of their interaction with cognate receptors, adsorption to the surface or uptake by the target cells and subsequent initiation of downstream signaling. In fact, during the past three decades experimental evidence has been presented for each of these putative roles of the lipolytic cleavage of GPI-AP.

Ad (i): Lipolytic cleavage and release from the cell surface of GPI-anchored cell adhesion molecules (CAM) leads to disruption of the adhesive state between cells in controlled fashion in the ciliate *Paramecium tetraurelia* (Müller et al. 2012a). The cleavage, presumably catalyzed by the two PLC isoforms PLC2 and PLC6, seems to be specific for a subset of GPI-AP rather than to cause the shedding of all GPI-AP from the cell surface. This is also reflected in the demonstrated preferential release of large surface antigens *vs.* smaller GPI-AP into the ciliate culture medium. Tissue non-specific alkaline phosphatase (TNAP), a member of the ecto-phosphatase family and expressed in plasma membrane lipid rafts in different tissues and cells such as brain and neurones, can be released into the ECM (Ermonval et al. 2015). Upon arrival of its soluble version at the neuronal microenvironment and dependent on certain (patho)physiological conditions prevalent in the central nervous system, it acts as a phosphomonoesterase on various soluble substrates, as an ectonucleotidase on nucleotide-phosphates and presumably as a phosphatase dephosphorylating phosphoproteins and phospholipids. GAS1 is associated with plasma membranes of cells in different organs, such as neuronal progenitors, or becomes secreted as a soluble version by various cell types, such as glomerular mesangial cells, and appears in the cerebrospinal fluid (CSF) and plasma of rats (Ayala-Sarmiento et al. 2016). On the basis of *in vitro* findings with a fibroblast cell line, that releases two distinct soluble versions of GAS1 into the medium supernatant, and experimentation with rats, it was suggested that membrane-associated and soluble GAS1 exerts differential effects, such as inhibition of proliferation upon overexpression and GPI-dependent membrane association in tumor cells and promotion of cell proliferation and survival in different organs and tissues upon release of the soluble version into body fluids. GPI-anchored CD14 represents a component of a receptor complex involved in the recognition of peptides, which accumulate in various neurodegenerative diseases, such as prion disease, Alzheimer's disease and Parkinson's disease. Lipolytic release of CD14 from cultured RAW264 and microglial cells upon activation of the cell-endogenous GPI-PLC by the sulfonylurea drug glimepiride leads to downregulation of the peptide-induced signaling and production of cytokines (Ingham et al. 2014). Thus agents, that modulate the efficacy of the membrane release of certain GPI-AP through specific control of the activity of a relevant phospholipase, such as the mammalian GPI-PLC, *via* engagement of the corresponding signaling pathway, such as inflammation, in relevant tissue(s), such as in the brain, may be useful as novel drugs, such as anti-inflammatory remedies, for the therapy of neurodegenerative diseases.

Ad (ii): The kinetic ligand-binding properties of the cAMP-binding protein, Gce1p, of the yeast *Saccharomyces cerevisiae* become drastically altered upon lipolytic cleavage of its GPI anchor (Müller and Bandlow 1993 and 1994). Other examples for the alteration of the functional characteristics of GPI-AP in course of anchor cleavage and/or membrane release are given above.

Ad (iii): On basis of the finding of infertility of ACE knockout male mice caused by defective sperm-egg interaction at the zona pellucida (Hagaman et al. 1998), it was proposed that fusion competence of sperm cells critically depends on the release of a subset of GPI-AP from their surface by the (eventually lipolytic) cleavage activity of tACE, which has been found associated with sperm. According to this view, shaving off the GPI-AP coat from the sperm cell would facilitate its binding to the zona pellucida of the egg (Kondoh et al. 2005).

Ad (iv): Mammalian Notum has previously been demonstrated to operate as negative regulator of Wnt signaling by causing the lipolytic cleavage of the GPI anchor of glypican-3, a co-receptor of the Wnt pathway (see above), and in consequence its release from the cell surface (Traister et al. 2007). This argues for a function of the released and soluble version of glypican-3 as competitive inhibitor or antagonist for the Wnt receptor-receptor interaction. Furthermore during almost three decades (1988-2015), (phospho)inositolglycan-peptides produced at the surface of donor cells by dual cleavage of GPI-AP within the anchor structure by the GPI-PLC and at a site a few amino acids upstream from the carboxy-terminus of the protein moiety by an (unknown) protease, as prepared *in vitro* from yeast Gce1p with bacterial PI-PLC and pronase (Müller et al. 1997) or obtained *in vivo* upon challenge of cells with insulin or glimepiride (Müller et al. 1993 and 1994a) have been demonstrated to exert considerable insulin-mimetic effects in insulin target cells, such as primary and cultured adipocytes and myocytes (e.g., Saltiel 1990, Field 1997). Unfortunately, the use of naturally occurring or synthetic (phospho)inositolglycan-peptides (Müller et al. 1997) for bypass of the insulin-resistant state, typical for insulin target cells during the development of T2D, has been hampered so far by (i) their metabolic instability due to the action of serum proteases and lipases, (ii) their putative hepatic clearance, (iii) the need for intraveneous injection due to their large size and polar character, which prevent oral bioavailability, and (iv) the high expenditure for their synthesis by recombinant or chemical means. Interestingly, it was demonstrated recently that injection of acylated inositolglycans of low molecular mass, which structurally resemble the glycan core of GPI-AP and exert insulin-mimetic effects in insulin target cells, provokes lowering of blood glucose levels, increases in liver and muscle glycogen content and improvement of glucose tolerance in rodent models of insulin resistance and prediabetes, the *db/db* mice and C57B6N mice fed a high-fat diet (Suzuki et al. 2014).

Ad (v): As already mentioned, a large body of experimental evidence has accumulated during the past three decades that insulin-responsive mammalian cells and tissues express a number of GPI-AP, the majority of them with their GPI anchor embedded in the outer leaflet of the plasma membrane bilayer, as well as certain GPI-PLC/D, which may trigger down-regulation of GPI-AP cell surface expression and simultaneously up-regulation of the soluble protein moiety released into the circulation. In fact, soluble versions of GPI-AP were detected in human plasma, such as 5′-nucleotidase, Thy-1, AP and CD16 receptors (Saltiel et al. 1986, Romero et al. 1988, Low 1989 and 1990, Thomas et al. 1990, Saltiel 1990, Romero and Larner 1993, Müller et al. 1994a, Varela-Nieto et al. 1996, Nosjean et al. 1997, Jones and Varela-Nieto 1999, Müller et al. 2005).

Additional relevant findings have been raised in more recent years: The GPI-AP CD59, a complement regulatory protein that is ubiquitously expressed at the surface of mammalian cells, specifically interferes with the formation of the cytotoxic pore-generating membrane attack complex (MAC) and protects the cell from complement-mediated lysis (Davis et al. 1989, Meri et al. 1991, Davis and Lachmann 1993). Interestingly, the complement regulatory function of human CD59 was found to become impaired in course of the development of diabetes (Acosta et al. 2000, Zhang et al. 2002), which is caused by the non-enzymic glycation of lysine at position 41 located at the active site of CD59 and by the resulting functional inactivation of the glycated CD59 (Davies et al. 2005). Since MAC and CD59 are often co-localized in the nerves, kidney and vasculature of diabetic patients (Hakulinen and Meri 1994), glycation and inactivation of CD59 may be regarded as candidate mechanism for the hyperglycemia-induced cell lysis and tissue damage which are typical for micro- and macrovascular diabetic late complications (Qin et al. 2004). Importantly, the soluble anchor-less protein moiety of human CD59 can be detected in human serum at concentrations of about 100 ng/ml (Landi et al. 2003) and in urine at about 4 μg/ml (Fletcher et al. 1994), apparently as result of its lipolytic release from the cell surface. Using a specific and standardized ELISA, glycated CD59 was also detected as soluble version circulating in human serum/plasma at levels which are significantly higher and discriminative for diabetes patients *vs*. healthy subjects and independently associated with the levels of glycated hemoglobin A1c (Ghosh et al. 2013).

CD59 is expressed as a GPI-AP at the surface of both tissue cells and cells circulating in interstitial spaces and body fluids and is therefore exposed towards glucose concentrations which are roughly identical with those of the venous blood. As a consequence it will be amenable to glycation and release into the circulation dependent on the exposure time as well as glucose concentration (Hakulinen and Meri 1995,

Cocuzzi et al. 2001). Interestingly, previous studies demonstrated the glucose-dependent stimulation of a GPI-PLC in yeast (Müller and Bandlow 1993) and rat adipocytes (Müller et al. 1994a), which could explain the observed positive correlation between soluble and membrane-released glycated CD59 in blood and hyperglycemia or diabetes. In any case, glycated CD59 released into blood seems to be of potential value for use as a biomarker for the prediction, monitoring, prognosis and stratification of vascular diabetic late complications, which can be measured with high precision, accuracy, sensitivity and limited expenditure.

Another example for the release and function of GPI-AP protein moieties is represented by glypican-4 (Gpc4). Gpc4 is a member of the family of GPI-anchored heparan sulfate proteoglycans and supports as a coreceptor a number of growth factors, such as Wnt, fibroblast growth factors and Hedgehog in mammals (De Cat and David 2001, Filmus et al. 2008, Fico et al. 2011). Gpc4 was reported to regulate insulin signaling *via* interaction with the insulin receptor (Ussar et al. 2012). Importantly, both membrane-associated GPI-anchored as well as soluble anchor-less Gpc4 managed to interact with the unoccupied insulin receptor and to stimulate insulin signaling, whereas binding of insulin led to abrogation of the interaction of Gpc4 with the occupied insulin receptor. Overexpression of the native GPI-anchored Gpc4 in or incubation of the recombinant anchor-less Gpc4 with 3T3-L1 adipocytes caused upregulation of insulin signaling, whereas depletion of Gpc4 blocked insulin receptor autophosphorylation and downstream signaling (Ussar et al. 2012). The physiological relevance of the cross-talk between GPI-AP and receptor tyrosine kinase signaling pathways was reinforced by demonstration of the inability of 3T3-L1 fibroblast with disrupted Gpc4 gene to undergo differentiation into mature adipocytes due to defective insulin action.

Interestingly, Gpc4 was detected in serum of mice and humans with levels being positively correlated with body fat mass and insulin resistance. The expression of the soluble version of Gpc4 in serum and its relationship to BMI and glucose tolerance (reflecting insulin sensitivity) may rely on its lipolytic release from the surface of donor cells. In fact, (G)PI-PLC was demonstrated to cleave the GPI anchor of Gpc4 (Brunner et al. 1994, Traister et al. 2007) and to be upregulated by insulin (Saltiel and Cuatrecasas 1988, Raikwar et al. 2010). Furthermore, serum levels of GPI-PLD were shown to be elevated in response to feeding a high-sucrose diet (Kurtz et al. 2004), but to be diminished in *ob/ob* mice (Bowen et al. 2001) as holds true for Gpc4. The strong correlation of serum Gpc4 levels with BMI in humans and the observation that Gpc4 is released from primary adipocytes strongly argues for adipose tissue as the major source of serum Gpc4 (Ussar et al. 2012). In conclusion, these findings indicate that Gpc4 acts as an insulin-sensitizing adipokine by direct interaction with the insulin receptor and accompanying activation and downstream signaling independent of whether being presented in the GPI-anchored or soluble lipolytically cleaved version. It may be speculated that the elevated insulin levels in course of insulin resistance and obesity induce upregulation of the cleavage of the GPI anchor of Gpc4 in adipose tissue by an insulin-dependent GPI-PL of C or D specificity. The resulting upregulated serum levels of soluble anchor-less Gpc4 would counteract the impaired insulin signaling as prevalent in the obese and (pre)diabetic state *via* a novel type of feedback loop (Ussar et al. 2012). Some aspects about GPI-PL and other GPI-AP-releasing enzymes as well as about the physiological relevance of the lipolytic release of GPI-AP were reviewed recently (Fujihara and Ikawa 2016).

## 2. Release of GPI-AP with Complete GPI Anchor

In contrast to the above releasing strategies of removal of the total or hydrophobic portion of the GPI anchornchor in total or its hydrophobic portion, at least, by proteolytic or lipolytic cleavage, GPI-AP may be released from the cell surface with their GPI anchors remaining complete and still harboring the diacylglycerol or phosphatidic acid moiety (Figure 4). In order to be compatible with the release, the (hydrophobic portion, at least, of the) complete GPI anchor attached has to be shielded from access of the aqueous milieu of the extracellular environment by one of several strategies for its masking and accompanying stabilization in ordered configuration by (i) insertion into the outer leaflet of the phospholipid bilayer of extracellular membrane vesicles (EV), such as microvesicles and exosomes, (ii) insertion into the phospholipid monolayer of (lipoprotein-like, surfactant-like, milk fat globule-like) particles, (iii) binding to the hydrophobic cleft of carrier proteins or (iv) assembly together with phospholipids and cholesterol into micelle-like complexes (GLEC).

### 2.1. Release *via* Membrane Vesicles

According to my knowledge, the AMP-hydrolyzing enzyme ecto-5'-nucleotidase or CD73, represents the first GPI-AP for which the release from the cell surface of eukaryotic cells into small membrane vesicles, during a process termed by the authors exfoliation, was reported (Trams et al. 1981), albeit at those times membrane anchorage of proteins solely by GPI structures still remained to be discovered. In fact, measurement of 5'-nucleotidase activity in the particulate materials obtained by sedimentation (at 310.000x*g* for 90 min) of the medium from monolayer cultures of various normal and neoplastic cell lines after incubation (for 24 h) revealed that as much as 70% of the total plasma membrane-associated activity was recovered in form of two distinct phospholipid- and sphingomyelin-containing membrane vesicle populations with average diameter of 500 to 1000 nm and 40 nm (Trams et al. 1981), subsequently termed microvesicles and exosomes, respectively. It should be kept in mind, that extracellular phospholipid membrane vesicles harboring subsets of typical transmembrane proteins, which may correspond to microvesicles and exosomes, were detected earlier and, for instance, shown to accumulate in the culture medium of chick embryo intestinal cells (Black et al. 1980). They have been the subject of intense research efforts during the last three decades (Figure 4).

*2.1.1. Exosomes*

Around the year 1980 small membrane vesicles of 50 to 200 nm diameter and constituted by a phospholipid/cholesterol-containing bilayer membrane were found to be engaged in the removal of the transferrin receptor from the surface of sheep reticulocytes in course of their maturation into erythrocytes *in vitro* during the receptor-mediated endocytosis of transferrin through packaging of the transferrin receptor into vesicles destined for secretion rather than recycling to the plasma membranes (Harding et al. 1983, Pan and Johnstone 1983, Pan et al. 1985). On the basis of these findings the hypothesis was created that those vesicles, meanwhile termed exosomes (Johnstone et al. 1987 and 1991), may operate as waste containers to get rid of superfluous, inactivated, defective or dangerous proteins from the interior as well as surface of eukaryotic cells (Johnstone 2003). Almost two decades later, four GPI-AP, acetylcholinesterase (AChE), lymphocyte function-associated antigen 3 (LFA-3), decay accelerating factor (DAF, CD55) and membrane inhibitor of reactive lysis (MIRL), the latter two are involved in the inhibition of membrane attack complex (MAC) formation on autologous cell surfaces (Kinoshita et al. 1986, Rollins et al. 1991), were demonstrated to be liberated from reticulocytes in course of their maturation *in vitro* into erythrocytes as components of exosomes, too (Rabesandrata et al. 1998). The identification of AChE, DAF, MIRL and LFA-3 at the surface of exosomes, which were isolated from human reticulocytes and analyzed by their specific coupling to antibody-coated beads combined with flow cytometry immunodetection, is compatible with efficient trafficking of GPI-AP to exosomes in general and a specific physiological role of exosome-associated GPI-AP upon their release into blood, such as the regulation of the assembly and activity of the MAC in case of DAF and MIRL.

Accordingly, the function(s) of exosomes would exceed that of a mere waste container in the removal of inactive or unwanted cell surface proteins, but encompass novel and unique ones, such as the intercellular (paracrine or endocrine) transfer of materials and/or information (Record et al. 2011, Thery 2011)(see below). The same conclusion was subsequently drawn from the detection of a multitude of other GPI-AP, including cellular prion protein (PrPc), 5'-nucleotidase (CD73), the glycolipid-anchored cAMP-binding ectoprotein and cAMP-specific phosphodiesterase (Gce1), Ly6 (CD59) and AP, as components of exosomes released from many eukaryotic blood and tissue cell types, such as platelets (Heijnen et al. 1999, Reed 2004), T cells (Peters et al. 1991, Blanchard et al. 2002), enterocytes (Karlsson et al. 2001, van Niel et al. 2001), B lymphocytes (Raposo et al. 1996), mast and dendritic cells (Zitvogel et al. 1998, Laulagnier et al. 2004), neurons (Fevrier et al. 2004, Parkin et al. 2004), primary rat adipocytes (Müller et al. 2008a-e) and some tumor cells (Wolfers et al. 2001, Iero et al. 2008). Meanwhile proteomic analyses of exosomes derived from various sources, such as dendritic cells and urine, have been performed (Thery et al. 2001, Huber et al. 2002, Pisitkun et al. 2004, Hogan et al. 2009). The differential patterns of the protein composition demonstrated unequivocally that exosomes are clearly distinct from apoptotic vesicles. Moreover, human tumor-derived and dendritic cell-derived exosomes were found to display distinct molecular profiles, which might be related to distinct biological roles (Wieckowski and Whiteside 2006).

Importantly, in addition to GPI-AP, a multitude of (bi- and polytopic) transmembrane and peripheral membrane proteins, such as MHC Class I and II antigens and their co-stimulatory molecules (Raposo et al. 1996, Zitvogel et al. 1998, Coppinger et al. 2004), the tetraspanin family members CD9, CD63, CD81 and CD82 (Escola et al. 1998), proteins engaged in membrane transport and fusion, such as GTPases and flotillin (Mathivanan et al. 2012), and in multivesicular body formation (Gibbings et al. 2009), such as TSG101 and Alix (Thery et al. 1999, Clayton et al. 2001), as well as lipid-handling proteins, such as phospholipases (Conde-Vancells et al. 2008, Subra et al. 2010, Vlassov et al. 2012), have been identified as constituents of the phospholipid bilayer of exosomes. These data argue for a broad range of polypeptides with pronounced heterogeneity in structure and function which are released *via* exosomes, in either spontaneous and constitutive fashion or, possibly in many cases, in response to exogenous or endogenous signals, only (Mathivanan et al. 2010).

The latter hypothesis gained further credit by a number of observations, among them the following most convincing ones:

i. In macrophages exposed to tobacco smoke extract, the expression, translocation from the nucleus to the cell surface *via* the nuclear exporter CRM1 and final release into exosomes of the high mobility group (HMG) proteins were upregulated considerably (Chen et al. 2016).

ii. Exosomes with the abundant GPI-anchored MHC molecules CD55 and CD59 and MHC Class I-like molecule CD1a became secreted from monocyte-derived dendritic cells, which was blocked or further stimulated upon treatment of the antigen presenting-cells with calcium ionophore or phorbol ester and the phosphatidylinositol 3-kinase inhibitor wortmannin, respectively (Clayton et al. 2001). The presentation of active GPI-anchored complement regulators at the surface of exosomes was suggested to mediate resistance towards the productive engagement by and rapid disruption of exosomes through MAC as consequence of the intrinsic property of vesicular structures to trigger complement activation leading to their rapid disrupture (Wassef et al. 1991, Liu et al. 1995). Thus the interference with complement-mediated disrupture of exosomes by expression of certain GPI-AP and consequently their persistence and *in vivo* activity in the extracellular milieu hint to a role of exosomal GPI-AP in immune function, such as in the presentation of (glyco)lipidic antigens and in complement regulation.

iii. GPI-anchored PrPc is exposed at the surface of peripheral blood cells and acts as precursor for the pathogenic conformational variant scrapie prion protein (PrPSc), which is responsible for the propagation and transmission of spongiform encephalopathies (Caughey et al. 2009). PrPc was found to be released from cultured THP-1 monocytes in exosomes in close association with heat shock protein Hsp70, but not with typical exosomal marker proteins, such as Tsg101 and flotillin-1 (Wang et al. 2011). Furthermore, GPI-anchored PrPc was identified in exosomes released from cultural cortical neurons in physical neighborhood to the GPI-AP ceruloplasmin and the monotopic membrane protein flotillin, which all are well known to reside predominantly in cholesterol-rich lipid rafts (Faure et al. 2006). Interestingly, the efficacy of the release of PrPc and ceruloplasmin *via* exosomes was demonstrated to depend on depolarization of the neurons, which argues for exosome-mediated intercellular transfer of GPI-AP at neuronal synapses.

iv. The NKG2D-ligands, members of two functionally related families of MHC Class I-related proteins, MICA/B and ULBP (Bahram et al. 1994, Bauer et al. 1999, Lopez-Cobo et al. 2016), and expressed as both transmembrane proteins and GPI-AP (Fernandez-Messina et al. 2012, Ashiru et al. 2013), were shown to be released from the plasma membranes of donor cells into the extracellular environment, such as urine, in course of proteolytic cleavage by metalloproteases (see above) or incorporation into exosomes (Gonzales et al. 2009). The corresponding receptor NKG2D, expressed at the surface of human natural killer cells and T cells, is known to act in the protection against different types of stress, such as propagation of tumor cells, bacterial and viral infections or autoimmune diseases (Gonzales et al. 2008), which involves upregulation of the efficacy of the immune response. Interestingly, tumor cells of a broad variety of cancers, such as hematological malignancies, gliomas and melanomas (Baragano Raneros et al. 2014), manage to escape the immune system through the release of excess of soluble, i.e., not cell-associated NKG2D-ligand (Gonzales et al. 2008). These are generated from either their transmembrane or GPI-anchored version by proteolytic cleavage (mediated by $Zn^{2+}$-dependent matrix metalloproteases of the MMP and ADAM families) or incorporation into exosomes, respectively. Remarkably, inhibition of the families leads to accumulation of the transmembrane NKG2D-ligand in exosomes. However, the site of action of the metalloproteases is not restricted to the cell surface, but apparently also encompasses the exosomal surface as revealed by proteolytic shedding of the

NKG2D-ligand from isolated exosomes. A number of metalloproteases were reported to be associated with exosomes, such as ADAM17 (Tellier et al. 2006), possibly as a result of targeting to lipid rafts (see above), which are present in exosomes. Moreover, tumor cells seem to manage the release of exosomes carrying a variety of MMP (Ginestra et al. 1997, Muralidharan-Chari et al. 2009) which has been linked to enhanced migration of tumor cells across solid tissues.

At present, it is assumed that in course of prolonged interaction of the soluble anchorless NKG2D-ligand protein moiety or the exosome-associated GPI-anchored and transmembrane (upon inhibition of MMP and ADAM) NKG2D-ligand with NKG2D, this receptor will become down-regulated, internalized and desensitized for NKG2D-ligand induced signaling and immune modulation (Groh et al. 2002). As already mentioned, one mechanism of NKG2D-ligand release critically depends on GPI anchorage for targeting to (placental, ovarian or prostate) exosomes (Ashiru et al. 2013). Strikingly, exosomal and GPI-anchored NKG2D-ligand is more efficious than the soluble and proteolytically cleaved NKG2D-ligand in binding to NKG2D and interference with downstream signal transduction to the immune effectors. This difference is presumably related to the intrinsic capability of GPI-AP, in general, for homo-oligomerization within exosomal lipid rafts (Ashiru et al. 2010, Labani-Motlagh et al. 2015). Thus the release of exosomes with embedded GPI-anchored NKG2D-ligand from malignant cells and the resulting impairment of effective NKG2D-dependent immune responses may represent one of the multiple mechanisms fostering the distribution of tumor cells through tissue invasion and creation of the appropriate tumor microenvironment. In accordance, the appearance of specific GPI-anchored NKG2D-ligands alone or in combination with the proteolytically cleaved versions in sera and tumors from cancer patients has been found to be positively correlated with poor prognosis and low survival (in the absence of treatment) in clinical studies during the past decade (Spear et al. 2013, Baragano Raneros et al. 2014).

A large body of experimental evidence strongly argues that exosomes are formed in the cytoplasm within multivesicular bodies (MVB) in course of inward budding of their membrane into their luminal space, followed by incision and fusion of the invaginated MVB membrane (Stoorvogel et al. 2002, Thery et al. 2002). This leads inevitably to entrapment of soluble components of the cytoplasm as well as incorporation of proteins of the membrane of the MVB into the interior and (outer and inner) surface, respectively, of so-called intralumenal vesicles (ILV).

The membrane constituents of the future ILV may arrive at the MVB membrane *en route* from the ER *via* the Golgi apparatus along the secretory pathway or, alternatively, from the plasma membrane or other organelles *via* the endosomal system and lysosomes along the endocytotic/degradative pathway. These dual biogenetic sources of the membrane components of the ILV may explain why newly synthesized GPI-AP, which had never or already been expressed at the cell surface, were reported to be efficiently released from eukaryotic cells in exosomes (Vidal et al. 1997). Nevertheless, the majority of GPI-AP become exposed at the extracellular leaflet of plasma membrane bilayer during their lifetime. Thus the major portion of GPI-AP leaving the cells *via* exosomes most likely is derived from those located at the cell surface prior to arrival at ILV. They may become constituents of the MVB membrane upon endocytosis and subsequent fusion-induced transformation of the cytoplasmic endocytic vesicles into MVB with their protein moiety facing the MVB lumen and their GPI anchor inserted into the luminal leaflet of the MVB phospholipid bilayer. This topology of the GPI-AP is maintained during the following formation of the ILV with the GPI-AP residing at their outer membrane leaflet directed towards the MVB lumen, which according to (the "Palade dogma" of) vectorial transport corresponds to the extracellular space.

Cos proteins in cooperation have been attributed a role in the organization and segregation of ubiquitinated transmembrane proteins prior to their incorporation into the membrane of ILV. Strikingly, albeit GPI-AP do not undergo ubiquitination, they are critically dependent on Cos proteins (in yeast) and tetraspanins (in mammals) for trafficking to the ILV, too (MacDonald et al. 2015). Moreover, GPI-AP engage the typical **E**ndosomal **S**orting **C**omplex **R**equired for **T**ransport (ESCRT) machinery and ubiquitin ligases (i.e., Rsp5-Sna3 complexes in yeast and MARCH ligases in mammals) that (ubiquitinated) transmembrane proteins do. It is conceivable that Cos proteins and tetraspanins act as cargo adaptors and carriers which in course of their abundant ubiquitination physically force GPI-AP upon their incorporation into the MVB membrane into areas specialized for the budding of future ILV. For the last step in exosome biogenesis, the MVB are targeted to and then fuse with the plasma membranes thereby releasing their contents of ILV, now termed exosomes, into the extracellular space (Denzer et al. 2000).

Importantly, the releasing process for a given GPI-AP can be quite efficient as observed with reticulocytes that release about half of their GPI-anchored AChE in exosomes during the transformation of reticulocytes into erythrocytes. Vesicles with considerable similarity to exosomes have been identified in seminal plasma and named prostasomes (Rooney et al. 1993). They are thought to assist sperm function and express the GPI-AP CD59, CD55 and CD52 at their surface. Interestingly, these GPI-AP can also be recovered with small and apparently non-vesicular aggregates in seminal plasma (Rooney et al. 1996, see below).

A very recent study shed new light on the relationship between exosomes and GPI-AP with regard to the involvement of GPI-AP in exosome biogenesis, in general, and the attachment to and displacement from the donor plasma membranes of exosomes, i.e., in the control of short-range *vs.* long-range action of exosomes, in particular (Edgar et al. 2016). In greater detail, it was demonstrated that inactivation of the vacuolar ATPase in

HeLa cells causes considerable upregulation of the formation of exosomes according to the amount of internalized cholesterol and CD63. Interestingly, under these conditions the newly generated exosomes were found to remain attached at the plasma membranes in a clustered state. And strikingly, this association seems to be mediated by the GPI-AP tetherin, since the corresponding gene knock-out resulted in significant lowering of the amount of cell surface-attached exosomes and concomitantly in considerably elevated release of exosomes into the medium. Compatible with a role of tetherin in the retention of exosomes at the cell surface, the phenotype of tetherin knock-out mice was reverted by its overexpression, however only, when the GPI-anchored wild-type version had been used (Edgar et al. 2016).

*2.1.2. Microvesicles*

In addition to vesicular release as constituents of exosomes, GPI-AP have been identified as components of so-called microvesicles (size 100 nm to 1 μm) that are generated by budding, incision, fusion and shedding of plasma membrane areas of eukaryotic cells (Martinez et al. 2004) and thereby released into various body fluids including blood (Morel et al. 2004, Simak and Gelderman 2006, Piccin et al. 2007), saliva (Berckmans et al. 2011), synovial fluid (Berckmans et al. 2002), seminal fluid (Franz et al. 2013) and urine (Bulut et al. 2009). In general, the microvesicle membrane resembles the phospholipid composition of the plasma membrane bilayer of the parental cell (Hugel et al. 2005, Bauer et al. 2009, Camussi et al. 2010). Integrins, selectins and CD40 ligands are being regarded as typical markers for microvesicles (Perez-Hernandez and Cortes 2015). On basis of the observed aggregation and accumulation of GPI-AP at nanoclusters within plasma membrane lipid rafts, which seem to occur (even) in the monomeric non-liganded or oligomeric cross-linked state (see above), the site of microvesicle budding has been suggested to be identical with lipid rafts (Suzuki et al. 2000a). According to this view, lipid raft-associated GPI-AP will automatically be recruited into and end up in microvesicles. In one of the first reports demonstrating the residence of a GPI-AP in microvesicles, the vesiculation of the plasma membranes of cultured bovine aortic endothelial cells upon incubation *in vitro* and of human umbilical vein or bovine aorta upon perfusion *ex vivo* was induced by chemical means using low concentrations of formaldehyde and dithiothreitol (Moldovan et al. 1987). Thorough analysis by a multitude of procedures including phase-contrast, transmission and scanning electron microscopy, freeze-fracture preparations, extraction of cholesterol by filipin, staining by Ruthenium Red, binding to cationized ferritin hydrazide, lectin agglutination assays and labeling of the externally exposed polypeptide domains with $^{125}$I-Iodine, the released vesicles seemed to maintain the biophysical and biochemical characteristics typical for the plasma membranes of the respective donor cells as well as to keep on their surface the polypeptide ectodomains expressed at the donor plasma membranes. Importantly, the isolated sealed microvesicles displayed the enzyme activity of the GPI-AP ecto-5'-nucleotidase (CD73) when assayed in the native (i.e., absence of detergent) state (Moldovan et al. 1987). This finding is compatible with a topology of the GPI-AP at the microvesicular membranes (facing the interstitial spaces or body fluids) identical to that of GPI-AP at the plasma membranes (facing the extracellular environment) as well as with the isolation and purification of microvesicles on basis of routine protocols leading to the "right-side out" native configuration. Taken together, those initial studies revealed that (subsets of) microvesicles are equipped with GPI-AP at their surface and can be released from the plasma membranes of endothelial cells in response to certain chemicals, such as crosslinkers.

In addition to unphysiological and experimental exogenous cues, the release of microvesicles is stimulated in response to a variety of naturally occurring cellular processes such as cell activation, signaling molecules such as cytokines, hormones or endotoxins and environmental stimuli such as stress conditions and mechanical (shearing) forces. Those factors are often reflected in the actual (patho)physiological situation which affect parameters of the cardiovascular system (e.g., hypoxia, elevated blood flow, lipoprotein-associated oxidative stress, blood pressure), metabolism (e.g., hyperglyceridemia, hypercholesterolemia, hyperglycemia, hyperinsulinemia) and aging (e.g., inflammation, vascular dysfunction) of cells and tissues (Mallat et al. 2000, VanWjik et al. 2002, Gasser and Schifferli 2004 and 2005, Distler et al. 2005, Eilertsen and Osterud 2005, Ayers et al. 2015, Lawson et al. 2016, Nielsen et al. 2016), blood coagulation (Warkentin et al. 1994, Satta et al. 1997), tumor development (Kim et al. 2003, Janowska-Wieczorek et al. 2005, Muralidharan-Chari et al. 2010), vascular dysfunction (VanWijk et al. 2002), acute sepsis (Nieuwland et al. 2000, Joop et al. 2001), T-cell activation (Kolowos et al. 2005), multiple sclerosis (Minagar et al. 2001) and virus infection (Mack et al. 2000). Nevertheless, microvesicle release also happens in apparently spontaneous and uncontrolled fashion. For instance, microvesicles enriched in the GPI-AP, including CD55 and CD59 (Bütikofer et al. 1989), are released from the membranes of erythrocytes upon their prolonged storage as well as of melanoma cells leaving behind GPI-AP-less erythrocytes (Long et al. 1993) and fibroblasts (Brasoveanu et al. 1997). The observed loss of DAF/CD59 has been interpreted as a hallmark of erythrocyte senescence and linked to the clearance of aged erythrocytes by autologous complement-mediated cell lysis, which becomes facilitated by the absence of those GPI-anchored inhibitors of reactive lysis (Bütikofer et al. 1989).

It is also conceivable that the release of GPI-AP together with microvesicles is controlled by stimulus-induced alterations of the phospholipid or cholesterol content of plasma membrane lipid rafts, in particular of those expressing GPI-AP. In fact, early findings showed that exposure of the surface of intact GPI-AP expressing cells, such as ROS cells, to phospholipids, such as dilaurylphosphatidylcholine (Kobayashi et al. 1984), or to cholesterol-binding agents, such as streptolysin-O, saponin and digitonin (Bergman and Carlsson 1994, Xie and Low 1995) leads to concentration-, time- and $Ca^{2+}$-dependent appearance of GPI-AP, among them AP and AChE, in the culture medium. Importantly, GPI-AP equipped with the complete GPI anchor were recovered together with cholesterol and sphingomyelin with the particulate fraction upon centrifugation at high speed. These pellet materials were comprised of vesicular structures of 100-200 nm diameter and enriched for the GPI-AP up to five-fold compared to total plasma membranes with regard to identical amounts of membrane protein (Xie and Low 1995). Moreover, meanwhile there is sound experimental evidence available that streptolysin-O specifically interacts with plasma membranes through the formation of complexes with cholesterol and concomitant self-aggregation, thereby leaving intracellular membranes largely intact (Bhakdi et al. 1993). Thus, the release of complete GPI-AP in response to reduced cholesterol content of plasma membranes presumably relies on shedding of microvesicles from plasma membranes rather than on exocytosis of exosomes (Xie and Low 1995).

In general, the release of microvesicles rather than that of exosomes is accompanied by increases in cytosolic $Ca^{2+}$ concentrations, resulting in the activation of a multitude of enzymes such as calpains, gelsolins, protein kinases and scramblases (Comfurius et al. 1990, Cocucci et al. 2009, Inal et al. 2013). Moreover, phosphatidylserine becomes exposed at the outer leaflet of plasma membranes in elevated amounts together with the GPI-AP in cells upon their specific activation or challenge with stress (Fujii et al. 2015). The accompanying diminution or even abrogation of the phospholipid asymmetry between the outer and inner plasma membrane leaflets with phosphatidylserine typically being enriched at the inner leaflet in healthy cells necessitates the inactivation of flippases along with the activation of scramblases, responsible for the asymmetric and symmetric phospholipid distribution, respectively (Segawa et al. 2014). So far, five out of the 10 members of the protein family TMEM16, a family of $Ca^{2+}$-activated chloride channels possessing eight transmembrane domains, have been identified in mouse platelets which apparently contribute to the control of microvesicle release and thereby to tissue factor-induced thrombin generation and blood coagulation (Fujii et al. 2015). It is assumed that ion channels in concert with enzymes, which become activated or inhibited due to the elevated cytosolic $Ca^{2+}$ concentration, trigger (i) the reorganization of the cytoskeleton, (ii) the loss of membrane asymmetry in course of relocation of phosphatidylserine to outer leaflet plasma membrane sites, where microvesicular budding takes place and (iii) the blebbing of plasma membranes with the formation of outside-out membrane curvature. The coordinated sequence of these steps may represent the basic mechanism of microvesiculation and thereby the predominant mode of GPI-AP release from the cell surface *via* membrane vesicles (Lynch and Ludlam 2007, Muralidharan-Chari et al. 2010, Chistiakov et al. 2015).

In addition to reticulocytes, erythrocytes and endothelial cells, a multitude of other eukaryotic cell types including leucocytes, hepatocytes, human tumor cells, mesenchymal stem cells and platelets (Mesri and Altieri 1999, Bruno et al. 2009, Herrera et al. 2010, Fong et al. 2011) have been demonstrated to release vesicular structures or plasma membrane fragments during cultivation or *in vivo* in the basal and/or stimulated state. Those most likely correspond to microvesicles and, when studied, harbor GPI-AP in addition to a subset of transmembrane and soluble proteins as well as cholesterol and (glyco)sphingolipids with enrichment towards plasma membranes of the donor cell with regard to total protein and phospholipid content, respectively (De Broe et al. 1977, Allan et al. 1980, Allan and Thomas 1981, Carr et al. 1984, Kobayashi et al. 1984, Masella et al. 1989, Bütikofer et al. 1989). Complete GPI-AP of diverse physiological functions associated with microvesicles were also detected in various body fluids, such as blood and serum in healthy probands (Carr et al. 1984, Bütikofer et al. 1989) and patients. For instance, elevated amounts of microvesicles carrying complete GPI-AP were measured in the blood from cancer patients (Carr et al. 1984, Masella et al. 1989). When studied by adequate experimental methods, it turned out to be difficult to discriminate whether the loss of microvesicle-associated GPI-AP from the relevant donor cells or tissues is the cause for or a mere consequence of the initiation or upregulation of accompanying (patho)physiological processes, as is true for the apparent correlation between elevated levels of plasma microvesicles, immunomodulation during pregnancy and pre-eclampsia (Redman and Sargent 2007).

Importantly, the exosome- and microvesicle-based mechanism for the release of GPI-AP with complete GPI anchor have in common that in the majority of cases and in most cell types investigated so far they are operating in a regulated rather than constitutive fashion and become activated by various stimuli, which are sometimes shared by exosomes and microvesicles, such as receptor agonists and environmental cues, among them oxygen deprivation, oxygen radicals, certain drugs (e.g., glimepiride), mechanical stress (e.g., shearing forces), metabolic state (e.g., high glucose or insulin) and viral infection (e.g., HIV-1). During the past decade, a number of reports have documented the release of exosomes and/or microvesicles, meanwhile often referred to as extracellular vesicles (EV) to avoid the difficulties in mutual differentiation and characterization on basis of the currently used experimental (biophysical and biochemical) procedures (György et al. 2011, Yánez-Mó et al. 2015, Rupert et al. 2016, Xu et al. 2016), from primary rat and cultured mouse adipocytes (Aoki et al. 2007, Müller et al. 2009a, 2010b and 2012b, Ogawa et al. 2010). In mouse and rat adipocytes those vesicular structures, analyzed under the term EV, harbor specific subsets of luminal proteins, transmembrane proteins and GPI-AP, among them Gce1 and CD73, which were demonstrated to cooperate in the degradation of cyclic adenosine monophosphate (cAMP) *via* AMP to adenosine (Müller et al. 2008b and d), the mRNAs coding for glycerol-3-phosphate acyltransferase-3 (GPAT3)(Müller et al. 2012) and fat-specific protein-27 (FSP27)(Puri et al. 2007 and 2008, Li et al. 2010), that drive lipid synthesis and lipid droplet biogenesis, as well as the miRNAs, miR-16 and miR-22, that have been implicated in the coordination of lipid metabolic pathways (Müller et al. 2012b) and adipogenesis (Xie et al. 2009). The release was found to become considerably upregulated upon challenge with certain physiological stimuli, such as hormones, palmitate or reactive oxygen species as well as the anti-diabetic sulfonylurea drug glimepiride (Aoki et al. 2007 and 2010, Müller et al. 2008b, e and f). Moreover, the release of CD73 and Gce1 from the adipocyte surface into the EV turned out to critically depend on the cell size with large rat adipocytes being significantly more efficient than small ones (Müller et al. 2010a, b, c and 2011b). Interestingly, incubation of small rat adipocytes with EV from large ones resulted in upregulation of lipid synthesis which was considerably more pronounced compared to the reverse combination of large adipocytes exposed to EV from small ones (Müller 2010a and b, Müller et al. 2011c). Consequently, it was suggested that adipocyte-derived EV are involved in the paracrine regulation of lipid metabolism between large "full" and small "empty" adipocytes within the same adipose tissue depot in course of the transfer of (some of) their constituent components to shift the burden of lipid loading (Müller 2010c, d and 2013).

In fact, the transfer of transmembrane proteins from donor to relevant acceptor cells and tissues *via* EV leading to (patho)physiological consequences had been demonstrated prior to that of GPI-AP, such as for the oncogenic version of the bitopic receptor tyrosine kinase EGFRvIII in course of tumor metastasis (Al-Nedawi et al. 2008) and for other bi- and polytopic transmembrane proteins (Valadi et al. 2007, Skog et al. 2008, Shao et al. 2015). These findings are compatible with the meanwhile broadly anticipated role of EV in intercellular information transfer and endo- or paracrine communication (Pap et al. 2009, Simons and Raposo 2009, Ahmed and Xiang 2011, Tkach and Thery 2016). Finally, in addition to EV, GPI-AP with complete anchor are known to be released into membrane fragments in apparently non-specific fashion or more specifically into apoptotic bodies or vesicles (> 1 μm) in course of cell death (Aupeix et al. 1997).

## 2.2. Release *via* Particles

As already discussed, the requirement of physical shielding the hydrophobic long-chain saturated fatty acyl residues of the complete GPI anchor of GPI-AP against the aqueous milieu of the extracellular environment is fulfilled by EV. Alternatively, it is conceivable that the diacylglycerol moiety of the GPI anchor of certain GPI-AP becomes embedded in phospholipid (mono- or bi)layers which surround certain non-vesicular lipid-filled particulate assemblies, such as surfactant-like particles (Eliakim et al. 1989), milk fat globules, nodal vesicular particles (Tanaka et al. 2005) and lipoprotein-like particles (Panakova et al. 2005)(Figure 4).

| | Typical Secretory Proteins | GLEC |
|---|---|---|
| **MECHANISM** | Classical „Palade" Pathway<br>Removal of Membrane Anchor | Plasma membrane shedding<br>Multivesicular bodies, unknown |
| **CELL TYPE** | Specific > General | General > Specific |
| **RANGE** | Endocrine > Paracrine | Paracrine > Endocrine |
| **DESTINATION** | Circulation<br>Receptors at Target Cells<br>Endocytic Uptake by Target Cell | Cell Surface of Donor Cells, ECM<br>Circulation, Cell Surface and<br>Cytoplasm of Acceptor Cells |
| **FUNCTION** | Signaling>Transport>Enzymatic | Enzymatic>Remodeling>Signaling |
| **NATURE** | Monomers (Multimers) | Protein-Lipid Complexes |
| **INFORMATION** | Single Protein Entity | Multiple Proteins + Lipids + RNAs<br>Overall Composition<br>Organization, Assembly State<br>Biophysical Properties |
| **EXAMPLES** | Hormones, Cytokines,<br>Transport Proteins | GPI-AP, Caveolins, HSPs<br>Extracellular Nucleotidases |
| **ANALYTICS** | ELISA, Proteomics (LC-MS) | Chip-based Sensing |

Figure 4. Compilation of critical distinguishing features between typical secretory proteins and GPI-AP. The most important differences with regard to their composition, structure, analytics and (patho)physiological roles for their putative use in the prediction, diagnosis, prognosis and stratification of complex stress-related (metabolic) diseases are indicated by green rectangles.

### *2.2.1. Surfactant-Like Particles*

Decades before the identification of the GPI glycolipid as mode for the anchorage of ectoproteins at the cell surface of eukaryotic cells and the recognition of the principal possibility of release of GPI-AP from plasma membranes and the cell surface, the presence of the GPI-AP intestinal AP in human lymph and serum had been reported (Keiding 1966). Subsequently, AP was also identified in the intestinal lumen and the mucosal surface of rats with pronounced increases in its concentration and abundance, respectively, upon high fat feeding (Young et al. 1981) or cholecystokinin administration (Götze et al. 1972, Dyck et al. 1973). In both the fasting and the fat-fed state intestinal AP harboring the complete GPI anchor appeared to be associated with the apical microvillar membrane surface of the intestinal lumen in course of its assembly into lamellar bodies (Young et al. 1981). These are morphologically very similar to surface-active and phospholipid-containing so-called surfactant-like particles (SLP) rather than to typical membrane vesicles with phospholipid bilayer structure of uniform nature.

Initially these SLP, considered as products of type II pneumocytes, were identified as membrane-bound lamellar structures in enterocytes of adult and suckling rats and could be labeled with phospholipid-staining reagents. They appear to be whorled structures or partially coiled membrane fragments when viewed by electron microscopy, and to contain saturated PC as the major phospholipid and, in small amounts, the surfactant-specific protein SP-B, a protein cross-reacting with collagenous protein-4, in combination with pulmonary surfactant as well as other brush-border enzymes (Eliakim et al. 1989). These structural and compositional characteristics in concert mediate the surfactant-like appearance and properties, i.e., the ability to lower surface tension. However, the lipid and protein composition of the SLP derived from the lumen and mucosal surface is distinct from that of brush-border membranes and fragments thereof, as is their morphological appearance (Eliakim et al. 1991).

In general, SLP are generated by budding, vesiculation and shedding of plasma membrane protrusions, which may carry a subset of the GPI-AP expressed by the donor cells (Eliakim et al. 1989). In particular, the absorption of dietary lipids is known to trigger the transient and selective endocytosis of AP from the brush border membrane into a low-density detergent-resistant, i.e., lipid raft-like membrane fraction (Hansen et al. 2007), as well as the secretion of SLP containing AP from enterocytes (DeSchryver-Kecskemeti et al. 1989). Consequently, it was proposed that dietary free fatty acids cause the release of GPI-AP from enterocyte lipid rafts into the SLP (Hansen et al. 2011). The underlying molecular mechanisms seem to be based on the negative membrane curvature necessary for the formation of endocytic vesicles, which is energetically more favorable in case of the insertion of free fatty acids with small polar carboxylic end group, derived from the ingested fatty acid-rich food. In consequence, physiological lipid absorption seems to rely on the requirement of endocytosis of GPI-AP, such as AP, with accompanying packaging of the SLP together with all the other GPI-AP and subsequent release of the fully assembled SLP.

Interestingly, SLP could be shown to be released at sites of the basolateral surface of intestinal epithelial cells into the lamina propria, where they become attached at the luminal surface and thereby deposited beneath the basal membrane of the enterocytes (Alpers et al. 1989). Subsequently, the basolaterally secreted SLP with incorporated and highly enriched AP move from the basolateral to the apical surface *via* passage across the intercellular tight junctions and then either enter the intestinal lumen upon detachment from the luminal surface or remain attached (DeSchryver-Kecskemeti et al. 1989, Eliakim et al. 1989). Previous data are consistent with the hypothesis that the time spent by secreted SLP in the lamina propria can be prolonged by strong interactions with proteins resident at the basolateral membranes, and in the intestinal lumen by weaker interactions with apical membrane components, such as heparin (Mahmood et al. 2002). These interactions may enable SLP to exert their functions in each relevant tissue compartment in precisely time-controlled fashion.

Importantly, exposure of microvillar plasma membrane vesicles, prepared from brush border membranes, to a lipidic preparation, which closely resembles the mixed micelles formed by dietary free fatty acids in the intestine, led to drastic reorganization of the lipid rafts within the intestinal plasma membranes (Hansen et al. 2011). This rearrangement was accompanied by considerable enrichment of AP at the lipid rafts, which could be explained by higher lateral mobility within plasma membranes as well as higher affinity of the GPI-AP for lipid rafts compared to those displayed by brush border transmembrane proteins. Taken together, these findings extend the possible range of the physiological roles of GPI-AP contained in small SLP from gas exchange in pulmonary epithelia to transepithelial transport of dietary fatty acids and triacylglycerol across and out of the enterocyte which is mediated by the release of intestinal AP and other GPI-anchored hydrolases in extrapulmonary epithelia (Mahmood et al. 1994). Interestingly, intestinal SLP were detected in rat and human serum (Eliakim et al. 1996), which was unexpected considering their release from the apical enterocyte surface into the intestinal lumen. Consequently, it has been suggested, that the presence and thus concentration of SLP in serum may reflect mucosal integrity, in general, and the intactness, i.e., sealed state of tight junctions between the enterocytes, from which serum SLP are derived, in particular.

### *2.2.2. Milk Fat Globules*

Milk fat globules (MFG) are constituents of mammalian milk (at about 4% by volume in humans)(Garza et al. 1993). In principle, they can be regarded as lipid droplets (LD) of 1-10 μm diameter which are surrounded by a phospholipid bilayer membrane. The structure of the LD of MFG closely resembles that of typical cytoplasmic LD of mammalian adipocytes and other triglyceride-storing cells, such as myocytes and hepatocytes (Cermelli et al. 2006, Martin and Parton 2006, Dugail and Hajduch 2007, Ducharme and Bickel 2008, Fujimoto et al. 2008, Le Lay et al. 2008, Le Lay and Dugail 2009, Ohsaki et al. 2009) with a monolayer of phospholipids and unesterified cholesterol surrounding a hydrophobic core of cholesterylester and triglycerides which is scaffolded by members of the family of apolipoproteins, inserted into the outer phospholipid monolayer (Patton and Huston 1986, Welsch et al. 1990). MFG are assembled and secreted into milk by mammary secretory cells lining the mammary gland (e.g., breast) duct epithelia. The biogenesis of MFG can be followed experimentally using breast cancer cell lines (MCF7, T47D), initially developed from mammary secretory tumor cells of breast cancer patients (Hakulinen and Meri 1994). The outermost phospholipid bilayer membrane is derived from plasma membranes of the secreting donor cells, presumably in course of shedding, which involves consecutive and coordinated blebbing, budding, incision and fusion processes, and apparently resembles the release of microvesicles from the plasma membranes. It is assumed that during shedding the MFG-specific, former cytoplasmic LD become enclosed into the lumen of the emerging MFG. Furthermore, it is tempting to speculate that the shedding of the MFG is triggered by the interaction of the LD phospholipid monolayer with the cytoplasmic leaflet of the plasma membrane bilayer, which could be mediated by corresponding proteinaceous components, ultimately residing in the released MFG, such as apolipoproteins.

In consequence, MFG display at their surface, i.e., outer leaflet of their phospholipid bilayer, a variety of transmembrane (glyco)proteins and GPI-AP typical for the cell surface of the donor cells of the mammary glands (Ceriani and Blank 1988). In accordance, GPI-AP, such as the inhibitor proteins of the MAC of complement CD59 (protectin), CD55 (DAF) and HRF (C8bp homologous restriction factor) were found associated with MFG with their polypeptide moieties being exposed at the surface and directed towards the milk fluid and their GPI anchors buried in the outer leaflet of the phospholipid bilayer (Sugita et al. 1988, Davis et al. 1989, Holguin et al. 1989, Okada et al. 1989, Hakulinen and Meri 1995). Due to the high content of carbohydrate structures which typically are coupled to the glycoproteins, glycolipids and GPI-AP expressed at MFG, the outermost layer of MFG can be regarded as a type of glycocalyx coat which structurally corresponds to the glycocalyx of the mammary secretory donor cells (Buchheim et al. 1988). The space between the inner leaflet of the phospholipid bilayer and the phospholipid monolayer of the MFG-specific LD seems to be filled with a proteinaceous gel-like layer of so far unknown composition and structure.

Interestingly, the surface of MFG secreted into human breast milk seems to display an unequal or heterogeneous distribution of CD59 as revealed by areas of intense fluorescence compared to darker regions during indirect immunofluorescence microscopy (Hakulinen and Meri 1995). In fact, the outer leaflet of the MFG phospholipid bilayer apparently suffers from intrinsic instability or fragility since loss of membrane constituents and glycocalyx materials including GPI-AP from the MFG into the alveolar lumen soon after their secretion from mammary glands has been reported (Patton and Huston 1986, Hakulinen and Meri 1995). Strikingly, this loss was found to be accompanied by the appearance of GPI-AP, such as CD59, equipped with the complete GPI anchor in aggregates of heterogeneous nature, which lack the other (not GPI-anchored) MFG protein constituents. It may be speculated that these aggregates represent CD59-expressing membrane vesicles released from the MFG into the environment, i.e., the alveolar lumen, in course of spontaneous blebbing, budding, incision and fusion of the MFG phospholipid bilayer. Alternatively, non-vesicular mechanisms, such as the assembly of CD59 into oligomers/multimers or the binding of CD59 to carrier proteins (see below) are conceivable as well.

With regard to the physiological role of the secretion of MFG exposing certain GPI-AP at their surface from mammary gland epithelial cells into milk it may be of relevance that their soluble hydrophilic counterparts lacking the GPI anchor were identified for a subset of those GPI-AP, such as CD59, in a variety of body fluids, such as tears, saliva and urine (Davies et al. 1989, Meri et al. 1991, Fletcher et al. 1993). In case of CD59, the naturally occurring soluble anchor-less version manages to specifically interact with terminal complement complexes, which holds true for the corresponding recombinant protein moiety of CD59, too (Sugita et al. 1994). However, this version exerts a rather moderate cytolysis inhibitory activity, only, in comparison to GPI-anchored CD59. In fact, only the latter potently interferes with the MAC due to efficient GPI anchor-mediated incorporation into plasma membranes and concomitant prevention of C5b-8-

triggered insertion of C9 into phospholipid bilayers as a result of a competitive mechanism (Meri et al. 1990, Rollins and Sims 1990, Lehto and Meri 1993). The demonstration that CD59 secreted into milk as GPI-AP with the complete GPI anchor is functionally active and of higher potency than its soluble (proteolytically or lipolytically cleaved) anchor-less counterpart argues for the physiological relevance of the release of GPI-AP with complete anchor *via* MFG (Lehto et al. 1995). Nevertheless, on basis of the current knowledge it cannot be excluded that the presence of GPI-AP in MFG of mammary milk simply represents the consequence of the biogenesis of MFG at the sites of lipid rafts, which by nature express GPI-AP (see above), but is not associated with a specific role of the GPI-anchored *vs.* the soluble versions in the milk. In other words, according to this possibility the expression of GPI-AP at MFG has to be regarded as mere result of their lipid raft residency and would not contribute to functions, exerted by GPI-AP with complete anchor, only.

### *2.2.3. Nodal Vesicular Particles*

Lipid-modified, in particular acylated proteins, among them the morphogens Wnt and Hedgehog, are contained in extracellular phospholipid-containing particles which apparently display similarity to SLP and MFG with regard to structure and biogenesis. Typically, the members of the Wnt and Hedgehog families operate as membrane-associated ligands for receptors of cell type-specific signal transduction pathways that control the patterning and proliferation processes during normal development and oncogenesis, respectively (Pasca di Magliano and Hebrok 2003, Reya and Clevers 2005). They are linked to cholesteryl and fatty acyl moieties at their amino- and/or carboxy-termini, which increases their affinity to the plasma membranes of the target cells with accompanying upregulation of their signaling capacity. Importantly, lipid-modified Wnt and Hedgehog proteins can be released from the producing cells and manage to function at distant target cells after their transport *via* the extracellular space, which apparently relies on distinct molecular mechanisms (Eaton 2006 and 2008). One of them relies on transport vehicles for Sonic Hedgehog, which prevent access of the aqueous milieu of the extracellular environment to its lipid modification, and have been termed nodal vesicular particles (NVP).

NVP represent extracellular parcels built up by several membrane layers, which initially were described for mice emerging at the surface of the ventral node in course of embryonic development (Tanaka et al. 2005). According to live imaging and electron microscopic studies, NVP are constituted by a multitude of lipophilic granules (as revealed by labeling with 3,3'-dioctadecylindocarbocyanine) that are surrounded by a phospholipid bilayer membrane. Furthermore, the images implied that this membrane is presumably derived from plasma membranes of the donor nodal cells in course of blebbing, budding, incision, fusion and shedding from plasma membrane apical microvilli. The nature of the lipophilic granules in the lumen of the NVP remains to be clarified, but on the basis of their ultrastructural appearance it is tempting to speculate about a close relationship to typical cytoplasmic LD or extracellular lipoprotein particles. In addition to the morphogens, NVP were shown to harbor fatty acid-binding proteins, such as the retinoid transport protein lipophorin in *Drosophila melanogaster* (Duncan et al. 1994, Trowell et al. 1994) and the cholesterol-binding protein prominin-1 in mice (Marzesco et al. 2005). Interestingly, the extracellular transport and distribution of lipid-modified and NVP-associated Sonic Hedgehog critically depend on the (leftward) flow of liquid in the interior of the ventral node driven by the nodal cilia.

The biogenesis of NVP is thought to rely on mechanisms similar to those operating for the synthesis, assembly and release of SLP and MFG as well as of membrane-enveloped mammalian viruses. First, lipophilic granules resembling cytoplasmic LD with regard to structure and origin are formed at the cytoplasmic face of the ER membrane in course of separation of their luminal and cytoplasmic leaflets of the phospholipid bilayer at specific sites of synthesis of triacylglycerol and cholesterylester. The deposition of those lipids at these sites leads to filling up of the interleaflet space in the ER membrane with an emerging neutral lipid core. Subsequently, the phospholipid monolayer covering the lipid core buds from the cytoplasmic leaflet of the ER membrane and following incision and fusion is shedded into the cytoplasm. Upon trafficking of the resulting LD-like lipophilic granules to the plasma membranes, possibly preferably to lipid rafts, interaction of the cytoplasmic bilayer leaflet of the latter with the phospholipid monolayer of the granules may induce blebbing, budding, incision and fusion of the plasma membranes, presumably under the control of protein-protein interactions between the two compartments. Finally, shedding of the fully assembled NVP from the donor cells, e.g., pneumocytes, epithelial cells from the intestine and mammary (breast) glands, endothelial nodal cells, causes their release into the extracellular space, i.e., the lung, intestinal lumen, milk and the neural tube lumen as well as nodal surface area, respectively.

Albeit to my knowledge the presence of GPI-AP in NVP has not been reported so far, the apparent similarities in structure and function between NVP and SLP as well as MFG as carriers for acylated (signaling) proteins predict a role of NVP in mediating the release and transport of certain GPI-AP in(to) extracellular compartments, as have already been demonstrated unambiguously for SLP and MFG. If this holds true for NVP, the incorporation of the GPI-AP should happen late at the level of the donor cell plasma membranes, as is the case for SLP and MFG, rather than early at the level of the ER membranes with the GPI-AP synthesized as "secretory" protein at the luminal face of the ER. This would result in budding of lipophilic granules from the cytoplasmic face of the ER which are devoid of GPI-AP at their phospholipid monolayer. At variance, the expression of the GPI-AP Gce1 and CD73 at the surface of LD in the cytoplasm of primary rat adipocytes has been amply documented upon their incubation with EV harboring these GPI-AP (Müller et al. 2008a, b and d, 2009b, 2010a). The experimental evidence available is compatible with translocation of Gce1 and CD73 from the EV *via* the adipocyte plasma membranes (in course of membrane fusion) to the LD rather than with loading of the GPI-AP onto the LD during (simultaneous) biogenesis of the two constituents at the ER (Müller et al. 2010c and d). The apparent translocation of GPI-AP from plasma membranes to cytoplasmic LD in adipocytes could represent a mechanism leading to the incorporation of GPI-AP onto the monolayer surface of lipophilic granules of SLP, MFG and NVP. It would be interesting to study whether the GPI-AP expressed at SLP, MFG and NVP are located either at their surface, i.e., at the outer leaflet of the phospholipid bilayer, or at their core, i.e., the phospholipid monolayer of the lipophilic granules, exclusively, or at the two sites simultaneously.

*2.2.4. Lipoprotein-Like Particles*

A variety of GPI-AP has been found associated with so-called lipoprotein-like particles (LLP), among them lipophorin, which is a constituent of LLP released from imaginal disc epithelial cells of *Drosophila melanogaster* together with lipid-modified Hedgehog (Panakova et al. 2005). The phospholipid monolayer of typical lipoproteins, which surrounds a core of neutral triacylglycerol and cholesterylester and is scaffolded by members of the apolipoprotein family (Olofsson et al. 1999), will represent an excellent environment for the embedding of the amphipatic GPI anchor of GPI-AP as well as of the fatty acyl and cholesteryl residues of lipid-modified proteins, such as Hedgehog and Wnt. This structural characteristics of LLP makes them to perfect candidates for stable expression and surface exposure of GPI-AP. In this regard it may be of relevance that cytoplasmic LD, which from a structural point of view can be regarded as the intracellular counterparts of extracellular lipoproteins, have been reported to express a subset of GPI-AP as well as the monotopic membrane protein caveolin-1 in primary rat adipocytes (Müller et al. 2010a). The intimate localization of Gce1 and CD73 in concert with caveolin at the immediate surface area of LD, which is presumably mediated by interactions between the GPI anchors, the single membrane leaflet-spanning domain of caveolin-1 and the LD phospholipid monolayer, was suggested to contribute to the cAMP-mediated coordinated regulation of the lipid synthesis and degradation at LD (Müller et al. 2011c).

Interestingly, the regulated association of caveolins with LD was also observed during the differentiation of 3T3-L1 adipocytes (Blouin et al. 2008) and found to be induced by cholesterol and fatty acids (Martin and Pol 2005, Pol et al. 2005, Le Lay et al. 2006). Furthermore, the insertion of the GPI-AP, bovine intestinal AP, into model monolayers consisting of PC at the air-water interface was demonstrated and reported to be affected by the degree of saturation of the phospholipid fatty acyl chains (Ronzon et al. 2002 and 2004, Kouzayha and Besson 2005). In detail, the interactions of the saturated acyl chains of the GPI anchor with palmitoyl chains of the PC monolayer were found to be much stronger than those with unsaturated oleoyl chains. In consequence, the dipalmitoyl- rather than the dioleoylphosphatidylcholine monolayer was almost completely covered by the protein moiety of AP (Kouzayha and Besson 2005).

Additional evidence for the operation of LLP as shuttle entities for the transport of lipid-modified proteins in hydrophilic environment was provided by the findings that several cell types known to release Wnt and Hedgehog proteins as well as GPI-AP are capable of producing their own lipoproteins (Gregorieff et al. 2005). Moreover, investigation of the presence of Hedgehog proteins as constituents of defined lipoprotein particles in human plasma by LTQ-Orbitrap XL analysis demonstrated the association of Indian Hedgehog (but not of Sonic and Desert Hedgehog) with very low density, but not with low and high density lipoprotein fractions (Queiroz et al. 2010).

On the basis of transcriptome and immunofluorescence analysis, primary white adipocytes were identified as the site of loading of Indian Hedgehog onto very low density lipoprotein particles, which are typically produced in the lumen of the ER of liver parenchyma cells (Jung et al. 2007, Omenetti and Diehl 2008). Importantly, Indian Hedgehog protein, which is expressed predominantly by white adipocytes rather than hepatocytes, was found to be located mainly at the surface of white adipocytes and inserted into the extracellular leaflet of their plasma membranes *via* its cholesteryl and fatty acyl moieties (Queiroz et al. 2010). This portion of Indian Hedgehog may be assembled into specialized LLP in course of incorporation into very low density lipoproteins, which subsequently have to be transported from hepatocytes as the predominant site of lipoprotein biogenesis to the adipocyte surface. Alternatively, it is conceivable that the very low density lipoproteins have to be taken up by the adipocytes and loaded with Indian Hedgehog at intracellular compartments, such as the ER and LD, relying on mechanisms similar to the reported translocation of GPI-AP from adipocyte plasma membranes to LD (see above).

Alternatively, LLP may be generated during the biogenesis of typical lipoproteins in course of separation of the luminal leaflet of the ER membranes with the inserted GPI-AP, cholesterol and apolipoproteins from the cytoplasmic leaflet. Following the filling-up of the interleaflet space with newly synthesized cholesterylester and triacylglycerol, the emerging LLP will finally bud into the ER lumen. Thus it seems likely that GPI-AP are incorporated into LLP as constituents of their phospholipid monolayer immediately after biosynthesis of the GPI-AP during their subsequent assembly into the LPP at the luminal leaflet of the ER membranes. Thereafter, the LLP will be directed into the classical secretory pathway *via* targeting to and inclusion into specialized transport vesicles which bud from the ER. In analogy to typical lipoproteins, vesicular ER-to-Golgi transport of LLP may require the action of liver-fatty acid binding protein CD36 and VAMP7 in addition to COPII proteins, such as Sar1B, which presumably form a scaffolding coat for vesiculation (Gillon et al. 2012). Upon fusion of Golgi-derived transport vesicles with the plasma membranes, LLP become released from the cells into the extracellular space, such as the bloodstream (Neumann et al. 2007), where the LLP-associated GPI-AP might fulfill their physiological roles. In addition to LLP, high density lipoproteins were demonstrated to act as transport vehicle for CD59 and to manage its transfer to erythrocytes *in vitro* (Väkevä et al. 1994). Taken together, LLP and other lipoproteins act as shuttles for lipid-modified proteins including GPI-AP in the bloodstream and concomitantly may facilitate their release from the donor cells.

## 2.3. Release *via* Carrier Proteins

Specific mechanisms supporting the release of GPI-AP equipped with the complete GPI anchor from the cell surface of eukaryotic donor cells seem to involve their oligo- or multimerization as homomeric structures in the absence of additional constituents or as heteromeric assemblies together with specific carrier proteins, which seem to replace phospholipids or cholesterol for masking of the fatty acyl chains of the anchor as is realized with EV and particles (Figure 5).

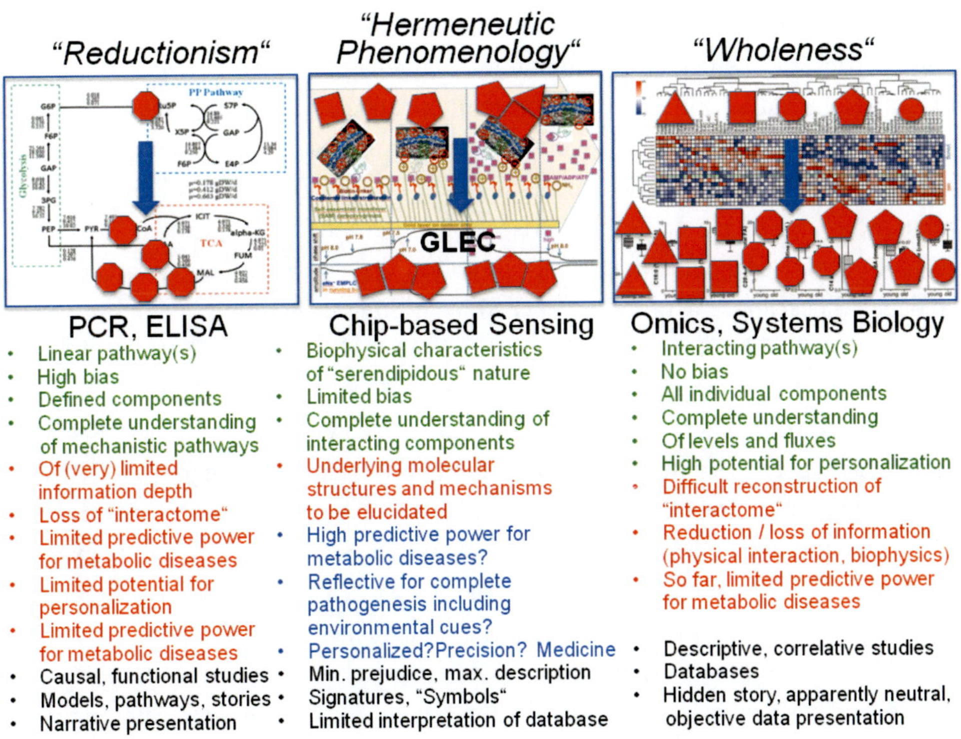


Figure 5. Compilation of critical distinguishing features between three principal modes of scientific thinking and acting. The advantages and disadvantages of "reductionism," "hermeneutic phenomenology" and "wholeness" are indicated in green and red, respectively. Putative beneficial options for the use of chip-based SAW sensing in the future are given in blue. The distinct modes of presentation of the experimental findings in typical scientific publications, which originate from critical differences in the understanding of the scientific discovery as well as documentation and writing processes (philosophy of science) are marked in black.

*2.3.1. Multimers*

Recent experimental evidence has raised the possibility, that NVP and LLP do not act as the sole transport vehicles for acylated morphogens and that other mechanisms for their release from donor cells and subsequent distribution *via* the extracellular space to the target cells and tissues operate in *Drosophila* (Zeng et al. 2001, Mann and Culi 2005). In fact, Hedgehog proteins have to be regarded as rather atypical intercellular messengers due to their modification with two very hydrophobic moieties, which upon exposure to the aqueous environment of the secreting cells would strongly limit their diffusion capacity. Nevertheless, Hedgehog has long been known to exert long-range signaling between a broad variety of tissues in specific fashion (Porter et al. 1996, Chen et al. 2002). To solve the challenge imposed by the release of amphipatic proteins with extended hydrophobic domains, such as Hedgehog, from the producing cells into the hydrophilic surroundings and by their subsequent controlled delivery to the various (Hedgehog) target tissues, the formation of homomeric (Hedgehog) oligo- and multimers has been proposed. Thereby the need for (complex) vesicles or particles in order to hide and protect the (Hedgehog) lipidic moieties could be circumvented. Accordingly, the hydrophobic membrane-anchoring lipidic moieties of several (Hedgehog) molecules seem to be packed together into a homomeric aggregate, whereby they become deeply buried in the aggregate core with the hydrophilic (Hedgehog) protein moieties remaining exposed at the aggregate surface. Overall, these aggregates display a detergent micelle-like configuration with the acyl chains taking over the structural and solubilizing role of detergent molecules (Feng et al. 2004, Briscoe and Therond 2005).

In fact, Sonic Hedgehog that is released into the aqueous incubation medium of cultured cell lines appears to be associated with soluble homomeric high-molecular-weight aggregates. Their formation strongly depends on the modification with cholesteryl (Gallet et al. 2006) and palmitoyl residues (Chen et al. 2004, Feng et al. 2004). Presumably, mutual interactions of the lipidic moieties drive the spontaneous assembly of several Hedgehog copies into micelle-like multimeric Hedgehog aggregates, which are compatible with the aqueous milieu of the extracellular space. This apparent self-assembly process might be facilitated by or even depend on the absence of exogenous phospholipids, which could be provided by the bilayer leaflets of the EV membranes or monolayer leaflets of the particles and enable the spontaneous insertion of the Hedgehog lipidic moieties. Importantly, biochemical and tissue-explant experiments revealed the involvement of both the amino- and carboxy-terminal lipids in the oligo-/multimerization processes, tissue distribution and long-range delivery of Sonic Hedgehog. These findings were confirmed by using mutant mice defective for the production of active Hedgehog palmitoylation enzyme Skinny Hedgehog and consequently expressing a non-palmitoylated version of Hedgehog (Chen et al. 2004).

However, RNA interference-mediated knockdown of the proteinaceous LLP constituent lipophorin in *Drosophila melanogaster*, which is known to operate as releasing as well as transport vehicle for the Wingless and Hedgehog morphogens as well as for certain GPI-AP (see above), led to altered distribution of and defective long-range signaling by Hedgehog (Panakova et al. 2005). These data suggest that multimeric morphogen aggregates cannot fully substitute for LLP in triggering the release, mediating the transport and guaranteeing the multiple functions of Wingless and Hedgehog and presumably of GPI-AP during development. It remains to be clarified whether proteins distinct from the lipid-modified morphogens are present in the high-molecular weight multimeric morphogen aggregates, i.e., whether these are actually of homomeric rather than heteromeric nature. The presence of additional proteins in the aggregates would fit to previous findings that the lipid-dependent release, assembly and distribution of morphogen aggregates are all necessary but not sufficient for the signaling capability of morphogens in relevant tissues (Feng et al. 2004).

Putative candidates for additional proteinaceous constituents expressed in morphogen aggregates represent GPI-AP. The long-chain saturated fatty acyl moieties of their GPI anchors could be packed together with the palmitoyl and cholesteryl moieties of the morphogens and be directed towards the hydrophobic interior of the aggregate, whereas the hydrophilic protein moieties of the GPI-AP could be exposed at the aggregate surface and mediate the contact with the aqueous interstitial fluids, blood stream or other body fluids. In fact, in seminal plasma the GPI-AP CD59, CD55 and CD52 were recovered with the pellet fraction upon high-speed centrifugation (Rooney et al. 1996). It remains to be elucidated whether these materials represent multimeric aggregates of these GPI-AP, exclusively, or, alternatively or in addition, vesicles or particles with the GPI-AP as one of their constituents. In this regard it may be of relevance that bile salts, present in bile and blood, foster the release of GPI-AP, such as AP, from the surface of exposed cells into high-molecular weight (> 500 kDa) multimeric aggregates in detergent-like fashion (Raymond et al. 1991). The structural relationship between the multimeric aggregates formed by either lipid-modified morphogens or GPI-AP or both in course of (patho)physiological response or detergent action remains to be elucidated.

### *2.3.2. Monomers*

As a further strategy in order to avoid access of the aqueous milieu to the hydrophobic portion of the GPI-AP upon their release from the plasma membranes of donor cells, the saturated long-chain fatty acyl moieties of their GPI anchor may be buried within a hydrophobic cleft or groove of certain (specialized) binding or carrier proteins. For instance, in response to elevated temperature the heat shock protein HSP70 was found to be translocated from the cytoplasm to the plasma membranes and released into the extracellular environment in a complex of unknown structure together with phospholipids in a non-vesicular configuration, which is compatible with the activation of macrophages (Vega et al. 2008). Furthermore, the stress-inducible HSP70 and the constitutively expressed cytoplasmic HSC70 heat shock proteins as well as the luminal

ER glucose-regulated protein GRP94 were reported to be actively released from various cell types *via* a non-classical pathway, that does not rely on vesicular structures, such as EV, but apparently depends on intact plasma membrane lipid rafts (Evdokimovskaya et al. 2010 and 2012)(Figure 5). Furthermore, heat shock protein Hsp60 was detected in saliva and serum of T2D patients (Yuan et al. 2011), however, the (vesicular, micelle-like or lipid-free) nature of its released version remains to be determined. To my knowledge, the association of GPI-AP in the monomeric state with typical fatty acid- or phospholipid-binding proteins, including serum albumin (van der Vusse 2009, Fujiwara and Amisaki 2013), or putative specific GPI-interacting polypeptides has not been reported so far, but at least remains a theoretical option for facilitating the release and distribution of GPI-AP with complete GPI anchor in case of non-availability of appropriate EV, particles or (other) GPI-AP (at a number sufficient for arrangement in multimeric aggregates).

## 2.4. Release *via* Micelle-Like Complexes

Three decades ago it has been shown that cell-free seminal plasma (SP) contains the GPI-AP CD59, CD55 and CD52 equipped with the complete GPI anchor as well as the transmembrane protein CD46 (Ronquist and Brody 1985, Hale et al. 1990, Hara et al. 1992 and 1993, Rooney et al. 1993). Initially, the major portion of CD59 in SP was found associated with a special type of EV called prostasomes (Rooney et al. 1993), which are thought to be derived from prostatic epithelial cells (Ronquist and Brody 1985). Subsequently, a minor portion of the GPI-AP was detected in the 200,000x*g*-supernatant of SP. However, a soluble and monomeric nature of the GPI-AP was excluded of basis of their equilibration at high buoyant density upon sucrose gradient density centrifugation and elution at high molecular mass upon size exclusion fast performance liquid chromatography (Rooney et al. 1996). In contrast, CD46 was recovered with prostasomes, exclusively. On basis of the failure of sedimentation during ultracentrifugation it was concluded that a fraction of GPI-AP with complete GPI anchor in SP is not embedded in vesicular structures, such as prostasomes. The buoyant density measured together with the observed partitioning into the detergent (TX-114) phase argued for the presence of (phospho)lipidic components in the GPI-AP-containing fraction. These apparently form a tight and possibly micelle-like complex together with the GPI-AP, which may guarantee stabilization and protection of the GPI anchor in the aqueous milieu (Rooney et al. 1996).

Two sources are conceivable for the GPI-AP released into the non-vesicular phospholipid-harboring (micelle-like) complexes, the EV membranes and the cellular plasma membranes. The release of the complexes from these two putative sources could rely on the same hypothetical mechanism encompassing multiple steps in sequence, (i) separation of the outer and inner phospholipid leaflets of the EV bilayer membrane, (ii) bulbing of the outer leaflet together with the inserted GPI-AP, (iii) incision and (iv) fusion of the phospholipid-GPI-AP monolayer into complexes with possibly micelle-like configuration, i.e., apolar core constituted by the fatty acyl chains and polar surface area constituted by the phospholipid head groups and the GPI-AP protein moieties. However, formation of those complexes from cellular plasma membranes might be more efficient than from EV membranes, given the susceptibility of the extended surface of whole cells towards (patho)physiological mechanical stress, such as shear and stretch forces, compared to the small size, compact shape and rigid configuration of EV.

More recently, it was found, that in lipid rafts prepared from atherosclerotic homozygous ApoE knockout mice under high-fat diet, the expression of HSP70, HSP90 and the atypical monotopic membrane protein caveolin-1, a resident and structural component of caveolae which binds cholesterol, was considerably upregulated in parallel with an increase in cholesterol (Fielding and Fielding 1996). Strikingly, these alterations in the lipid raft composition were correlated to the release of the heat shock proteins and caveolin-1 into serum (Figure 5). Strikingly, treatment of the mice with the cholesterol-lowering drugs of the statin class caused almost complete reversal of these increases in both expression and release of the heat shock proteins and caveolin-1 (Uyy et al. 2013). Together with previous reports, that caveolins and GPI-AP often biochemically cofractionate along with lipid rafts in a variety of cell types (Sargiacomo et al. 1993, Müller et al. 2001b, Razani and Lisanti 2001) and the assumption that they colocalize also *in vivo* despite being embedded on opposite leaflets of the plasma membrane phospholipid bilayer, it is tempting to speculate that GPI-AP and phospholipids in concert with caveolin-1 and cholesterol form non-vesicular micelle-like complexes. They may include a subset of soluble proteins, such as heat shock proteins, by mere "passive unspecific engulfment" of passenger components, albeit targeted specific enclosure cannot be excluded at present. Lipid rafts may be regarded as their site of origin in those animal and human tissues, such as adipose, liver and muscle, which are prone to a variety of stress conditions, such as high-fat diet induced hyperlipidemia or systemic inflammation in golden Syrian hamsters (Haraba et al. 2011), tumor load in mice and patients (Zhang et al. 2006) or hyperglycemia and hyperinsulinemia in rats (see below).

During the past five years this view has gained additional credit through the chip-based detection of extracellular complexes built up by GPI-AP, phospholipids and cholesterol, now termed GLEC, in serum of rats, which enable their differentiation according to metabolic phenotype and genotype (Müller et al. 2018). Initially, the rationale for studying the possibility of the release of GPI-AP with complete GPI anchor from eukaryotic cells into GLEC together with phospholipids in a non-vesicular arrangement was based on the following considerations and assumptions: (i) GPI-AP are particularly prone to spontaneous or regulated release from the cell surface due to sole anchorage at the extracellular leaflet of the plasma membrane phospholipid bilayer *via* their covalently attached GPI moiety. (ii) The release of the amphiphilic GPI-AP equipped with the complete GPI moiety (in contrast to that of the soluble protein moiety of GPI-AP as a result of lipolytic or proteolytic cleavage of the GPI anchor or protein moiety, respectively) necessitates its embedding into an amphipatic environment, which may be created by vesicles (EV), particles (SLP, NVP, LLP, MFG), carrier proteins (caveolin-1) or micelle-like structures. (iii) The rate of the release of GPI-AP from the surface of metabolically relevant cells into the circulation and/or their specific structural and biophysical characteristics is determined by certain environmental (chemical and mechanical) stress factors, such as elevated levels of plasma lipids, glucose, insulin, reactive oxygen species and high surface tension, shear and stretch forces, respectively). These factors are known to be heavily affected by the metabolic state of the donor cells and organisms, such as the (pre)diabetic state in humans.

The following experimental findings can be interpreted in favor of this rationale and for the operation of GLEC rather than vesicles, particles or carrier proteins in the release of complete GPI-AP in response to the metabolic state: (i) Certain GPI-AP with complete GPI anchor, such as CD73 and Gce1, have been reported to be released from cultured and primary rodent adipocytes in response to metabolically relevant stress factors, such as high levels of palmitate, $H_2O_2$, and anti-diabetic drugs (Aoki et al. 2007, Müller et al. 2008d and f, 2009a and 2012b, Ogawa et al. 2010). (ii) GPI-AP, but only those harboring the complete GPI anchor, are known to be susceptible for transfer from donor to acceptor cells *in vitro* and *in vivo*, thereby putatively transmitting functional materials and/or biological information within or between tissues (Rifkin and Landsberger 1990, Rooney et al. 1993, Kooyman et al. 1995, Anderson et al. 1996, Dunn et al. 1996, Ilangumaran et al. 1996, Miller 1998, Sloand et al. 1998). (iii) The level of the GPI-AP CD73 in plasma was shown to be correlated with insulin sensitivity in diabetic mice and human probands (Lunkes et al. 2004, Guzman-Flores et al. 2015, Oyarzún et al. 2015). (iv) Elevated plasma levels of GPI-AP displaying the complete GPI anchor have previously been measured for patients with ovarian cancer, glioblastoma brain tumors and lower grade colon adenocarcinomas, known to be exposed to metabolic stress as a consequence of their disease (Dolezal et al. 2014). (v) Plasma phospholipids analyzed by untargeted lipidomics were demonstrated to predict early neurodegeneration during presymptomatic Alzheimer's disease, that is presumably associated with metabolic disturbances closely linked to diabetes (Chatterjee et al. 2015, Wurtman 2015, Casanova et al. 2016) as well as to identify antecedent memory impairment in older adults (Mapstone et al. 2014). (vi) Phospholipids in complex with membrane proteins were reported to be released from vascular endothelial cells into culture medium upon cultivation and into plasma from mice following oxidative stress or high fat diet (Zhou et al. 2001, Mancek-Keber et al. 2015). (vii) Vesicle- and particle-like structures carrying GPI-AP were identified in plasma (Rooney et al. 1993 and 1996, Väkevä et al. 1994, Neumann et al. 2007). (viii) A GPI-specific lectin was used as a specific and reliable tool to detect GPI-AP in human plasma (Dolezal et al. 2014), which however fails to differentiate between GPI anchors having lost (due to lipolytic cleavage) or retained their diacylglycerol or phosphatidic portions. Together, these data convincingly demonstrate alterations in response to certain (patho)physiological states in the plasma levels of GPI-AP, which are constituents of GLEC rather than typical secretory proteins, such as hormones or transport proteins. The most important differences between GLEC and typical secretory proteins are summarized (Figure 6). Consequently, plasma GLEC-associated GPI-AP may be regarded as potentially useful biomarkers for the prediction, diagnosis or monitoring of certain human diseases (see below) as has already been demonstrated successfully for a number of serum proteins (lacking a GPI anchor).

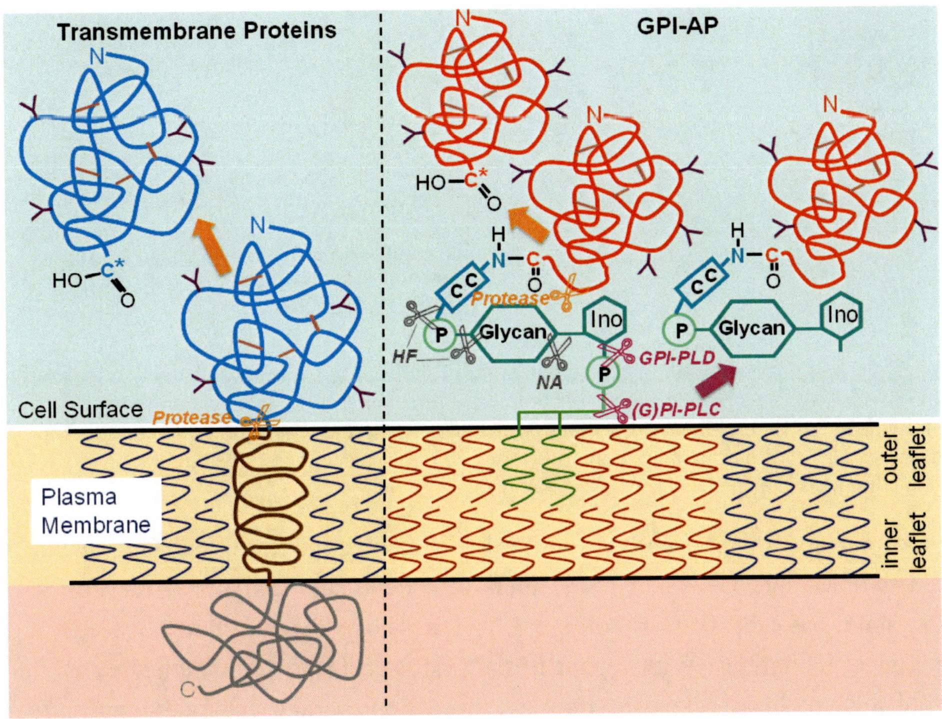


Figure 6. Release of membrane proteins from the cell surface by transmembrane or GPI anchor cleavage. The cleavage sites for proteases within transmembranous domains of transmembrane proteins and for GPI-specific PLC or PLD (GPI-PLC/D) with specificity for the GPI anchor of GPI-AP as well as for the chemical cleavage procedures of hydrogen fluoride dephosphorylation (HF) and nitrous acid deamination (NA) within GPI-AP are indicated. Upon protease action, the extracellular protein moiety of transmembrane proteins and GPI-AP without any residue left from the (glycan core of the) anchor will be released into the extracellular space. Dependent on the site of the proteinaceous cleavage, the protein moiety released will be truncated at its carboxy-terminus by a single (C) or a few amino acids (C*). Nevertheless, the loss of (single or a few) carboxy-terminal amino acids could affect the functionality of the truncated protein. Alternatively, GPI-PLC or GPI-PLD cleavage will cause the liberation of the complete protein moiety with the phosphoinositolglycan or inositolglycan moiety, respectively, of the GPI anchor remaining attached. Finally, chemical dephosphorylation or deamination of the GPI anchor will result in release of the complete protein moiety harboring the terminal ethanolamine residue only or in conjunction with the partially degraded glycan core of the GPI anchor, respectively. Ino, inositol; C-C, ethanolamine.

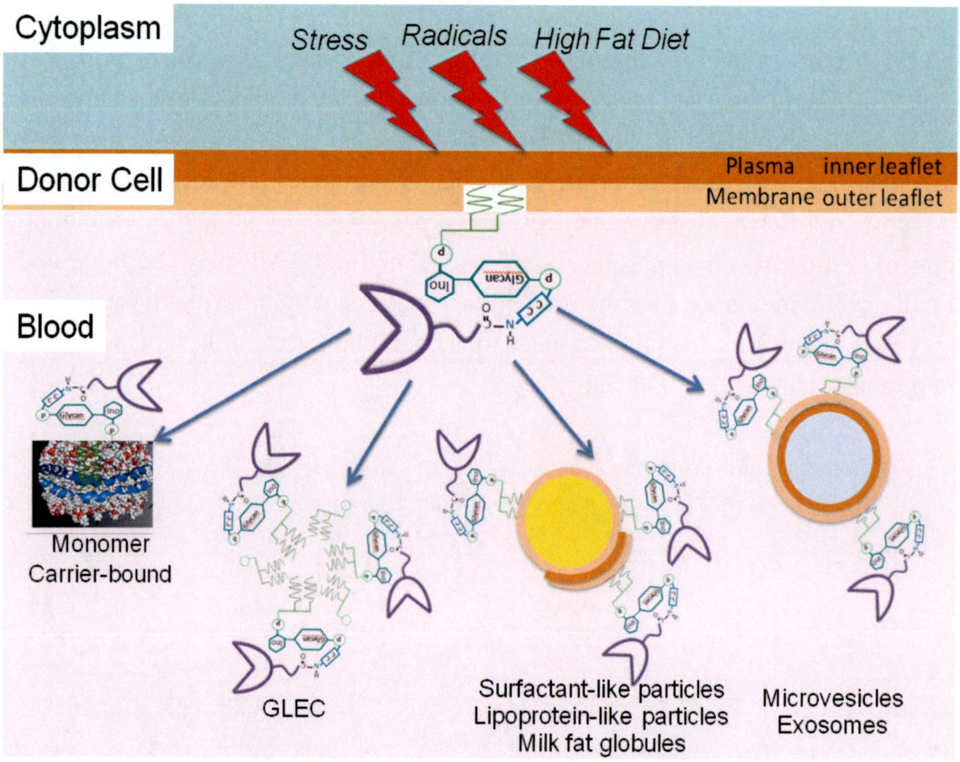


Figure 7. Release of GPI-AP with intact anchor from the cell surface. The four distinct entities which harbor GPI-AP with the complete anchor in complex with either specific (carrier) proteins or phospholipids in typical multimeric, micellar, particulate or vesicular configuration and are released from the outer leaflet of plasma membrane bilayer of donor tissue cells into the circulation in response to certain stimuli are depicted.

Unfortunately, the presence of GPI-AP displaying the complete GPI anchor and being assembled into GLEC in body fluids of rodents and humans in the healthy and disease state has not been studied so far, possibly due to both conceptual and technological limitations. With regard to the first aspect, the "reductionistic" as well as the "holistic" path of scientific thinking in extremes may mask the more fruitful hermeneutic phenomenological approach to the "natural things and phenomenons" (Figure 7) as stated by Edmund Husserl (1996), Martin Heidegger (1977 and 2009) and Ludwig Wittgenstein (1970, 1984a and b). With regard to the second aspect, the typical analysis of the serum secretome (by ELISA, Western blotting, 2D-PAGE, mass spectometry) necessitates complex fractionation and sample preparation procedures with most likely deleterious consequences for GLEC (e.g., loss as pelleted or floating materials in course of centrifugation or disintegration in course of sample solubilization) and may be of inadequate sensitivity and resolution power. To overcome these hurdles, a chip- and microfluidic channel-based sensor has been introduced recently for the specific detection and biophysical characterization of GLEC, that operates even in the presence of complex matrices, such as serum, and with lipid-containing samples. It relies on the interference of GLEC with the propagation of horizontal surface acoustic waves (SAW)(Gronewold 2007, Andrä et al. 2008, Saitakis and Gizeli 2012, Fogel et al. 2016), which are generated at the gold surface of chips equipped with microfluidic channels (Gronewold et al. 2005, Voiculescu and Nordin 2012, Ding et al. 2013). The sensor chip will detect specific capturing of the GLEC mediated by the GPI-binding molecules, aerolysin (Nelson et al. 1997, Diep et al. 1998) or α-toxin (Gordon et al. 1999, Shin et al. 2005), and monitor eventually associated phospholipids through binding of the phosphatidylserine-binding protein annexin-V (in the presence of $Ca^{2+}$)(Kuypers et al. 2007)(Figure 8).

Time-resolved right-ward phase shifts (i.e., declines in frequency) and amplitude reductions of the SAW reflect alterations in mass loading of the GLEC (reflecting upregulation in their number, size or spec. mass) to the chip surface and in their viscoelasticity, respectively (Figure 8), as validated by the author with EV from primary rat adipocytes, which *per definition* represent a subspecies of GLEC (Müller et al. 2018). Pairwise comparison of SAW signatures produced by the GLEC in pooled or individual serum samples in the presence of phosphoinositolglycans, which compete for binding of GLEC to α-toxin according to certain (minor) structural deviations (Frick et al. 1998), enabled the differentiation according to differences in body weight of (lean and diet-induced obese) rats of the same genotype as well as according to differences in genotype of rats of similar body weight (between insulin-sensitive lean Wistar and ZF or insulin-resistant obese Wistar and ZDF or insulin-resistant obese ZF and ZDF rats)(Müller et al. 2018). Loss of the differentiating SAW signatures in course of elimination of the GPI-AP, phospholipids or cholesterol as well as exposure toward physical stress (centrifugation, ultrasonication, vibration) is compatible with serum GLEC as a non-vesicular configuration of their constituents GPI-AP, phospholipids and cholesterol, held together by weak secondary interactions.

According to the following working hypothesis, GPI-AP in serum GLEC may be correlated to the metabolic state as defined by the genotype and nutritional state and can be regarded as phenomenological biomarker for the prediction of (metabolic) stress-related disorders, such as T2D: The rate of spontaneous as well as signal-induced release of GPI-AP from the extracellular phospholipid leaflet of plasma membranes into the GLEC could be dependent on the membrane rigidity/fluidity, particularly at membrane areas enriched in GPI-AP and cholesterol, such as lipid rafts (Sargiacomo et al. 1993, Müller et al. 2005). These plasma membrane areas of high local stability could facilitate GLEC release *via* shedding of "open" fragments of the phospholipid leaflet or shaping, bulbing, incision and fusion of "closed" phospholipid monolayer structures enriched in GPI-AP and cholesterol in micelle-like configuration. The impact of the cholesterol content and thereby stability of lipid rafts on their functionality has previously been underscored. Cholesterol deprivation from lipid rafts was correlated to impairment of insulin signal transduction mediated by the lipid raft-associated insulin receptor and suggested as cause of insulin resistance in rat adipocytes (Le Lay et al. 2001, Dagher et al. 2003, Dugail et al. 2003). Interestingly, it has been proposed that during enlargement of adipocytes in the presence of high concentrations of fatty acids and insulin, the forced biogenesis of cytoplasmic LD with accompanying incorporation of cholesterol into the surrounding phospholipid monolayer (see above) drives the translocation of cholesterol from the cytoplasmic leaflet of plasma membrane lipid rafts to the LD. The transient deprivation of cholesterol from the lipid rafts with their concomitant destabilization may impair the release of GLEC. According to this view, hyperinsulinemia and obesity, the key stimuli for LD biogenesis, would be negatively correlated to the release of GLEC, which critically depends on the presence (i.e., stabilization by cholesterol) of lipid rafts (Figure 9).

In addition to the efficacy of the release of GLEC, their serum level seems to be determined by the expression of the GPI-PLD in serum which represents a candidate enzyme for the remodeling or elimination of GLEC. It cleaves the phosphatidate moiety from the GPI-AP anchor leaving its inositolglycan portion at the polypeptide moiety (Figure 3). Importantly, *in vitro* GPI-PLD activity strictly depends on the presentation of the GPI anchor within detergent micelles, i.e., on the solubilization of plasma membranes with non-ionic detergents, and *in vivo* on its localization in lysosomes after re-uptake by endocytosis from the serum together with the released GPI-AP (Hari et al. 1996, Stadelmann et al. 1997). In contrast, GPI-PLD is inactive towards GPI-AP if embedded in the (plasma) membranes of intact cells or vesicles (Brodbeck 1998). Thus, only GPI-AP presented in micelle-like configuration, such as GLEC, may be acceptable as substrates for GPI-PLD. Since a portion of GPI-AP released into serum in response to metabolic stress is apparently associated with micelle-like structures, such as GLEC, rather than with EV, particles or carrier proteins, upregulated GPI-PLD may be responsible for the removal of excess of GPI-AP from serum (Figure 10). The accumulation of GPI-AP with complete GPI anchor in blood may exert deleterious effects, that could encompass impairment of the integrity and functionality of blood cells in course of spontaneous insertion of the amphipatic GPI anchors into their plasma membranes to up to (partial) solubilization of cellular membranes and cell lysis. Compatible with this "detoxifying" role of GPI-PLD, serum GLEC of rats represent substrates for serum GPI-PLD as demonstrated by the recent findings that pretreatment of adipocyte supernatants and rat sera with GPI-PLD caused considerable changes in the maximal phase shifts and amplitude reductions as detected by chip-based sensing (Müller et al. 2018).

It would be interesting to study whether the single nucleotide polymorphism identified in the GPI-PLD gene and found associated with T2D patients in a recent genome-wide-association-study (von Toerne et al. 2016) is responsible for an increase in the amount and/or activity of GPI-PLD. This would explain the decline in phase shift as a consequence of elevated degradation by GPI-PLD and thereby reduced number, spec.

mass or size of the GLEC in serum of hyperinsulinemic (lean and obese ZDF) rats compared to normoinsulinemic (Wistar and ZF) rats (Müller et al. 2018).

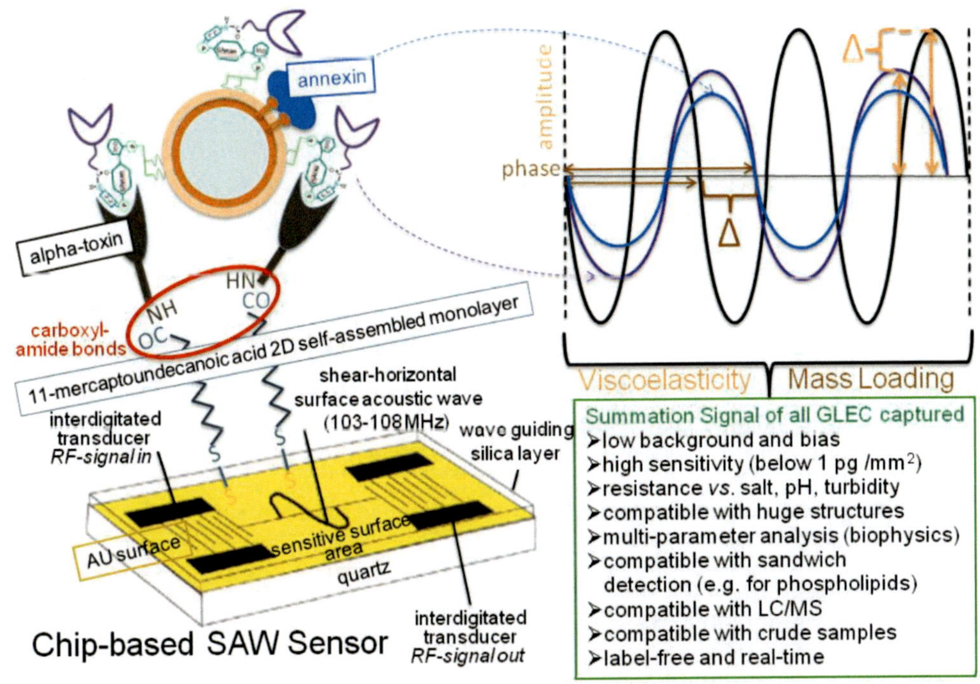


Figure 8. The principle of chip-based SAW sensing of serum GLEC. The specific capturing of the GLEC by the SAW chip-based sensor is accomplished by the GPI-binding protein α-toxin upon its covalent coupling to the chip surface. GPI-AP act as receptor for the bacterial α-toxin of *Clostridium septicum*. Ample evidence has accumulated during the past decade that the highly conserved glycan core of the GPI anchor is the major binding determinant for α-toxin. Nevertheless, it cannot be excluded at present that GPI anchors of different GPI-AP may differ in their relative affinity for α-toxin to a limited extent, which could be due to structural variation of the protein moiety of the GPI-AP. The covalent coupling of α-toxin to the gold surface of long-chain 2D self-assembled CM-dextran monolayer chips was performed using the conventional EDC/NHS-based protocol and monitored by measuring the phase shift in course of the reaction. The signals generated by the sensor and recorded in real-time reflect the loading of mass onto the chip surface and, in addition, depend on the (bio-)physical properties of the contacting sample fluid, including and predominantly its viscoelasticity. The phase shifts and amplitude reductions were corrected for unspecific effects caused by the sample buffer and non GPI-mediated (i.e., unspecific) interaction of sample components with the chip by subtraction of the "blank" channel (i.e., those lacking α-toxin) values measured upon sample injection. Any covalent (e.g., coupling of α-toxin) or secondary (e.g., capturing of GPI-AP) interaction of molecules or complexes with the chip surface will lead to right-ward shifts in phase and/or reductions in amplitude of the shear-horizontal SAW propagating along the chip surface. Those changes reflect alterations in mass loading and/or viscoelasticity, respectively, exerted by the interacting materials. Consequently, the coating with α-toxin *per se* and the capturing of complete GPI-AP and GLEC can be monitored by chip-based SAW sensing. The nature of the GLEC was shown by detection of phospholipids in course of sequential binding "in sandwich" of the $Ca^{2+}$-dependent phospholipid- (i.e., phosphatidylserine)-sequestering protein annexin-V (in the presence of $Ca^{2+}$). The sequential right-ward shifts in phase (mass loading of GLEC and annexin-V) and reductions in amplitude provoked by capturing of GLEC and annexin-V (blue curves) *vs.* control chips (black curves) are indicated by (hatched, brown and orange) arrows and (brown and orange) triangles, respectively. The advantages of the chip-based SAW sensing resulting in a summation signal for all GLEC captured are summarized in the green rectangle.

Extensive shaving of the GPI-AP coat from the surface of the GLEC would also provide an explanation for the seemingly contradictory decline in amplitude in parallel to the decrease in phase shift in course of development of the insulin-resistant and diabetic state. Certainly, a portion of the amplitude reduction can be attributed to capturing of the GLEC by the chip surface, since they are of viscous nature to some degree and thereby cause dampening of the maximal SAW amplitude in correlation to mass loading by the GLEC. However, the viscoelasticity of GLEC is presumably strongly affected by the GPI-AP surface coat with the less extended and intact, the lower/higher the GLEC elasticity/viscosity and, consequently, the higher the amplitude reduction provoked by the GLEC. Thus, hyperactivation of serum GPI-PLD in the insulin-resistant diabetic state should lead to considerable increases in viscosity and reductions in amplitude compared to GLEC with intact GPI-AP coat from normal serum. Thus the parallel declines in amplitude and phase shift when measuring serum from insulin-resistant or diabetic *vs.* normal rats can be reconciled best by a single mechanism, the upregulation of serum GPI-PLD activity leading to major losses of the GPI-AP coat (up to a single copy of GPI-AP, which has to ensure their capturing) and reductions in number, spec. mass and/or size of the GLEC, in parallel.

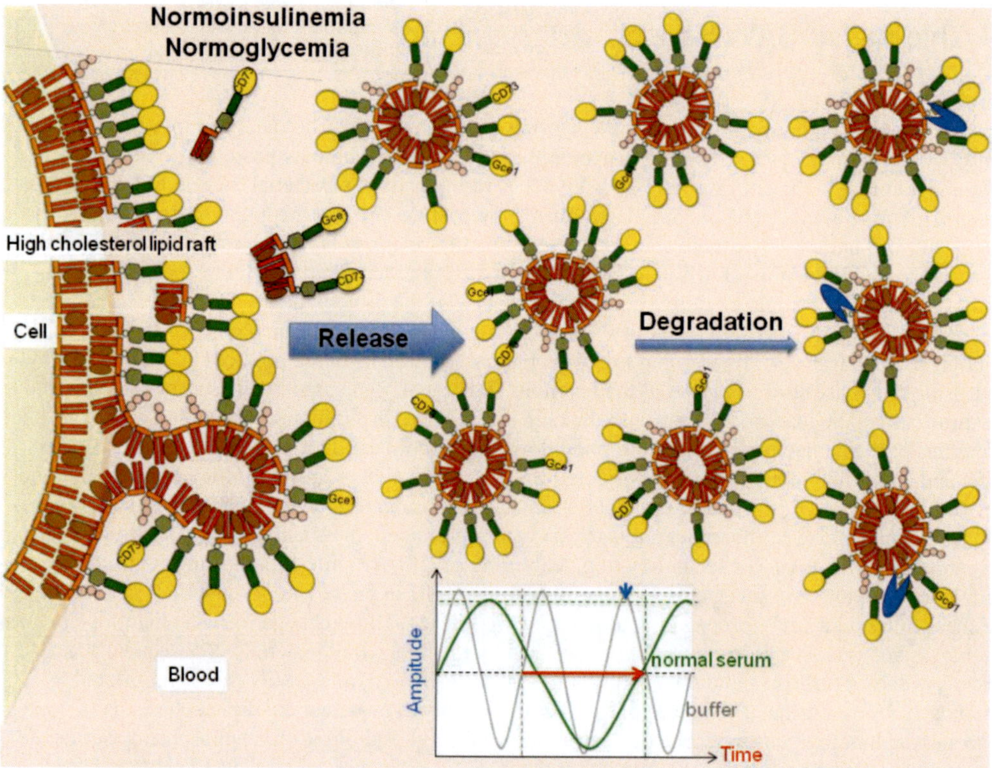


Figure 9. Model for the release and degradation of serum GLEC in normal rats with normal levels of glucose and insulin (normoglycemic and normoinsulinemic state) and their analysis by chip-based SAW sensing. See text for details and Figure 10 for explanation of the symbols.

According to this explanation, GLEC from healthy rats exhibit an intact and homogenous GPI-AP coat, whereas those from insulin-resistant or diabetic rats are of incomplete and heterogeneous nature in course of digestion by GPI-PLD to varying degree. No doubt, delineation of differences in the relative amount, size and spec. mass of GLEC between normal and hyperinsulinemic or hyperglycemic serum would be difficult using conventional methods, such as immunoblotting or "omics." In any case, their application would necessitate pre-fractionation, which seems to be critical due to the reported structural heterogeneity and instability of the GLEC (Müller et al. 2018). The putative interrelationship between spontaneous release into and enzymatic degradation in serum of the GLEC, the possible roles of lipid raft cholesterol and serum GPI-PLD activity, and the consequences of alterations in number, spec. mass or size and viscoelasticity of GLEC on phase shift and amplitude have recently been summarized in a working model (Müller et al. 2018).

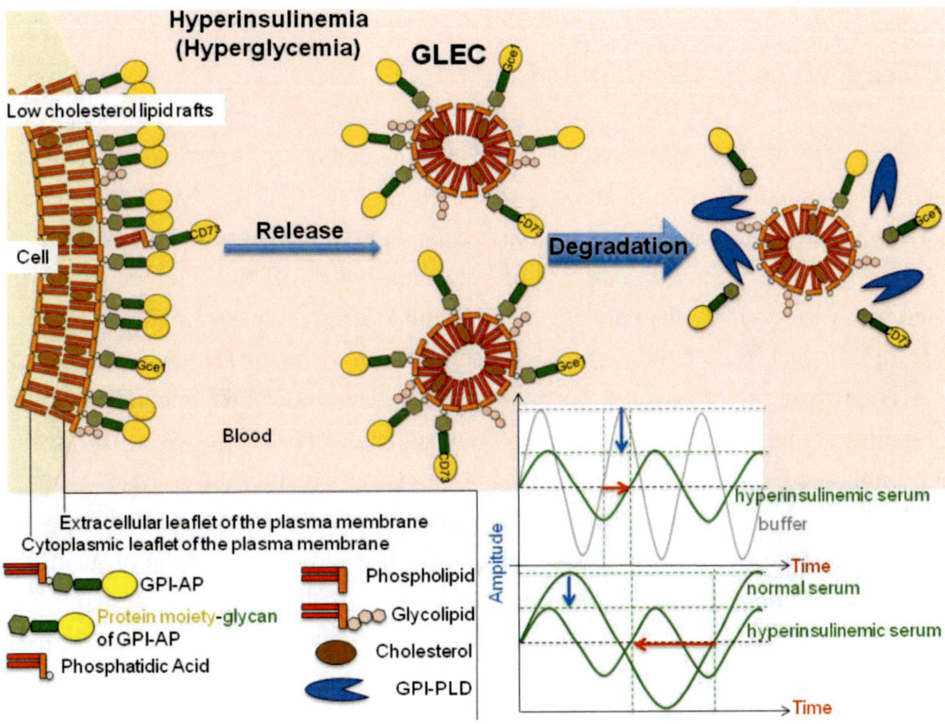


Figure 10. Model for the release and degradation of serum GLEC in insulin-resistant/diabetic rats with elevated levels of glucose and insulin (hyperinsulinemic and hyperglycemic state) and their analysis by chip-based SAW sensing. See text for details.

Accordingly, in the healthy (normoglycemic normoinsulinemic) state GPI-AP are continuously and constitutively released from plasma membranes of a wide variety of tissues into the blood stream due to the inherently loose association with the outer phospholipid leaflet in concert with a spontaneous mechanism or due to the controlled biogenesis of GLEC at cholesterol-stabilized GPI-AP-harboring lipid rafts (Figure 9). In the diabetic (hyperglycemic hyperinsulinemic) state both processes are counteracted by (metabolic) stress factors and flux of cholesterol from the plasma membranes to cytoplasmic LD in course of insulin-driven triacylglycerol synthesis (Figure 10). The released GPI-AP may act as a type of vacuum cleaner adsorbing phospholipids and cholesterol from the environment (blood) by chance for mediation of GPI-AP solubilization and stabilization despite the presence of the intact amphipatic anchor moiety. According to this view, GLEC do not represent unique homogenous complexes characterized by a specific morphology and composition, such as realized in organelles or protein-protein complexes. Rather GLEC may represent configurations/constellations of their constituent components, which are formed by heterogeneous "fields" of molecules. In this regard they can be interpreted as "symbols," "signatures," "local milieu" or "Gestalten" reflecting distinct metabolic states. Consequently, the "Gestalten" could reflect distinct (metabolic) genotypes (wild-type/ZF/ZDF) and phenotypes (lean/obese).

## 2.5. Release *via* Direct Membrane Contact

The transfer of GPI-AP between neighbouring cells, i.e., between cells in direct physical contact, which necessitates (i) the release of the GPI-AP with complete GPI anchor from the plasma membranes of donor cells and (ii) the subsequent insertion of the GPI-AP *via* their GPI anchor into the plasma membranes of acceptor cells, has been reported in a variety of *in vitro* and *in vivo* systems during the past three to four decades (Figure 11). In fact, the phenomenon of intercellular transfer of GPI-AP with complete GPI anchor was described even before the initial discovery of GPI anchors. During the investigation of the phospholipid exchange between cells and liposomes it was observed, that GPI-anchored AChE and a number of other (trans-) membrane proteins manage to leave the erythrocyte membranes and to appear at the liposomes, which apparently reflects their release from donor cells and subsequent insertion into acceptor membranes, respectively (Bouma et al. 1977). This apparent cell-to-membrane transfer of an erythrocyte GPI-AP turned out to be reversible. Later CD59 was recognized to be transferred from vesicular prostasomes (see above) of the SP to erythrocytes and other cells (Rooney et al. 1993) as well as from erythrocytes to endothelial cells in mice made transgenic for this GPI-AP (Kooyman et al. 1995 and 1998).

More recently it was found that EV released from donor rat adipocytes are able to transfer their constituent GPI-AP, Gce1 and CD73, to plasma membranes of acceptor adipocytes, which following challenge with palmitate, glimepiride or hydrogen peroxide become translocated to cytoplasmic LD resulting in inhibition of lipolysis and stimulation of esterification (Müller et al. 2010a and b). Small adipocytes were found to be more efficient than large ones in acting as acceptor cells (Müller et al., 2010c) as well as in being susceptible towards regulation of lipid metabolism by the EV (Müller et al., 2010a and b). In contrast, large adipocytes turned out to operate as donor cells with higher efficacy compared to small ones. Esterification stimulation and lipolysis inhibition by palmitate, glimepiride and hydrogen peroxide was demonstrated to be more prominent in mixed populations of small and large rat adipocytes compared to those consisting of either small or large adipocytes (Müller 2010c, Müller et al. 2010d) and to be completely abrogated by depletion of Gce1- and CD73-harboring EV from the incubation mixtures.

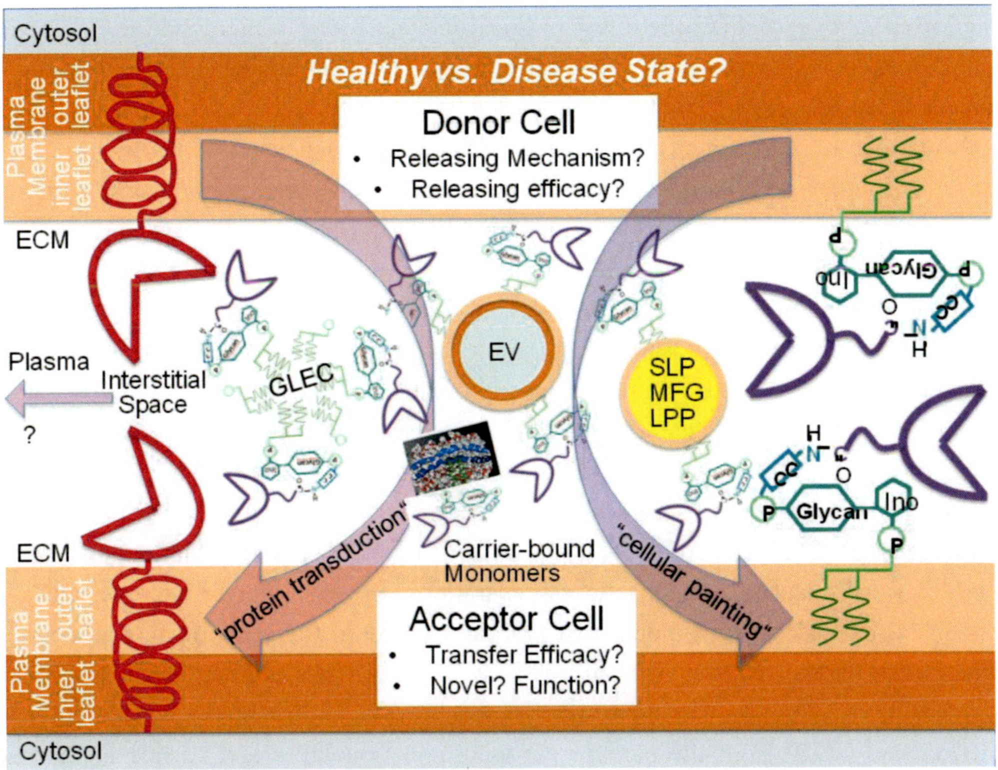


Figure 11. Model for the so-called "protein transduction," cellular painting" or "cell surface engineering." GPI-AP with complete anchor located at the outer leaflet of the plasma membrane bilayer of tissue donor cells are released into the ECM or interstitial space, from where they can be transferred to the surface of neighboring tissue acceptor cells, either directly or *via* the circulation (resulting in transient appearance in the plasma) following transport across vascular endothelial cells. Following insertion of the GPI-AP in the outer leaflet of the plasma membrane bilayer, which is mediated by its GPI anchor, the GPI-AP may acquire either its original or a novel function for the acceptor cells. The molecular mechanisms and structural intermediates involved in the release from the donor plasma membranes and passage across the aqueous milieu of the interstitial space/plasma (as carrier-bound multimers, SLP, LLP, MFG, GLEC or EV) as well as the insertion into the acceptor plasma membranes and the relevance of the transfer process for (patho)physiological processes, such as the development of stress-related diseases (e.g., T2D, obesity), remain to be determined. Furthermore, it cannot be excluded that the transfer of GPI-AP relies on the direct and intimate physical contact of the plasma membranes of donor and acceptor cells rather than on the generation of hydrophilic structural intermediates of the GPI-AP harboring the complete anchor. Finally, the transfer of transmembrane proteins from donor to acceptor cells has been reported (Eisenberg-Bord et al. 2016, Jain and Holthuis 2017).

Together these findings provide additional strong evidence for the release of GPI-AP equipped with the complete GPI anchor from membranes (of EV) and their subsequent insertion into the plasma membranes of intact acceptor cells (adipocytes). However, they leave open the underlying molecular mechanisms encompassing two fundamental options, (i) the involvement of extrinsic soluble mediators/vehicles for the transport of the GPI-AP between the donor and acceptor membranes/cells across the aqueous milieu of the extracellular space, such as multimeric or micelle-like complexes or carrier proteins (see above) and (ii) the direct physical contact between the donor and acceptor membranes/cells without engagement of extrinsic mediators/vehicles, which do not permanently reside at or leave the membranes/cells in course of the GPI-AP transfer.

A number of experiments have been performed which shed some light on the mechanisms underlying the transfer of GPI-AP from the phospholipid bilayer of vesicles or liposomes or from the phospholipid monolayer of (lipoprotein) particles or from detergent micelles to the surface of eukaryotic cells during the last three decades. In addition, considerable evidence for the (patho)physiological relevance of GPI-AP transfer were gathered in course of those studies: Paroxysmal nocturnal hemoglobinuria (PNH) is a chronic disease with severe hemolytic anemia, which is caused by a general defect in the biosynthesis of GPI anchors (Rosse 2001). Erythrocytes from patients with PNH are extremely sensitive to complement-mediated lysis, because they lack GPI-anchored CD55 and CD59, which in concert inhibit the formation of the MAC as MAC inhibitory proteins (Rooney et al. 1992). When erythrocytes from PNH patients, that are deficient in GPI-AP, were incubated with HDL preparations or erythrocyte microvesicles, a pronounced increase in the amount of CD55 and CD59 at the erythrocyte surface was monitored in parallel with significantly reduced susceptibility to complement-mediated lysis (Sloand et al. 1998). These findings argued for efficient transfer in functional state of CD55 and CD59 from the particle phospholipid monolayer and vesicle membrane outer leaflet, respectively, to the extracellular leaflet of the plasma membrane phospholipid bilayer. Pretreatment of the microvesicles and HDL with bacterial PI-PLC led to total abrogation of GPI-AP transfer to the GPI-AP-deficient cells (Sloand et al. 1998). This demonstrated that the apparent transfer of CD55 and CD59 critically depends on the expression and, most likely, on the tight insertion of the GPI fatty acyl chains into the plasma membrane phospholipids rather than on the unspecific adhesion of the GPI-AP protein moieties to other plasma membrane surface proteins or phospholipids.

Moreover, over decades ample experimental evidence has been accumulating that GPI-AP, in addition to transfer from vesicles, particles and mixed detergent micelles to cells, can undergo transfer from cells to cells without loosing their GPI anchor (Kooyman et al. 1995, Anderson et al. 1996, Ilangumaran et al. 1996, Liu et al. 2002)(Figure 11). Intercellular transfer of GPI-AP to cells deficient in GPI-AP was demonstrated in cell culture, animal models and even in patients. For instance, ectopically expressed CD59 was found to be transferred from erythrocytes to vascular endothelial cells (Kooyman et al. 1995, Rooney et al. 1996). Furthermore, cell-to-cell transfer of the GPI-AP Thy-1 was detected in chimeric murine embryoid bodies composed of normal and PNH cells (Dunn et al. 1996). In parasitemic patients, VSG was recognized to be transferred from trypanosomal membranes to erythrocytes (Rifkin and Landsberger 1990), finally leading to anemia *via* destruction of the erythrocytes by immune response towards VSG, which reinforces the maintenance or acquisition of functionality, such as antigenicity, of the transferred GPI-AP.

Thus, a considerable body of experimental data hints to the possibility of the release of GPI-AP from the surface of donor cells as a consequence of the direct physical contact with the surface of acceptor cells with simultaneous or subsequent insertion into the acceptor cell plasma membranes. This transfer mechanism does not rely on the involvement of (detectable) soluble (catalytic) mediators or facilitators, transport vehicles or GPI-AP variants, which freely diffuse in the intercellular or intertissue spaces or body fluids. Initially, the operation of a transfer mechanism of this type was hypothesized in the course of studies about the redistribution of plasma membrane proteins, including GPI-AP, upon incubation of mammalian cells with artificial phospholipid vesicles or liposomes (Dunnick et al. 1976, Bouma et al. 1977, Cook et al. 1980, Ott et al. 1981, Sunamoto et al. 1992, Nakamura et al. 1994, Okumura et al. 1994, Suzuki et al. 1995, Kogure et al. 1997). The seminal observation was the appearance of the GPI-AP AChE in the outer membrane leaflet of protein-free sealed dimyristoylphosphatidylcholine liposomes upon their incubation with human erythrocytes which is indicative for the possibility of efficient release from cellular membranes and concomitant insertion into liposomal membranes of GPI-AP (Bouma et al. 1977). Importantly, the erythrocyte transmembrane protein band 3 was recovered with the liposomes with considerably lower efficacy. These findings obtained with a precisely defined system consisting of donor (cellular) and acceptor (liposomal) membranes were supported by the subsequent demonstration of the asymmetric insertion of GPI-AP into liposomal membranes (Villar et al. 1999, Nosjean and Roux 2003) as well as the spontaneous partitioning of AP into model lipid rafts (Milhiet et al. 2002).

Interestingly, the transfer of GPI-AP from donor to acceptor membranes or cells was recognized to be correlated with the reverse translocation to and accumulation in the donor (i.e., erythrocyte) membranes of acceptor (i.e., liposomal) PC, in parallel (Ott et al. 1981). In fact, the resulting measured increase in the morphological index of the erythrocytes seems to be prerequisite for the initiation of AChE release from the erythrocytes. Unexpectedly, an increase in the membrane fluidity of the liposomes apparently does not play a major role (Suzuki et al. 1999) in contrast to previous conclusions (Cook et al. 1980). Moreover, the generation of a considerable membrane defect resulting in a non-bilayer phase of irregular structure as a consequence of the accumulation of exogenous phospholipids was proposed as the driving force for the release of GPI-AP into liposomal phospholipid layers in course of direct physical contact.

Such a mechanism may rely on weakening of the interaction between the phospholipids and GPI anchors in the erythrocyte membrane (Suzuki et al. 1999) as a consequence of the generation of high "curvature energy" driven by curvature-related packing stress (Yang et al. 1997).

Finally, the role of the interaction between spectrin and membrane proteins in the selective release of GPI-AP *vs.* transmembrane proteins in the presence of liposomes was addressed in studies using human erythrocyte ghosts (Suzuki and Okumura 2000b). Polymerization of spectrin and tryptic removal of the cytoplasmic portion of band 3 transmembrane protein were demonstrated to drastically reduce and increase, respectively, the release of band 3 from erythrocyte membranes. These findings hint to the critical impact of lateral diffusion of plasma membrane proteins, which is determined by their interaction with the cytoplasmic gel-like protein mesh and the plasma membrane-underlying cytoskeleton (Zhang et al. 1991) as well as their glycosylation state (Karnchanaphanurach et al. 2009), on the release of membrane proteins from the plasma membranes. High lateral diffusion coefficients have been measured for GPI-AP in general (Hartel et al. 2016), presumably caused by lack of contact with cytoskeletal elements as exclusive residents of the outer membrane leaflet, as for instance is true for erythrocytes most likely due to escape of the GPI-AP from the spectrin mesh in their crenated membranes. On the basis of these data and considerations, a "fence selection model" has been proposed for the selective release of GPI-AP *vs.* transmembrane proteins upon direct touch with acceptor membranes (Suzuki and Okumura 2000a and b).

*Chapter 5*

# IMPLICATIONS FROM THE STRUCTURE, FUNCTION AND MEMBRANE RELEASE OF GPI-AP FOR POTENTIAL CLINICAL AND PHARMACEUTICAL APPLICATIONS

The available experimental data unambiguously demonstrate the release of a multitude of GPI-AP from cells of a large panel of eukaryotic organisms which typically relies on a variety of complex molecular mechanisms and the formation of a multitude of distinct structural arrangements of the GPI-AP into EV, particles, GLEC or multimers. Less clear so far remains the functional or (patho)physiological relevance of the release of GPI-AP into the extracellular medium *in vitro*, and interstitial space or body fluids *in vivo* which may encompass (i) the removal of waste (inactivated or unwanted GPI-AP), (ii) the alteration of specific cell surface characteristics (brought about by the GPI-AP), (iii) the intercellular (paracrine or endocrine) transfer of materials or information (represented or encoded by the GPI-AP) from donor to acceptor cells and (iv) the biogenesis of extracellular structures and complexes with specific (local or systemic) function (exerted by the GPI-AP in concert with the other constituents of the structures). Moreover, at present it seems possible that certain GPI-AP become released to a certain (minor) degree as a consequence of spontaneous or induced (e.g., by exogenous environmental cues such as UV irradiation or endogenous cell-derived signals such as deformation of the cell shape) alterations in the biophysical properties of the plasma membranes, such as fluidity, surface tension, local curvature, stretching, shearing. It remains to be studied whether the release of GPI-AP from eukaryotic cells can be used for any medical or pharmaceutical application. In principal, the use of GPI-AP as disease biomarkers (Müller 2012b) or for novel biomaterials (Müller 2011a) is conceivable. In particular, vesicular (EV), particulate (SLP, LPP, NVP, MFG), multimeric and micelle-like (GLEC) assemblies constituted by GPI-AP equipped with the complete GPI anchor and specific components, such as phospholipids and cholesterol, may offer novel opportunities for the prediction, diagnosis, prognosis and stratification of common (e.g., metabolic) diseases (Müller 2012a), for oral delivery of protein drugs (Müller 2010a and b), such as insulins (Heinemann and Jacques 2009, Kapitza et al. 2010), and protein engineering or transfection of cell surfaces or target cells, respectively, by both recombinant and non-recombinant means (Müller 2011a).

## 1. GPI-AP AS BIOMARKER FOR COMMON DISEASES

In cases of known (patho)physiological function(s) of a GPI-AP it is tempting to speculate about the consequences of its putative release from the surface of the expressing cells into the corresponding body fluids caused by one or the other structures and mechanisms described above. Those considerations, in particular in case of relevance for (patho)physiological processes in humans, as is true for the following prominent examples for GPI-AP with known critical roles in the pathogenesis of human diseases, may justify a closer look at the presence and the levels of released versions of the GPI-AP in relevant body fluids under normal (healthy animals and probands) und pathological conditions (animal models of disease, patients). The observation of GPI-AP release and measurement of levels of soluble GPI-AP should motivate the evaluation of their potential use as biomarkers, in particular in comparison with traditional biomarkers for the delineation of putative advantages of GPI-AP biomarkers.

Traditional phenotypic biomarkers for common complex diseases, such as cardiovascular and metabolic disease, encompass blood glucose, glycated hemoglobin HbA1c, plasma LDL-cholesterol, blood pressure, body weight, BMI etc. and - in combination - reflect the complex interactions between the multitude of susceptibility genes (according to current estimations at least one third of all genes are involved in the regulation of metabolism) and the multitude of environmental influences which include the individual life stage and style and form a very dense and complex network. To account for the environmental impact, so-called phenotypic biomarkers have to be determined, however unfortunately at a rather late stage of the pathogenesis, since the earlier the time points of their measurement, the less predictive they are. As a consequence they fail to support disease prediction prior to disease onset or at early phases of the pathogenic development (Faerch et al. 2016). Moreover, the traditional phenotypic biomarkers typically do not allow stratification of common complex diseases into the multiple subtypes according to distinct underlying pathogenetic mechanisms and the resulting (late) consequences, such as diabetic late complications (nephropathy, retinopathy, neuropathy, microangiopathy, macroangiopathy) in T2D. In conclusion, phenotypic biomarkers often fail to support individualized preventive and therapeutic efforts (Müller 2012b).

As an alternative, so-called novel molecular biomarkers, predominantly soluble serum proteins, such as cytokines (e.g., TNF-α), adipokines (e.g., leptin), incretins (e.g., GLP-1) and others (e.g., cross-reactive protein) were determined in a number of current research clinical studies. Interestingly, in combination those novel peptidic biomarkers yielded prediction values approaching but not exceeding those achieved with combinations of traditional phenotypic biomarkers (Bahceci et al. 2007, Müller 2012b). Not surprisingly, the highest prediction probabilities have been reported so far for combinations of traditional phenotypic and novel molecular biomarkers. However, because of the partial overlap of the pathogenic pathways reflected by those biomarkers for a given disease, the predictive values upon their combination did not reach the sum of their individual contributions.

It may be argued that the future increase in the number of susceptibility genes identified for common complex diseases in course of genome-wide association studies will lead to considerable improvement of the predictive power of combinations of polymorphic genotypic biomarkers affecting multiple target tissues and pathogenic pathways. In fact, the required considerable improvement of the predictive power is frequently thought to rely on the inclusion of novel (single nucleotide polymorphic) genotypic biomarkers (Johansen et al. 2015, Müller et al. 2016, Wang et al. 2016). However, the path from genotype to phenotype with the underlying gene-gene and gene-environment interactions, genomic plasticity and epigenetic modifications is of extreme complexity and apparently does not obey simple cause-effect relationships and linear chains of events. Therefore, it remains questionable whether the complete genetic profiling of a given patient for (in theory all) disease susceptibility genes *per se* will provide information about the operation of a so-called complete causal mechanism (i.e., a chain of successive and interconnected physiological and molecular processes that will be necessary and sufficient to induce and drive the pathogenesis) for complex common diseases, such as cardiovascular diseases and T2D, and thereby enable their prediction with the desired probabilities (i.e., higher than 0.90). These considerations nourish the current hope that the limitations of phenotypic and genotypic profiling could be overcome by novel "complex" biomarkers, such as GPI-AP with complete anchor in specific complex arrangements, thereby combining the advantages of the traditional phenotypic, molecular (peptidic) and genotypic disease biomarkers.

## 1.1. Specific GPI-AP with Cleaved or Complete Anchor

By nature, released GPI-AP with lipolytically or proteolytically cleaved GPI anchor have to be regarded as candidates for molecular (peptidic) biomarkers with their soluble protein moieties corresponding to typical secreted serum polypeptides with regard to their putative (limited) information content. In contrast, GPI-AP released from the cell surface with the complete GPI anchor remaining attached may represent novel "complex" biomarkers of higher information content. This assumption is based on the demonstrated presence of additional biomolecules, such as phospholipids, cholesterol, other (membrane) proteins, in characteristic structural arrangements, such as EV, particles, (heteromeric) multimers and GLEC and the possibility that the composition and configuration as well as the accompanying biophysical characteristics of the complexes undergo changes of differing extent dependent on the (individual) pathogenic development. The resulting variations could reflect differences in the (patho)physiological (e.g., metabolic) state of the releasing cells or donor tissues of the proband/patient.

However, the co-expression of GPI-AP and other components in a complex does not necessarily imply an increase in biomarker "quality" compared to that exerted by the individual entities in "sum." A surplus may arise through the coordinated interaction of the distinct (biophysical) properties and (enzymic, signaling) functions of the individual entities assembled into the complex. If (some of) those hypotheses and assumptions would turn out to be true in the future on the basis of significant correlations in course of longitudinal and cross-sectional research and clinical studies, GPI-AP equipped with the complete GPI anchor and assembled into complexes will be useful for the prediction of the initiation, diagnosis of the onset and prognosis of the progression as well as endpoint of a common complex disease, including the desirable stratification between affected individuals. However, at present severe methodological constraints of the reported studies often do not allow the clear-cut differentiation as to whether the measured serum levels of GPI-AP reflect released protein moieties lacking the GPI anchor (and other entities) and thus can be regarded as peptidic biomarkers, only, or are related to complexes harboring GPI-AP with complete anchor and additional lipidic (and proteinaceous) constituents and thus have to be classified as novel "complex" biomarkers. In fact, a study dealing with the differentiating determination of monomeric (hydrophilic) GPI-AP *vs.* multimeric (amphipatic) GPI-AP in complex in body fluids of probands and patients has not been presented so far.

#### *1.1.1. CD59*

A typical example for the wide distribution in distinct body fluids of a GPI-AP and its occurrence as amphipatic and/or hydrophilic version displaying and lacking its GPI anchor, respectively, is provided by human CD59 (protectin), a complement regulatory glycoprotein of 18-20 kDa, which is expressed at the surface of red blood cells, leucocytes (Stefanova et al. 1989, Sawada et al. 1990), platelets (Sims et al. 1989, Morgan et al. 1993) as well as epithelial and endothelial cells of various origin (Hamilton et al. 1990, Nose et al. 1990, Brooimans et al. 1992). It interferes with the complement-mediated cell lysis by insertion into the forming MAC C5b-8, thereby competing with C9-binding to phospholipid membrane bilayers and its polymerization (Davis et al. 1989, Rollins and Sims 1990). In addition, CD59 is detectable at varying concentrations in human body fluids, such as plasma (Meri et al. 1991, Björge et al. 1993), sweat and tears (Meri et al. 1991), colostrum and milk (Björge et al. 1996), cerebrospinal fluid (Roddy et al. 1994), amniotic fluid (Rooney et al. 1992) and SP (Rooney et al. 1993).

A soluble anchor-less and glycated version of CD59 generated by the action of an (unknown) PLC was measured in human serum and urine at concentrations of 100 ng/ml and 4 μg/ml, respectively, applying a newly introduced and thoroughly validated ELISA with optimized precision, accuracy, reproducibility and sensitivity, suitable for clinical use (Ghosh et al. 2013). The use of this assay for the determination of the glycated version of CD59, formed by non-enzymic glycation of lysine41 in course of blood glucose at high concentrations, in the serum of normal and diabetic probands demonstrated that the level of soluble glycated CD59 (i) is significantly elevated in diabetic patients compared to healthy controls, (ii) is independently associated with and positively correlated to the level of glycated hemoglobin HbA1c and (iii) enables the identification of patients with diabetes with high sensitivity, specificity and predictive power. On basis of a putative contribution of soluble glycated CD59 to the pathogenesis of diabetic late complications as causative biomarker (assuming functional impairment due to its glycation), the determination of the serum levels of glycated CD59 should be useful for the prediction and management of diabetes.

In addition to the recognition of a correlative relationship between serum CD59 and diabetes, the concentrations of unglycated CD59 as measured using Western blotting, dot blotting or ELISA turned out to be rather low in normal human serum or plasma under basal conditions (7.8 $\pm$ 6.3 ng/ml)(Ratnoff et al. 1993, Väkevä et al. 2000), but to be elevated in plasma of patients after myocardial infarction (27.3 ng/ml)(Väkevä et al. 2000), during development of melanomas (Durrant and Spendlove 2001, De Nardo et al. 2002) and in the urine of patients with idiopathic glomerulonephritis (Lehto et al. 1995). This may be related to the measured reduced levels of GPI-anchored complement inhibitors during early complement activation in human and experimental diabetic retinopathy (Zhang et al. 2002). Strikingly, these values increased by 5- to 9-fold upon addition of the detergent octylglucoside to the sample diluent (Landi et al. 2003). This may be explained with the detergent causing solubilization of vesicular membranes or phospholipid monolayers or dissociation of multimers or micelle-like complexes or unmasking of the antibody epitope of the GPI-AP in course of its liberation from a carrier protein (e.g., albumin), in other words with the disruption of structures which mediate between the anchor and the aqueous milieu. In fact, experimentally prepared radiolabeled GPI-anchored CD59, but not soluble anchor-less urinary CD59 was shown to be incorporated into HDL particles upon incubation with human serum as analyzed by high resolution gel filtration and anti-apoA-I affinity chromatography (Väkevä et al. 1994).

Interestingly, only a very minor portion of CD59 managed to insert into the LDL fraction. This unambiguous differential behavior of CD59 in the *in vitro* assembly of LLP with insertion into the corresponding outer phospholipid monolayer seems to be mimicked *in vivo* since immunoaffinity purification and immunoblotting analysis revealed that HDL, but not LPL prepared from normolipidemic human serum harbor CD59, albeit at low amounts (Väkevä et al. 1994).

Thus it is well conceivable that GPI-AP liberated from associated phospholipids or proteins (i.e., other GPI-AP, carriers) can be measured under these optimized conditions with greater sensitivity and accuracy compared to GPI-AP in the native state. In addition, this detergent effect argues for the presence of CD59 with complete GPI anchor in complexes of the above structures in human plasma and may even be used as indicator for complex formation of GPI-AP in body fluids, in general. In contrast, the ELISA performed in the absence of detergent may preferentially or even exclusively measure the proteolytically or lipolytically cleaved soluble hydrophilic versions of GPI-AP. This view is compatible with the finding that in SP the portion of total CD59 measured with the detergent-free ELISA (about 10%) was recovered almost quantitatively from the aqueous phase upon partitioning between a water/Triton X-114 solution (Rooney et al. 1993 and 1996) and thus obviously reflects the soluble anchor-less portion of CD59. In other words, the detergent-free ELISA failed to measure GPI-anchored and complex-associated CD59.

### *1.1.2. Alkaline Phosphatase (AP)*

Instead of using antibody-based assays, such as ELISA, for the detection of GPI-AP in body fluids, such as blood and urine, the measurement of catalytic activity in case of enzyme nature of the GPI-AP protein moiety may provide evidence for the release of the GPI-AP, irrespective of whether harboring or lacking the GPI anchor, from cells or tissues into the circulation. For instance, feeding of adult rats with high-fat (corn oil) diet for seven hours resulted in a 2- to 3-fold increase in membrane-free intestinal AP activity in serum (Eliakim et al. 1991). This was qualitatively mimicked by subcutaneous injection of bethanecol or cholecystokinin with peak values reached after 7.5 min and 60 min, respectively. Importantly, the serum intestinal AP did display the characteristics typical for a soluble rather than membrane-associated entity (Eliakim et al. 1991). Taken together, the data obtained with rats imply that bethanecol and cholecystokinin trigger elevations of AP activity in intestinal mucosal surface scrapings of the same order of magnitude as associated with particulate materials which appears in the blood immediately thereafter (Eliakim et al. 1991). These data are compatible with the release of intestinal AP from the enterocytes into the intestinal lumen and in parallel (or shortly thereafter) into the circulation by means of particles.

In fact, the times course for the marked elevation of intestinal AP activity in serum turned out to be related to that of particle-bound AP activity as had been reported several decades ago for proximal intestinal light mucosal scrapings and distal luminal washings in several species, such as rats after high-fat feeding (Keiding 1966, Inglis et al. 1967, Warnock 1968, Kleerekoper et al. 1970, Young et al. 1981). Moreover, at those times vesicles were isolated from the medium of mouse and human explant cultures or from the lumen of adult intestine (De Broe et al. 1977, Berteloot et al. 1981), which resembled brush border membranes from a biochemical point of view (i.e., identical hydrolytic activities) and according to a buoyant density (1.19) similar to plasma membranes (De Broe et al. 1977). Based on these findings, shedding of vesicles from plasma membranes was suggested to be responsible for the release of AP into the intestinal lumen. However, this shedding process does neither explain the presence of AP in serum nor answer the question, whether it plays a regulatory role in normal intestinal physiology or is involved predominantly in the degeneration, renewal or restructuring of the intestinal microvilli. In fact, early data obtained with cultured chick embryonic intestinal cells indicated that plasma membrane vesiculation contributes to growth and development of microvillar structures (Black et al. 1980).

In an effort to characterize the pathway responsible for the trafficking of intestinal AP to the serum and to confirm or exclude the possibility of lumenal membrane vesiculation as critical step herein, the release of AP was investigated in an ileal-like cultured cell line derived from a human colonic carcinoma Caco-2. A tightly sealed monolayer of differentiated post-confluent cells of this continuous cell line, grown on a semi-permeable membrane, enabled the analysis of the surrounding medium fluids which reflect either the apical or the basolateral surface (Sussman et al. 1989). Interestingly, newly synthesized intestinal AP protein and activity were recovered from both the apical (25%) and basolateral (75%) serum-free medium. Moreover, a considerable portion of the released AP remained associated with membranes as manifested by its sedimentation (105.000x*g*, 1 h) and partitioning into the detergent-enriched phase upon TX-114 partitioning. This behavior allowed the exclusion of proteolytic or lipolytic release from the cell surface. Strikingly, the GPI-AP carcinoembryonic antigen (CEA) was recovered from the serum-free medium in parallel and in a hydrophilic and soluble version, exclusively (Sack et al. 1988). On the basis of these cell culture studies it was suggested that intestinal AP becomes released from the basolateral surface of enterocytes and subsequently moves through diffusion into the apical lumen *via* passage across the tight junctions and into the serum compartment *via* transport through the lamina propria (Alpers et al. 1989).

Consequently, the next experimental steps addressed the identification of (a) morphological correlate(s) for such a pathway using the *in vivo* high-fat fed rat as a model on basis of its high rate of secretion of intestinal AP. Luminal washings of the rat intestine yielded membranous lamellar particles with surfactant-like properties, as revealed by lowering of the surface tension in a pulsating bubble assay, similar to the SLP described above, from which the intestinal AP became liberated by treatment with bacterial PI-PLC (DeSchryver-Kecskemeti et al. 1989). Immunogold labeling of osmicated and non-osmicated sections demonstrated that intestinal AP is colocalized with these SLP in the cytoplasm of enterocytes (DeSchryver-Kecskemeti et al. 1991). Most importantly and in agreement with the proposed structure of SLP (see above), these particles were found to enclose LD-like structures and to accompany those along their trafficking throughout the cytoplasm of the enterocytes. Subsequent more detailed analysis of these SLP revealed that the constituting membranes are enriched with PC which is predominantly equipped with saturated fatty acids (Eliakim et al. 1990). After fat feeding the SLP and intestinal AP appeared in the intestinal lumen and blood with comparable kinetics, arguing for their colocalization (Young et al. 1981). The buoyant density of these SLP was found to be similar to that of typical rat pulmonary surfactant particles, but to be significantly lower than that of purified plasma membrane vesicles. These SLP expressed the typical digestive enzymes of apical brush border membranes, however, in very different ratios as well as total amounts. Importantly, intestinal AP was reported to be considerably enriched in those SLP *vs.* the apical enterocyte membranes.

Interestingly with regard to the putative stratification of AP-related pathogenic pathways and diseases on the basis of compositional or structural differences between the SLP, those SLP associated with AP in serum can be differentiated into subclasses. This was accomplished by sucrose gradient density centrifugation with buoyant densities ranging from 1.18 to 1.20, similar to those of the enterocyte basolateral membranes (Bond and Butler 1987), which reflects differences in protein composition. These gradient-purified SLP harbor plasma membrane enzymes, only, with considerable enrichment of AP compared to the typical marker protein leucine aminopeptidase. The observations that the AP-containing SLP are absent from the bile and that initially the majority of newly synthesized apical proteins are trafficked to the sinusoidal (basal) membrane domain argue for this domain operating as the site of shedding of the SLP into serum. Importantly, in about 50% of human probands AP activity in serum can be recovered with SLP, which considerably contributes to the total serum activity. On basis of the relative enrichment of the sinusoidal membrane fragments when assaying for AP activity, a small portion of the SLP-associated enzyme seems to be released, only.

These results resemble data for the GPI-AP 5'-nucleotidase (CD73) obtained with various normal and cancer cell lines. Membrane vesicles recovered from the culture medium were found to be enriched with 5'-nucleotidase activity *vs.* other plasma membrane enzymic activities (Trams et al. 1981), arguing for the selective release of GPI-AP in course of membrane vesiculation at specific plasma membrane sites or subdomains, candidates of which represent lipid rafts (see above). Interestingly, GPI-AP embedded into either membrane vesicles or SLP after their release can be detected in normal and cholestatic serum as well as in culture medium, supplemented with serum, at high and constant levels despite the presence of large amounts of GPI-PLD in serum (see above). Apparently, serum GPI-PLD does not play a major role in releasing GPI-AP from relevant native membrane vesicles and SLP *in vivo*. This is in line with the observation that cleavage of the GPI anchor by GPI-PLD *in vitro* depends on their presentation in detergent-like milieu or mixed detergent micelles (see above).

### *1.1.3. GPI-Anchored High Density Lipoprotein Binding Protein 1 (GPIHBP1)*

GPIHBP1 is a GPI-AP that is expressed in capillary endothelial cells as the predominant binding site for lipoprotein lipase (LPL) and engaged in the transcellular transport of LPL after its synthesis in and secretion from adipose cells from the surrounding adipocyte and subendothelial interstitial spaces across the capillary endothelial cells into the capillary lumen where LPL remains attached to the surface of the capillary endothelial cells (Adeyo et al. 2012). LPL is avidly bound to heparan sulfate proteoglycans on the surface of capillary endothelial cells (Berryman and Bensadoun 1995, Lookene et al. 1997), from which it can be dissociated into the capillary lumen by the highly sulfated proteoglycan heparin (Korn 1955). Here LPL hydrolyzes triglycerides assembled in triglyceride-rich plasma lipoproteins (chylomicrons and very low density lipoproteins). However, since LPL, in contrast to endothelial lipase, which is expressed and secreted by vascular endothelial cells (Yasuda et al. 2010), is synthesized by parenchymal cells, such as myocytes and adipocytes, it has remained a mystery for decades how LPL gains access to the capillary lumen following its synthesis and secretion into the subendothelial interstitial spaces.

As the name already suggests, GPIHBP1 is capable of specific binding high-density lipoproteins (HDL), which represents the function used for its initial discovery by expression cloning (Ioka et al. 2003). GPIHBP1 harbors a GPI anchor for association with the cell surface, from which it can be released by action of bacterial PI-PLC. The subsequent development of GPIHBP1 knockout mice and GPIHBP1 overexpressing cell lines strongly suggested that GPIHBP1 (i) is critical for the lipolytic processing of triglyceride-rich lipoproteins (Beigneux et al. 2007), (ii) binds LPL avidly (Beigneux et al. 2009), (iii) interacts indirectly with triglyceride-rich lipoproteins which seems to be mediated by LPL (Gin et al. 2011), (iv) is involved in the delivery of lipoproteins to peripheral tissues (Young et al. 2011) and (v) is required for the transendothelial transport of LPL (Beigneux et al. 2007, Weinstein et al. 2008). Strikingly, in the GPIHBP1 knockout mice LPL was found to be mislocalized to the interstitial spaces surrounding the LPL-producing parenchymal cells and to be absent from the capillary lumen (Davies et al. 2010). Direct evidence for a role of GPIHBP1 in the transport of LPL was derived from the measurement of transendothelial transport using confluent monolayers of cultured endothelial cells, which express GPIHBP1. Strikingly, a monoclonal antibody directed against GPIHBP1 was found to be transported from the basolateral to the apical medium compartments (Davies et al. 2010). Convincingly, the release of GPIHBP1 from the basolateral surface of the endothelial cells by GPI anchor cleavage with bacterial PI-PLC led to significant impairment of the transendothelial transport of the antibody. Since

GPIHBP1 heterozygous mice displaying about 50% GPIHBP1 expression level have normal plasma lipoprotein levels, pathological GPIHBP1 deficiency seems to be a recessive syndrome (Weinstein et al. 2012). Half-normal levels of GPIHBP1 apparently manage to mediate the transendothelial transport of LPL to a sufficient degree in both mice and humans (Franssen et al. 2010, Coca-Prieto et al. 2011, Rios et al. 2011).

Of great relevance for the (patho)physiological situation in humans was the observation that GPIHBP1 deficiency in homozygous knockout mice with complete mistargeting of LPL after its initial secretion from striated muscle and adipose tissue leading to residence in the interstitial spaces (Olivecrona et al. 1997, Wang and Eckel 2009) was associated with severe hypertriglyceridemia and chylomicronemia (Davies et al. 2010). Subsequently, this phenotype was also recognized in humans to be caused by mutations in either GPIHBP1 (Beigneux et al. 2009, Olivecrona et al. 2010, Charriere et al. 2011) or LPL (Voss et al. 2011) which both lead to impairment of their mutual interaction. Interestingly, homozygous GPIHBP1 knockout mice were found to suffer from lipid- and macrophage-rich atherosclerotic lesions in the aortic root and coronary arteries (Weinstein et al. 2010). The manifestation of this phenotype in humans with GPIHBP1 or LPL mutations remains to be demonstrated.

Taken these findings together, an updated model for the lipolysis of triglyceride-rich lipoproteins by LPL in concert with GPIHBP1 was presented (Young et al. 2011): GPIHBP1, GPI-anchored at the basolateral surface of capillary endothelial cells, "picks up" with high affinity soluble LPL from the subendothelial interstitial space following its detachment from GPI-anchored heparan sulfate proteoglycans at the myocyte and adipocyte surface. Subsequently, GPIHBP1 actively shuttles LPL from the basolateral to the apical surface of the vascular endothelial cells with exposure towards the capillary lumen. The molecular mechanism underlying this shuttle remains to be clarified, but may depend on the lateral movement of the GPIHBP1-LPL complex along the basolateral and apical plasma membranes *via* passage across the tight junctions between neighboring endothelial cells and between the apical and basolateral plasma membrane domains. Alternatively, the transendothelial transport of GPIHBP1 in concert with LPL from the basolateral domain following its initial expression across the cytoplasm of the endothelial cells for exposure at their apical surface. The underlying transcytotic process may involve caveolae, arising from small invaginations, subsequent budding into the cytoplasm, incision and fusion of the plasma membranes to yield vesicles, which are covered by the atypical monotopic membrane and coat protein caveolin-1 (Root et al. 2015). Caveolae have been implicated to mediate transcytosis of fluids, small molecules and proteins in endothelial and epithelial cells (Cheng and Nichols 2016). Following arrival at the apical surface of the endothelial cells, LPL remains bound to GPIHBP1 initially, but then becomes detached upon interaction with triglyceride-rich particles (Lookene et al. 1997) or in response to fatty acids generated by lipolysis of the triglyceride-rich particles (Adeyo et al. 2012).

In conclusion, the GPI-AP GPIHBP1 plays an important role in the complex pathway of intravascular processing of triglyceride-rich lipoproteins as is also reflected in the consequences of the demonstrated artificial and naturally occurring defects herein for the lipid metabolism and cardiovascular system during human health and disease. This raises the interesting question about the (patho)physiological effects of the putative release of GPIHBP1 from the basolateral or apical surface of the capillary endothelial cells into the subendothelial interstitial space and capillary lumen, respectively, involving one of the molecular mechanisms and structures described above. Albeit the appearance of GPIHBP1 harboring or lacking the complete GPI anchor in the circulation of mice or humans has not been reported so far, preliminary data indicate that this GPI-AP is a minor constituent of GLEC released into the plasma of diet-induced and genetically obese rats and obese and diabetic humans (Müller, G.A., Herling, A.W. and Lechner, A. unpublished data). Thus it may be attractive to study whether the pathogenesis of metabolic diseases, such as T2D and obesity, is correlated to the levels of circulating GPIHBP1 in any, i.e., soluble, vesicle-associated, lipoprotein-associated, micelle-like or multimeric version.

## 1.2. Total GPI-AP

The release of a subset of GPI-AP or even of a single GPI-AP member, such as CD59, AP and GPIHBP1, only, rather than of all members of the GPI-AP family may rely on the type of the molecular mechanism engaged. Removal of the GPI anchor by proteolytic or lipolytic cleavage through a specific protease or phospholipase or interaction of the GPI anchor with a specific carrier protein may favor the release of specific GPI-AP, whereas insertion of the GPI anchor into vesicles, particles, (heteromeric) multimers or micelle-like complexes may support the release of all GPI-AP irrespective of the nature of their protein and anchor moiety. The formation of EV and SLP, LLP, NVP or MFG with GPI-AP as major constituents may be compatible with the release of both specific GPI-AP, subsets of GPI-AP and total GPI-AP, depending on the GPI-AP expression profile of the releasing cells and donor tissues. Thus it may be useful to introduce analytical methods for the unbiased determination of the GPI-AP in total rather than for the measurement of single and specific GPI-AP members in body fluids for the evaluation and validation of those GPI-AP as biomarkers which become released in course of operation of the seemingly "unspecific" mechanisms of multimerization and formation of micelle-like complexes.

### *1.2.1. Expression Level of GPI-AP-Synthesizing Components*

The expression level of GPI-AP seems to be tightly controlled given the rather low mRNA and protein expression levels for a number of enzymes and transport or assembly components engaged in the biosynthesis of the GPI anchor, among them the catalytic and non-catalytic subunits of the multi-subunit complex, which constitute the GPI transamidase GPIT in normal cells and tissues (Zhao et al. 2012). However, some GPIT subunits, among them PIGU (Jiang et al. 2007, Shen et al. 2007, Knaus et al. 2018), PIGT, GPAA1 (Ho et al. 2006, Wu et al. 2006, Jiang et al. 2007, Nagpal et al. 2008), PIGK (GP18)(Nagpal et al. 2008) and PIGS (Nagpal et al. 2008), are known to become upregulated with regard to both mRNA and protein expression during the development of various cancers in humans (Wu et al. 2006, Jiang et al. 2007, Nagpal et al. 2008). In particular, in breast carcinomas chromosomal amplifications of the genes encoding the non-catalytic subunits GPAA1 and PIGT of the GPIT occur during malignant transformation and lead to their significantly elevated expression (Wu et al. 2006, Zhao et al. 2012). Importantly, this upregulation was found to be associated with significant increases in the protein levels of total GPI-AP as well as in tumorigenicity and cancer progression.

#### *1.2.2. Expression Level of GPI-AP in the Circulation*

It was recognized that tumor cells release GPI-AP, among them CEA and mesothelin, from their surface into the circulation with higher efficacy than normal cells. The range of those GPI-AP, i.e., specific ones or in total, and the underlying molecular mechanisms and structures, remain to be elucidated as is true for the (putative soluble, vesicular, particulate, multimeric, monomeric, micelle-like) configuration of the tumor-expressed GPI-AP which enter the circulatory system. The question of the release of GPI anchor-harboring *vs.* lacking GPI-AP was tackled on basis of the GPI anchor remnants left at the released GPI-AP entity. The proteolytically released hydrophilic version of a GPI-AP lacking any anchor building blocks, including the terminal phosphoethanolamine residue (i), can be discriminated from (ii) the lipolytically released hydrophilic version lacking the anchor diacylglycerol (through PLC action) or phosphatidate (through PLD action) residue and (iii) the released amphipatic version harboring the complete anchor embedded in vesicular, particulate, multimeric, heteromeric or micelle-like complexes, since only the latter two (ii, iii) display the (phospho)inositolglycan core at their carboxy-terminus.

The experimental discrimination between (i) and (ii) can be achieved with the aid of the lectin α-toxin from *Clostridium septicum*, a member of the aerolysin-like pore-forming toxins, that interacts with GPI-AP with high specificity and affinity (Gordon et al. 1999). α-Toxin was used for the efficient capturing and identifying of GPI-AP by mass spectrometry (Zhao et al. 2012). Importantly, the capturing was not impaired upon lipolytic cleavage of the GPI-AP by bacterial PI-PLC and successful for GPI-AP with very divergent (carboxy-terminal) amino acid sequences. This argues for predominant or even exclusive interaction of α-toxin with the glycan core of the GPI anchor and makes involvement of the protein moiety of the GPI-AP rather unlikely. Moreover, the analysis of serum proteins by mass spectrometry-based proteomics is typically accompanied by considerable hurdles since it necessitates the elimination of abundant polypeptide species and tedious steps of protein fractionation. Thus, the use of α-toxin for the isolation and enrichment of GPI-AP could facilitate their detection upon release into serum and concomitantly enable their classification into proteolytically cleaved versions (i) and those versions with lipolytically cleaved (ii) or complete (iii) GPI anchor.

Consequently, in an investigation to evaluate the potential of GPI-AP as biomarkers for certain human cancers, α-toxin was used to measure the level of GPI-AP harboring the glycan core in plasma from patients suffering from several types of tumors (Zhao et al. 2012, Dolezal et al. 2014). Interestingly, the plasma of patients with cancers, that are typically associated with increased mRNA and protein expression of GPIT subunits, displayed significantly elevated α-toxin-binding compared to that of plasma from patients with no malignant disease. Thus the high and very low amounts of total GPI-AP in cancer and control plasma, respectively, as revealed by α-toxin signals suggest that GPI-AP are released (presumably from tumor cells) into plasma as glycan core-containing protein moieties by one or the other of the mechanisms discussed above. In fact, the released glycan core-displaying GPI-AP may be useful as biomarkers for the prediction, detection and stratification of certain cancers. In particular, the GPI-AP FERMT3 and FLNA captured by α-toxin could be relevant with regard to the development of breast carcinomas (Dolezal et al. 2014). Strikingly, based on the criterion of α-toxin-binding, these two GPI-AP were detected in more than 90% of the cancer patients, whereas only very few probands with non-malignant tumors were tested as positive for the presence of FERMT3 and FLNA in serum.

Surprisingly, in cancers with very high expression of GPIT mRNA and protein, such as ovarian cancer, albeit displaying elevated levels of α-toxin-binding in plasma compared to control plasma, the levels of α-toxin-binding were found to be lower than those measured for other cancers (Dolezal et al. 2014). Apparently, in addition to the expression level of the GPIT subunit genes, other factors seem to be directly involved in the release of those GPI-AP with the glycan core still attached, which as a consequence remain detectable by α-toxin. Alternatively, the GPI anchor glycan core of released GPI-AP may be modified in ovarian rather than in other cancers by factors, such as an endogenous protease or GPI-specific phospholipase, in such a way that it fails to be recognized avidly by α-toxin. Interestingly, high levels of α-toxin-binding to membrane-associated amphiphilic GPI-AP prepared from ovarian cancer tumors as well as from the surface of cells from patient ascites were measured (Dangaj et al. 2011). Together these data hint to considerable structural differences between the GPI anchors of GPI-AP from ovarian tissue, ascites and plasma as explanation for the relatively low α-toxin-binding capacity in the plasma from ovarian cancer patients.

A problem often associated with the development of assays for the detection of serum biomarkers relies on the formation of complexes between the protein of interest and certain serum components, which causes masking of the epitopes recognized by the detecting (e.g., ELISA) antibody. In fact, it cannot be excluded that GPI-AP assembled in complexes in serum as described above may escape the traditional detection techniques, such as ELISA, at least under native conditions. This potential limitation would be bypassed in course of isolation of native serum GPI-AP on basis of the interaction of α-toxin with their glycan core. Apparently, this interaction is not susceptible towards masking by serum proteins since according to my knowledge no reports about endogenous GPI-binding serum proteins have been reported so far. At variance, successful interaction of α-toxin with GPI-AP in the context of complexes such as GLEC (see below; Müller et al. 2018) and EV (Müller, data not shown) has been demonstrated recently.

It is tempting to speculate about the putative survival advantage of the release of GPI-AP (as versions with the glycan core still attached) from the surface of tumor cells on the basis of the (patho)physiological roles, which have been suggested so far for GPI-AP release, in general, such as in waste disposal, intercellular communication, control of antigenic variability and regulation of cell shape and, in particular, in the control of cell growth, differentiation and migration (Lee et al. 2001, Simon and Kusch 2013). Strikingly, the highest plasma levels of GPI-AP with the glycan core still attached were reported for patients with colon and brain cancer (Dolezal et al. 2014). In particular, patients with glioblastoma brain tumors and lower grade colon adenocarcinomas are characterized by the highest levels of α-toxin-binding of their plasma constituents. Along these lines it may be argued, that the upregulated release of GPI-AP from the corresponding cancer cells could support the increased need for intercellular communication as reflected in synaptic activity of the brain and for diversification of cell surface antigens in colonic epithelial cells, respectively.

### *1.2.3. GPI-AP Associated with Exosomes and Microvesicles (EV)*

Over the past decades experimental and clinical evidence is increasing that the presence and accumulation of EV harboring as major protein components monotopic (e.g., caveolins), bitopic and polytopic (e.g., tetraspanins) transmembrane proteins (Hemler 2005) as well as GPI-AP (e.g., CD73, Gce1) in body fluids, such as blood, liquor, saliva, mucus, urine and interstitial fluids (e.g., pleural fluids, ascites) are related to the pathogenesis of a variety of (common complex) diseases. In the European community, (cardio)vascular, metabolic and cancer diseases are the leading causes of reduction of life quality, disability and death with major impact on health care costs. In particular, metabolic overload is a paramount problem for people in Western and industrial countries in which the caloric intake exceeds the dietary needs, accompanied by increased risks for secondary disorders and complications, including obesity, hepatic steatosis, chronic inflammation, hypertension, T2D, atherosclerosis, atherothrombosis and neurodegeneration. It is estimated that more than 50% of the Western countries can be regarded as obese according to WHO standards. Moreover, obesity is the entry gate of T2D and most of the above mentioned disorders (De Ferranti and Mozaffarian 2010, Pereira-Lancha et al. 2012). Albeit many of the secondary disorders are cell type- and organ-specific, there is overwhelming information that the blood compartment and other body fluids contain many components that coordinate distant paracrine and endocrine processes during disease initiation and progression (Kolberg et al. 2009, Schulze et al. 2009, Herder et al. 2011, Younus and Rodgers 2011). Among those, metabolites of energy metabolism, such as carbohydrates and fatty acids (Lyons and Basu 2012), cytokines, such as adipokines, myokines and hepatokines (Reilly et al. 2005, Garaulet et al. 2007, De Luca and Olefsky 2008, Luo et al. 2010, Cui et al. 2011) and miRNAs (Hunter et al. 2008, Taylor and Gercel-Tayler 2008, Bonauer et al. 2010, Michael 2010) are well established culprits for (cardio)vascular, metabolic and tumor diseases as well as associated low-degree inflammation, which in combination ultimately lead to myocardial infarction, stroke, dementia, and failure of kidney, heart, liver and brain function (Pradhan et al. 2001, Duncan et al. 2003, Ridker et al. 2003, Holvoet et al. 2004, Holvoet et al. 2008, Doria et al. 2008, Heilbronn and Cambell 2008, Conen et al. 2009, Staiger et al. 2009, Heneghan et al. 2010, Dignat-George and Boulanger 2011, Martinez and Andriantsitohaina 2011, Siljander 2011, Samuel and Shulman 2012).

Experimental evidence is accumulating that the levels of EV and of their (protein) components including GPI-AP in plasma may be related to (cardio)vascular (Chargaff and West 1946, Wolf 1967, Nomura et al. 1998, Boulanger et al. 2001, Nomura et al. 2001, Freyssinet 2003, Bernal-Mizrachi et al. 2003, Horstman et al. 2003, Jimenez et al. 2001 and 2003, Preston et al. 2003, Brodsky et al. 2004, Ferreira et al. 2004, Tan et al. 2005, Esposito et al. 2006b, Setzer et al. 2006, Agouni et al. 2008, Freyssinet and Toti 2010) and metabolic diseases, in particular T2D and obesity (Nomura et al. 1995 and 1998, Diamant et al. 2002, Omoto et al. 2002, Sabatier et al. 2002, Esposito et al. 2006a, Goichot et al. 2006, Koga et al. 2006, Leroyer et al. 2008, Helal et al. 2010, Tramontano et al. 2010, Esposito et al. 2011), endothelial (Horstman and Ahn 1999, Horstman et al. 2003) and other common complex diseases (Horstman 2007, Mitchell 2009, Rabinowits et al. 2009, Simpson et al. 2009, Qazi et al. 2010). These findings can be explained most easily by the biological function of EV in the orchestration of signaling and transport processes, immunomodulation, tissue remodeling as well as the interaction of cells during angiogenesis, cell proliferation and apoptosis/survival (Baj-Krzyworzeka et al. 2006 and 2007, Deregibus et al. 2007, Schorey and Bhatnagar 2008, Skog et al. 2008, Al-Nedawi et al. 2008 and 2009a, Yuan et al. 2009, Castellana et al. 2010, Liu et al. 2010, Kogure et al. 2011, Yang et al. 2011, Montecalvo et al. 2012). As a consequence, control and patient cohorts have to be recruited at different stages of (cardio)vascular (stroke, myocardial infarction) and metabolic diseases encompassing obesity without and with insulin resistance, T2D and the full blown metabolic syndrome.

Concentration, composition (also with regard to GPI-AP), cellular origin and biological function of circulating EV are critical factors in (cardio)vascular and metabolic diseases (VanWjik et al. 2003, Al-Nedawi et al. 2009b). With regard to all these parameters, EV are of considerable heterogeneity. Since 70-90% of the circulating EV are derived from senescent or activated platelets, the transcriptome and proteome of the platelet and platelet-derived EV has to be investigated thoroughly in order to facilitate the discrimination between tissue cell-derived and platelet-derived EV. To further support this objective in addition to extensive *ex vivo* platelet analysis, an *in vitro* model for human cord blood $CD34^{+}$ progenitors capable of megakaryocytic differentiation was established to differentiate pro-platelets, shedded immature and mature platelets and EV (Liu et al. 2003 and 2004). Moreover, an *in vitro* system reflecting platelet senescence in human haemapheretic platelet preparations was introduced and analyzed for the composition of platelets and shedded EV over five days with multi-omics platforms. It was found that stored platelets double EV-shedding, whereas Apo-A-I/HDL antagonizes it (Liu et al. 2003).

Five subclasses of platelet-derived EV were identified, each having unique profiles for its protein components, encompassing GPI-AP (e.g., Apo A-I, ApoE, clusterin) and ligands for lipid-signaling G protein-coupled receptors (e.g., lysophosphatidic acid, lysophosphatidylcholine, sphingosine-1-phosphate, ceramide), DNA/RNA-binding proteins and miRNAs. In conclusion, EV with their complex and variable composition that is expressed in their overall signature rather than in single or multiple parameters seem to reflect those alterations in gene expression and function of the relevant tissues and cells being involved in the pathogenesis of metabolic diseases and thus could be useful for their prediction, diagnosis, prognosis, therapy monitoring and stratification. In this regard, EV with the embedded GPI-AP can be considered as a type of (patho)physiological "mirror" rather than classical biomarker since they combine complete sets of cellular proteinaceous and nucleic acid components with functional and (patho)physiological roles for common complex diseases. In fact, it seems plausible that all the typical criteria to be fulfilled by biomarkers according to the rules defined by the EMA and FDA (Lathia 2002, Colburn 2003, Morel et al. 2004, Naylor 2003 and 2004, Zolg and Langen 2004, Ciesla et al. 2011, Lee et al. 2011, Zineh et al. 2011, Müller 2012b) could be met by EV in course of their ongoing detailed investigation for approval and application in the future.

#### 1.2.3.1. EV-Associated GPI-AP and Liver Diseases

Non-alcoholic fatty liver disease (NAFLD) is a condition characterized by excessive lipid accumulation in the liver which strongly correlates with insulin resistance and its phenotypic manifestations, such as obesity, T2D, dyslipidemia, arterial hypertension (Abbasi et al. 2004, Zelber-Sagi et al. 2011). It is the most common chronic liver disease in Western countries and encompasses a non-progressive probably benign hepatological form of steatosis and a progressive form of non-alcoholic steatohepatitis (NASH), which significantly increases liver-related as well as cardiovascular mortality, the incidence of T2D and the overall mortality (Ratziu et al. 2012). Currently, the diagnosis is entirely based on histology as revealed by liver biopsies. This is an invasive and cost-intensive procedure and not appropriate for the screening of large numbers of patients at risk for NAFLD. Therefore there is a high medical need for non-invasive diagnosis of NAFLD with the objectives of patients with (i) the progressive form of NASH in order to monitor and propose early therapeutic interventions and (ii) (very) early stages of NASH in order to transfer knowledge about diagnostic and therapeutic options to those suffering from advanced stages of the disease.

Both NAFLD and NASH are often associated with obesity, T2D and asymptomatic elevations in serum levels of liver transaminases. The progression of these hepatic diseases is accompanied by the development of ER stress (Lauressergues et al. 2012) and inflammation involving chemokines such as MCP-1, cytokines such as TNF-alpha and metalloproteases. In comparison to NAFLD patients, NASH patients in general are older, more obese and more often have high serum liver enzymes and suffer from T2D and the metabolic syndrome. To date the diagnosis and prognosis of NAFLD/NASH are being performed using ultrasound analysis of the liver for the detection of fatty infiltration, which however does not allow assessment of the degree of inflammation and fibrosis. Therefore, diagnosis of lipid in the liver is easily made by ultrasound but diagnosis of NAFLD or NASH cannot be performed without liver histology. Several biomarker-based approaches have been proposed to enable the non-invasive diagnosis of liver fibrosis. These encompass panels of serum markers, genetic markers of disease progression, plasma lipidomic signatures, newer imaging methods and breath tests. For instance, a patented algorithm (Fibro-Meter, BioLiveScale, Angers, France) was evaluated and demonstrated improved diagnostic performance for the stages F2-F4 compared to the fibrosis score and AST-to-platelet ratio index (Poynard et al. 2012). This marker panel index is based on relatively simple and readily accessible parameters, such as age, glucose, liver transaminases, ferritin, platelets and body weight, but still requires further validation. Plasma lipidomic signatures were also used for the characterization of NAFLD/NASH patients. Increased lipogenesis as well as desaturase and lipooxygenase activities compared to healthy patients were measured. Moreover, impaired peroxisomal metabolism of polyunsaturated fatty acids and non-enzymatic oxidation were found to be associated with the progression to NASH. This strategy may provide information that serves as a basis for future hypothesis-driven studies of the role of metabolic pathways in the pathogenesis of NASH.

Even if the combination of various serum markers of liver fibrosis and the results from transient elastography measured by the fibroscan technology have been suggested to predict the development of NASH and fibrosis, liver biopsy remains the accepted gold-standard for the differentiation of NASH and NAFLD (Munteanu et al. 2011). There is still an unmet medical need for the discovery of non-invasive and precise diagnostic tools for the patients suffering from these liver diseases (Boursier et al. 2011).

Tissue remodeling which occurs during NAFLD/NASH may induce the formation of EV containing GPI-AP. Those EV are present in the circulation of healthy individuals at relatively low concentrations. Numerous clinical studies have reported increased plasma EV levels associated with (cardio)vascular risk factors, such as hypertension, smoking, obesity and prediabetes as well as frank (cardio)vascular and metabolic diseases, such as T2D (see above). Concerning liver diseases, an increase of the level of plasma EV was reported in patients with hepatitis C and hepatocellular carcinoma. However, it remains unknown whether liver-derived EV displaying GPI-AP were detectable in the plasma of NASH patients. It is commonly accepted that NASH pathogenesis represents a progressive process involving steatosis and inflammation. Therefore it is reasonable to assume that these conditions may result in the release of GPI-AP in association with EV from liver cells, Kupffer cells or (recruited) macrophages. So far, there are no clinical data available linking EV with fatty liver diseases. Nevertheless, in apolipoprotein E2 knock-in mice, which represents an experimental model for atherosclerosis and NASH (Shiri-Sverdlov et al. 2006), the presence of EV was observed both in the atherosclerotic lesions of the vasculature and in the liver.

In future projects correlations between the levels of EV displaying GPI-AP in the plasma and the specific stages of NAFLD/NASH have to be delineated by measuring liver cell-specific EV in order to provide and validate biomarker candidates in appropriate clinical cohorts and ultimately to elucidate a relationship between the appearance of specific subsets of GPI-AP-harboring EV and the pathogenesis (with regard to stage and severity) of NASH. The objectives should be as follows: (i) Optimization and routine use of cell culture protocols for the release of EV from human hepatoma, stellate, Kupffer and liver endothelial cell models including cultures of human primary hepatocytes, (ii) identification and validation of markers for GPI-AP-harboring EV originating from hepatic, stellate, Kupffer and liver endothelial cells *in vitro* and *in vivo*, including activation status markers, by using several omics- (e.g., flow cytometry, immune blotting, protein/lipid/DNA arrays, nucleic acid probes) and signature-based approaches, including data mining of public literature, patents and databases, (iii) the measurement and comparison of the total concentrations of EV, their compositions and cellular origins in plasma from patients with different stages of NAFLD/NASH, (iv) the identification of subsets of EV expressing GPI-AP with available biomarkers or imaging methods for monitoring the development of fibrosis and simple steatosis in NAFLD *vs.* NASH with fibrosis and (v) selection of liver cell-derived EV in the circulation for their use as candidate biomarkers in clinical studies.

From a methodological point of view, patients suffering from chronic liver disease caused by alcohol misuse, viral infection, autoimmune response or hemochromatosis rather than by NAFLD have to be excluded. All patients should be subjected to a liver biopsy, to detailed hepatological and metabolic phenotyping as well as ultrasound and hepatic elastometry. Based on the results of the liver biopsy the patients will be categorized into (i) patients with steatosis alone, (ii) patients with steatohepatitis (NASH), (iii) patients with NASH and no or minimal fibrosis and (iv) patients with NASH and advanced fibrosis/initial cirrhosis. Differences in the number, size, composition and cellular origin of GPI-AP-expressing EV are expected according to the absence/presence/degree of NASH and the extent of fibrosis and steatosis. If needed, patient sub-groups have to be selected in order to improve putative correlations.

#### 1.2.3.2. EV-Associated GPI-AP and Immune Diseases

There is increasing evidence that for certain acute lung diseases circulating leucocyte-derived EV that express GPI-AP can be used for their prediction as well as prognosis of the outcome (Guervilly et al. 2011, McVey et al. 2012). For instance, acute respiratory distress syndrome (ARDS) as the most severe clinical manifestation of acute lung injury occurs in systemic inflammatory response syndromes as well as in sepsis. Evidence from clinical and experimental studies is converging that leucocytes play a pivotal role during the acute phase of ARDS. The local and systemic pro-inflammatory responses, which accompany ARDS and sepsis, apparently are coordinated by EV, which display GPI-AP at their surface and are released from leucocytes, platelets and endothelial cells in response to their interaction (Bastarache et al. 2009). Those EV are commonly regarded as sensitive biomarkers for the assessment of the activation or apoptotic status of cells in systemic inflammatory response syndromes and sepsis. In fact, high levels of leucocyte-derived EV have been identified in the plasma of ARDS patients and associated with better outcome in ARDS.

#### 1.2.3.3. EV-Associated GPI-AP and Lipid Diseases

Recently gained experimental evidence demonstrated that the transfer of proteins and nucleic acids from releasing (i.e., donor) to target (i.e., acceptor) adipocytes *via* EV (Aoki et al. 2007, Ogawa et al. 2010) is associated with (patho)physiological consequences, such as the upregulation of the esterification of fatty acids into triacylglycerol and the downregulation of lipolytic fatty acid release from target adipocytes (Müller 2009, Müller et al. 2010a and b, 2011a-c). This apparent phenotype switching was triggered by the mere incubation of (preferentially small) acceptor adipocytes with EV released from (preferentially large) donor adipocytes which harbor the GPI-AP, Gce1 and CD73, as well as the mRNAs coding for fat-specific protein 27 (FSP27) and glycerol-3-phosphate acyltransferase (GPAT3) and the miRNAs, miR-16 and miR-222 (Müller et al. 2012b)(Figure 12). Importantly, the release of those EV from large primary rat adipocytes as well as differentiated human adipocytes in response to physiological (palmitate, $H_2O_2$) and pharmacological (anti-diabetic sulfonylurea drug glimepiride) stimuli was shown to be considerably reduced in the presence of 5-aza-2'-deoxycytidine

(5-Aza-CdR)(Müller et al. 2012b). This inhibitor of cytosine methylation as well as the specific inhibitor of histone lysine methyltransferases, BIX01294, are known to induce substantial remodeling of heterochromatic domains. The blockade of EV release by 5-Aza-CdR or BIX01294 did not correlate with an alteration in the apoptotic rate, but was accompanied by impairment of the $H_2O_2$- (but not insulin-) induced stimulation of esterification and inhibition of lipolysis in large (but not small) primary and differentiated human adipocytes (Müller et al. 2012b)(Figure 12).

In contrast, the simultaneous presence of 5-Aza-CdR and BIX01294 had almost no effect on the palmitate-, glimepiride- and $H_2O_2$-induced release of those EV and the regulation of lipid metabolism. These findings argue for the modulation of the induced release of EV harboring GPI-AP, mRNAs and miRNAs, that are specific for the control of lipid metabolism, from rat and human adipocytes by DNA and histone methylation in interdependent fashion (Müller et al. 2012b). Furthermore, it has been proposed that the EV-mediated transfer of lipogenic and anti-lipolytic information between large and small adipocytes in response to certain physiological and pharmacological stimuli may be inherited by epigenetic mechanisms (Musri et al. 2010a and b, Müller et al. 2012b). Consequently, it is tempting to speculate that interference with the epigenetic control of the information transfer between adipocytes as well as between adipocytes and other relevant cells (e.g., monocytes, macrophages, pericytes) could modulate the complex molecular mechanisms through which environmental (e.g., special nutritional, hormonal, stress) conditions, especially in early life, lead to a biochemical memory effect that influences the susceptibility towards lipid disorders, including metabolic syndrome, T2D and obesity (Stöger 2008, Campion et al. 2009, Cho et al. 2009, Ling and Groop 2009, Musri et al. 2009b, Heerwagen et al. 2010, Zalou et al. 2018).

The idea that epigenetic modifications could be manipulated by small molecules to modulate the expression of multiple genes and effectively reverse pre-programmed susceptibility for metabolic diseases is highly attractive as exemplified by the growing evidence for successful pharmacological interference with epigenetic mechanisms (Martin and Zhang 2005, Jenuwein 2006, Cole 2008, Spannhoff et al. 2009): (i) The selective p300 histone acetyltransferase inhibitor C646 was identified through virtual library screening and showed inhibition of cell growth and histone acetylation. (ii) Entinostat is the most advanced example of the 2-aminobenzamide chemotype, aimed to bypass pharmacokinetic and toxicological issues which are associated with hydroxamic acids, and is currently in phase II trials for the treatment of several malignant tumors. (iii) Tubastatin and PCI-34051 represent recent advances in targeting the histone deacetylase subtypes 6 and 8, respectively. (iv) Recently identified BET bromodomain inhibitors have demonstrated the accessibility of this protein family for small-molecule modulation and show promising anti-tumor and anti-inflammatory properties. (v) Advances in structure-based design have led to the discovery of two series of JMJD2 demethylase inhibitors, which exert effects on cellular H3K9me3 levels and anti-proliferative activities, respectively. (vi) The availability of a protein crystal structure for the methyltransferase G9a allowed the design of highly potent and selective inhibitors.

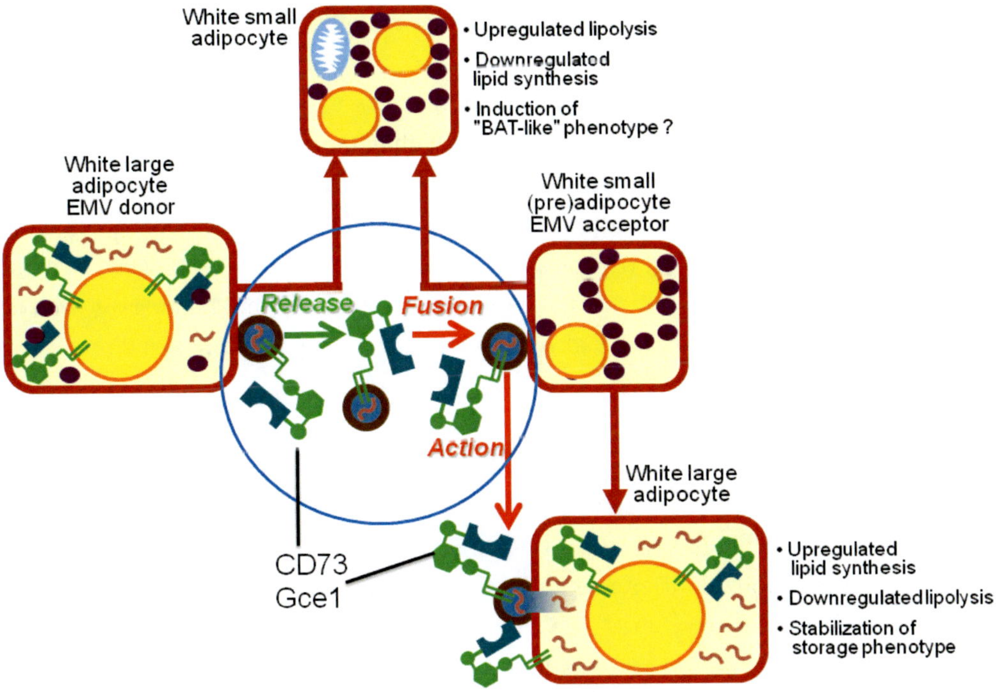


Figure 12. Model for the involvement of inter-adipocyte transfer and communication of the GPI-AP Gce1 and CD73 for the regulation of lipid metabolism, storage and cell size within adipose tissue depots. The model is based on a polarized structure of adipose tissue depots with large mature adipocytes harboring a single or a few LD only, which are capable of very pronounced esterification in parallel with high lipolysis and located in immediate neighborhood to blood vessels, where they form a gradient together with nascent adipocytes of small and very small size. Prolonged delivery of free fatty acids by the blood vessels and their resulting efficient esterification into LD will cause size increase of the adipocytes and exhaustion of their lipid storage capability with accompanying upregulation of the lipolytic release of fatty acids into the interstitial space. This will trigger the maturation of (very) small precursor adipocytes lacking visible LD in the deeper adipose tissue layers that are capable of esterification and lipolysis to a very moderate degree, only, by one or several types of intercellular signaling. Upon induction of novel and specific cell-to-cell contacts, as are already formed between the large adipocytes, or the (vesicular) secretion of adipokines (Rondinone 2006), such as acylation stimulation protein ASP (Maslowska et al. 2005), or soluble factors, such as 15-keto-PGE2 (Chou et al. 2007, Yu et al. 2013), the targeted cell adhesion molecules or adipokine receptors at the cell surface or nuclear hormone receptors, respectively, induce signaling for stimulation of the esterification and/or inhibition of lipolysis, which in concert finally lead to upregulated LD biogenesis in the growing small adipocytes. Alternatively or in addition, upon challenge with high concentrations of fatty acids originating directly from the blood vessels or from lipolytic release in response to physiological or pharmacological stimuli, such as hydrogen peroxide or the anti-diabetic sulfonylurea drug glimepiride, the large adipocytes release the GPI-AP Gce1 and CD73 into EV. Following passage through interstitial tissue spaces and eventually thereafter through the circulation, the EV interact with plasma membranes of (very) small adipocytes for subsequent transfer of Gce1 and CD73 to the surface of LD, where those enzymes degrade (c)AMP. Again, the resulting parallel upregulation of esterification and downregulation of lipolysis lead to enlargement of the LD and size of the adipocytes.

Thus, the recent progress in the discovery of drug-like small molecule inhibitors for the majority of the classes of histone modifiers, including "writers," "readers" and "erasers" of acetyl and methyl marks, convincingly demonstrates the attractiveness of these protein families for drug development. The availability of selective inhibitors and modulators for use as chemical probes as well as of test systems for analyzing the intercellular communication *via* EV between adipocytes as well as between adipocytes and immune cells *in vitro* will be essential in unraveling the involvement and function of individual players in the epigenetic control of the metabolic syndrome and the understanding of its underlying "histone code." This will be prerequisite for the initiation of future drug and biomarker discovery projects aimed to modulate the relevant "writers," "readers" and "erasers" as critical and novel targets for the modulation as well as biomarkers for the monitoring of EV-driven actions for the future therapy and prediction, respectively, of metabolic disorders.

#### 1.2.3.4. EV-Associated GPI-AP as Biomarkers

The predictive power of classical biomarkers for the prediction of common complex diseases, such as metabolic diseases (lipid disorders, T2D, obesity) is currently not sufficient for the individual differentiation of the pathogenesis and outcome, as well as of the therapeutic options and the individual monitoring of the therapeutic success. Monitoring of multi-parameter patterns may add a novel layer of stratifying quality. EV equipped with GPI-AP in addition to a variety of transmembrane proteins, mRNAs, miRNAs and phospholipids are known to be released from various tissues, such as adipose, liver, and cell types, such as immune cells, into multiple body fluids, such as plasma, saliva and urine (Cheruvanky et al. 2007, Conde-Vancells et al. 2010, Miranda et al. 2010, Keller et al. 2011) in differential fashion in response to normal and diabetogenic/obesogenic conditions. EV with their complex and variable composition that is manifested in their "overall-signature" rather than in single or multiple parameters seem to reflect those alterations in gene expression and function of relevant cells and tissues being involved in the pathogenesis of common complex diseases and thus could be useful for their prediction, diagnosis and therapy monitoring.

#### 1.2.3.5. Information or Material Transfer by EV?

The research area of EV has gained enormous credit in attention by the scientific community as well as funding during the past two decades after more than five decades of minor interest after their initial description (Chargaff and West 1946) and following the first attribution of a physiological function, which was waste disposal (Johnstone 2003). Meanwhile EV are frequently and typically regarded as carriers of biological information between cells, tissues and organs in multicellular organisms and this (hypothetical) function is often interpreted as the only and/or most physiologically relevant one. In fact, it is tempting to state that meanwhile the number of review articles dealing with EV as carrier/mediator/vehicle of information appears to exceed that of research studies about any aspect of their appearance and structure and, in particular, the number of experimental findings demonstrating the transfer of biological information by EV in unequivocal fashion. What is the state of data-based knowledge concerning intercellular information transfer by EV?

The transfer of biological information (a typical metaphor from IT sciences adopted by life sciences in rather unreflected way) in the only useful narrow meaning of the terms "transfer" and "information" implies the induction of the process of (re-)programming of a target cell by a donor cell for the adoption of a new phenotype or one to several specific physiological function(s), that may be related or unrelated to that/those of the donor cells, with the help of messengers, such as molecules, waves or currents, which from a structural and functional point of view have nothing to do with the transferred information and the machinery of its decoding and realization by the target cells. Thus biological information has to be regarded as the operator (e.g., ligand) of a switch (e.g., receptor) controlling a physiological circuit (e.g., for growth, differentiation, metabolism), which can be achieved by single entities (e.g., hormone, cytokine, ion, miRNA, light) and does not require a multi-component complex as its material/molecular basis, i.e., for pressing the button. The validity of this definition for the term "information" is exemplified best by the action of hormones, neurotransmitters or action potentials. For instance, the hormone insulin is secreted by the pancreatic ß-cells (donor cells) in response to elevated blood glucose levels in order to "instruct" the adipose, muscle and liver (target) cells to take up glucose for storage as glycogen or triacylglycerol. Insulin apparently encodes the information "High Blood Glucose," sensed by the ß-cells, which is subsequently decoded by the insulin receptor of the target cells. Importantly, insulin as the information carrier is completely unrelated with regard to both structure (peptide) and mode of action (hormone which binds to a cognate receptor for downstream signaling) to the transferred information "High Glucose" and the instruction for glucose uptake. In other words, the information carrier insulin as substance/molecule does not form and act as a component of the information decoding machinery of the glycogen and lipid biosynthetic pathways in the target cells.

Can this definition of biological information transfer be regarded as being fulfilled for EV on basis of the experimental evidence presented so far? As far I am aware of, at present only a few studies have been devoted to demonstrate (patho)physiological effects exerted by isolated, purified and adequately characterized EV, which display the typical characteristics of sealed phospholipid membrane vesicles with embedded membrane proteins and enclosed mRNAs/microRNAs, rather than by the individual constituents such as miRNAs (Hu et al. 2012). Among them are the seminal demonstrations that (i) a truncated constitutively active and oncogenic version of the epidermal growth factor, EGFRvIII, becomes transferred from tumor cells to normal cells *via* EV causing their malignant transformation (Al-Nedawi et al. 2008, 2009a and b), (ii) EV released from glioblastoma cells transfer RNAs and proteins which foster tumor development (Skog et al. 2008) and (iii) the GPI-AP Gce1 and CD73 together with the mRNA for GPAT3, which are involved in the control of lipolysis and lipogenesis, respectively, are transferred from large to small adipocytes *via* EV leading to upregulation of lipid synthesis in the latter (target) cells (Müller et al. 2011a and b). Importantly, in each of these cases, the EV actually transfer materials, i.e., molecular components, which following uptake are directly and stoichiometrically involved in the donor cell-instructed processes in the target cells, such as receptor downstream signaling for cell division by EGFRvIII or GPAT3-catalyzed lipid synthesis and its (c)AMP-dependent regulation by Gce1 and CD73. In other words, the transferred EV components EGFRvIII, GPAT3, Gce1, CD73 exert their authentic receptor and enzymic functions, respectively, rather than encode information, which in indirect and non-stoichiometric fashion presses the switches for oncogenic transformation and lipid storage, respectively.

On the basis of these considerations it is strongly recommended that the (patho)physiological role of EV should be interpreted as material transfer rather than information transfer as long as the experimental evidence available does not support the term information for the load of EV in the scientifically strict and only useful application. Moreover, the putative value of the concept of EV, in general, and of those equipped with GPI-AP, in particular, as novel "complex" biomarkers for the prediction, diagnosis, prognosis and therapy monitoring of common complex diseases critically relies on the expression of a multitude and variety of distinct physical materials by EV to be transferred from donor to target cells *via* body fluids for direct participation in multi-step pathways and/or in the coordination of distinct pathways. In contrast, the transfer of a single information encoded by EV, such as for tumor transformation or lipid synthesis, does not necessitate (extensive) variation of the composition or structure of the corresponding EV, making them unattractive for use as biomarkers. This limitation is exemplified best by the rather low predictive value of typical information carriers, such as insulin and growth factors. In fact, the unambiguous demonstration of the compositional and/or structural diversity of EV released from disease-relevant cell types or tissues, in general, and of the association of certain compositional and structural subtypes/variants of EV with the development or state of a disease, in particular, has not been presented so far.

Finally, it is reasonable to assume that material transfer exerted by EV is physiological and efficient for short distances, only, in order (i) to guarantee its specificity with regard to the material's destination, (ii) to limit the waste of materials and energy for its production in case of broad distribution (*via* the circulation) over the organism and eventual failure of reaching the envisaged destination, (iii) to avoid deleterious effects caused by materials upon transfer to undesirable locations and eventual unforeseen uptake by non-target cells, (iv) to bypass the restrictions for the path from donor cells to acceptor cells in distinct tissue depots or organs *via* the circulation which depends on passage of the EV across vascular endothelial cells of the underlying vessels, possibly involving transcytosis or transient opening of tight junctions.

In fact, the operation of a paracrine rather than endocrine mode of EV action was clearly reported for the adipose tissue depot (Müller et al. 2010b and d). EV, equipped with specific GPI-AP, mRNAs and miRNAs which are all engaged in triacylglycerol synthesis, become released from large adipocytes and then transferred to small neighboring adipocytes for uptake and direct participation in lipid synthesis (Figure 12). This apparent paracrine material transfer was interpreted as shift of the burden of lipid loading from large (filled-up) adipocytes to small (empty) adipocytes through the coordination of the triacylglycerol synthesizing capacity within a tissue depot. (Müller 2011a). A paracrine transfer of exosomal miRNA was also reported between T-cells and antigen-presenting cells within immune organs (Mittelbrunn et al. 2011) and of both mRNAs and proteins between embryonic stem cells and blood cell progenitors within hematopoietic tissues (Ratajczak et al. 2006).

At variance, during certain pathophysiological situations, such as malignant transformation, "spill-over" of EV from malignant tissues into the blood may lead to the distribution and subsequent transfer of oncogenic materials to hitherto benign tissue cells. This would explain the very low number of studies at present, which provide unequivocal evidence for the presence of tissue-derived EV in mammalian plasma. No doubt, the major portion of plasma EV is derived from blood cells (see above). This distribution ratio, i.e., the (presumably) very low concentration of tissue-derived EV, presumably reduces the usefulness of plasma EV as biomarkers for (common complex) diseases, which are caused by functional impairment of tissue cells. However, at present systemic appearance and function of EV cannot be ruled out completely. One of the rare experimental studies addressing this issue demonstrated the fusion of tissue-factor-bearing microvesicles with activated platelets and the accompanying initiation of coagulation (Del Conde et al. 2005).

### *1.2.4. GLEC-Associated GPI-AP as Biomarkers*

"Reductionistic" approaches for the identification of novel biomarkers are based on the measurement of the levels of a single or a few dominant and defined protein, lipid or metabolite species (e.g., by PCR or ELISA methods) and provide the complete but biased understanding of the only and seemingly linear metabolic pathway thought to be responsible (Figure 7). In contrast, "holistic" approaches rely on the determination of changes in the levels and fluxes of all relevant components amenable to untargeted "Omics" technologies and lead to the complete but unbiased understanding of the underlying network of interacting pathways without focusing on a specific one. Unfortunately, so far both approaches have demonstrated only limited predictive power for T2D, which does not support stratification and individualized therapy of lifestyle diseases. This failure may, in part, be due to the limited information depth intrinsic to both "reductionism" and "holism" which is presumably caused by loss of the "interactome" and of (biophysical) properties intrinsic to complexes, respectively.

A novel "hermeneutic phenomenological" approach (Figure 7) may lead to biomarkers of higher predictive power reflecting the intimate interplay between susceptibility genes and environmental cues in a more direct and precise fashion than previous "reductionistic" and "holistic" approaches. It relies on the demonstration and biophysical characterization of extracellular complexes in plasma, which harbor GPI-AP and phospholipids and possibly additional proteinaceous and lipidic components, the so-called GLEC (see above), and may be released from almost each cell type through non-classical secretory mechanisms (Figure 9).

The rationale of this "hermeneutic phenomenological" approach is as follows: (i) GLEC are released from metabolically relevant tissues (e.g., adipose, muscle, liver, ß-cells, endothelial cells) into the circulation in response to metabolic stress (e.g., high levels of glucose and fatty acids), as is prevalent during obesity and T2D. (ii) GLEC differ in level, type (structure, composition) and biophysical properties (viscoelasticity, rigidity) between distinct subtypes of frank T2D (prerequisite for biomarkers for disease stratification) and/or between obese subjects characterized by varying life styles, life stages and disease states along the pathogenesis of T2D (prerequisite for biomarkers for prediction). (iii) GLEC in serum can be detected and characterized with regard to level, type and biophysics using chip-based sensors. (iv) Reliable and significant correlations between the level, type and/or biophysics of GLEC in plasma and progressive pre-diabetic states and (subtypes of) overt diabetic states of obese rodents and human subjects will be delineated in future animal and experimental clinical studies. Thereby the project will clarify whether the analysis of GLEC using a chip-based sensor provides (i) high predictive power for T2D and (ii) in-depth information for its stratification into subtypes relevant for individualized therapy. Positive outcome will foster subsequent larger scale studies with patients to demonstrate that GLEC meet the biomarker criteria for prediction and stratification of T2D.

For the following reasons GLEC are thought to be relevant for the prediction and classification of T2D: (i) Certain GPI-AP, such as CD73 and Gce1, have been reported to be released from metabolically relevant cell types, such as adipocytes, in response to metabolically relevant stress factors, such as high levels of saturated fatty acids, reactive oxygen species and anti-diabetic drugs (Müller et al. 2009a and b). Likewise, elevated levels for a number of GPI-AP have been measured in plasma from cancer patients (Zhao et al. 2012, Dolezal et al. 2014). (ii) GPI-AP rather than typical transmembrane proteins are prone to rapid and efficient release from the plasma membrane of donor cells in course of mechanically or chemically induced stress. (iii) Phospholipids in complex with GPI-AP have been detected in the supernatants of cultured cells as well as in rodent and human serum and may represent constituents of mixed phospholipid micelles, such as GLEC, lipoprotein-like particles, such as SLP, NVP, LLP, globule-like particles, such as

MFG and EV, such as microvesicles, exosomes, which are released by ill-defined non-classical (e.g., blebbing, shedding, bicelle formation) or classical (e.g., exocytosis) secretory mechanisms (De Broe et al. 1977, Väkevä et al. 1994, Camussi et al. 2011). (iv) Plasma phospholipids analyzed by untargeted lipidomics have been shown to predict early neurodegeneration during preclinical and presymptomatic Alzheimer's disease (Ray et al. 2007). (v) GPI-AP are known to be susceptible for transfer from donor cells to acceptor cells *in vitro* (Medof et al. 1996) and *in vivo* (Kooyman et al. 1995) in a functional state, thereby putatively transmitting biological materials within or between tissues (e.g., CD73 as anti-inflammatory and immune-suppressive molecules). Interestingly, the level of the GPI-AP CD73 in plasma was shown to be correlated with insulin sensitivity in diabetic mice and human probands (Lunkes et al. 2003, Antonioli et al. 2013 and 2015). (vi) phospholipids in complex with membrane proteins (e.g., caveolins) were reported to be released from vascular endothelial cells in culture and into plasma from mice following oxidative stress and high fat diet (Yuan et al. 2011, Gumulec et al. 2012).

#### 1.2.4.1. Pathogenesis and Heterogeneity of T2D

Western and developing countries are faced with a dramatic increase in the number of patients suffering from metabolic diseases. Currently, in Germany more than 50% of people are overweight (i.e., BMI > 25) and 5-10% of them will develop T2D during lifetime. Numerous studies have established the huge impact of the lifestyle (in particular, nutrition and exercise) on the pathogenesis of T2D. This effect can be explained, in part, by excessive deposition of triacylglycerols in tissues, which are not primarily destined for energy and lipid storage (e.g., muscle, liver, pancreatic ß-cells). The resulting lipotoxic effect is assumed to impair the insulin sensitivity of the affected muscle and liver cells (*via* interference with the insulin signaling cascade) as well as to reduce the number of ß-cells (*via* induction of ß-cell apoptosis) and the glucose-induced insulin release (*via* interference with nutritional signaling in ß-cells and between ß-cells, brain and intestine). Thus, the hallmarks of T2D, diminished insulin action and secretion, can be explained both by a dysbalance of energy intake and energy expenditure (Samuel and Shulman 2012, Staiger et al. 2009, Duncan et al. 2003). However, longitudinal clinical studies have revealed that the onset of frank T2D can be prevented or delayed by appropriate reduction of caloric intake and normalization of the body weight within the period of reversible insulin resistance prior to irreversible damage of the pancreatic ß-cells in many but not all cases.

In addition of the role of obesity, the apparent involvement of multiple susceptibility genes and environmental cues affecting multiple tissues together with the clinical outcome of diabetic late complications, which all considerably differ between individual T2D patients, strongly argues for the heterogeneity of T2D. Importantly, the current molecular and clinical findings hint to a "garden variety" of pathophysiological causes as well as consequences for T2D rather than to "hot spots." Moreover, the late complications are improved by intensified insulin therapy and tight control of blood glucose levels to partial degree only, which differs between individual patients, too. Therefore, a sound stratification of "the" T2D into subtypes according to the different underlying pathophysiological mechanisms and consequences will be helpful for the prediction and individualized therapy of T2D. This necessitates the use of biomarkers enabling differentiation of the T2D "garden variety" into (at least major) subtypes. Ideally, prediction of increased susceptibility of individuals towards T2D and the accompanying clinical consequences with sufficient power, i.e., measurement of the individual rather than population risk, may facilitate decisions for adequate changes in the individual lifestyle.

At present, the risk assessment for T2D in non-obese and obese subjects lacking any symptoms and in those probands displaying first signs of the disease as well as its definite diagnosis typically after 5 to 15 years of pathogenic progression is based on the measurement of the levels of classical and easily accessible serum parameters, such as glucose, intermediary metabolites, such as 2-hydroxybutyrate, as well as surrogate entities, such as glycated hemoglobin $HbA_{1c}$. These so-called traditional biomarkers in combination enable the prediction of T2D with a probability of about 0.65 to 0.75 (Müller et al. 2012a, O'Neill et al. 2016). The combination with information about physical body parameters (BMI, waist-to-hip ratio, sex) as well as about the general health state, lifeworld and lifestyle of the probands (e.g., blood pressure, smoking, fitness, food consumption as evaluated in the *Deutsche Risikoabschätzung*) leads to a further increase in the predictive power to about 0.85 to 0.90 (Barber et al. 2014, Lotta et al. 2015). In future, functional assays, such as glucose and insulin tolerance tests, and non-invasive imaging procedures for metabolic processes and organs (Shin and Camilleri 2013, Stefan et al. 2016), magnetic resonance imaging of the ß-cell mass and positron emission tomography of the brown adipose tissue mass will further improve predictive power. However by nature, these phenotypic biomarkers have to be determined at a rather late stage of the pathogenesis, since the earlier the time points of their measurement, the less predictive they are. As a consequence they fail to support individualized preventive efforts (Faerch et al. 2016).

Certainly, the value of a predictive biomarker increases with (i) the time point of its initial appearance and first possibility of its technical measurement prior to disease onset, (ii) the detection in easily accessible body fluids, such as serum, and (iii) the detectable differences between individuals with regard to lifeworld, disease onset and disease outcome (Müller 2012b). These criteria may be fulfilled by novel "phenomenological" biomarker, such as GLEC for the prediction of T2D. The assumption of the release of GPI-AP with complete GPI anchor from cells into complexes together with phospholipids and cholesterol as GLEC is based on the following rationale: (i) GPI-AP may be particularly prone to spontaneous or regulated release from the cell surface due to sole anchorage at the extracellular face of the plasma membranes *via* their covalently attached GPI moiety. (ii) This release of the amphiphilic GPI-AP equipped with the complete GPI moiety (in contrast to that of the soluble GPI-AP as a consequence of lipolytic or proteolytic cleavage of the GPI or protein moiety, respectively) necessitates its embedding into amphiphilic structures, the GLEC, which may be comprised of mixed phospholipid micelles, surfactant- or lipoprotein-like particles or vesicles. (iii) The rate of the release of GLEC from the surface of metabolically relevant cells into the circulation and/or their morphological and biophysical characteristics are correlated to metabolic stress, exerted by chemical and mechanical factors (such as elevated levels of plasma lipids, insulin or reactive oxygen species, shearing forces, cell deformation through compression or stretching), as is prevalent during early stages of T2D pathogenesis. (iv) Gender-specific differences in the levels and the types of plasma GLEC detected by chip-based sensing for obese diabetic subjects are expected for both the prediction of T2D and its stratification into subtypes on basis of the known slightly yet significantly increased probability for T2D in case of being descendant of a diabetic father and unaffected mother compared to the inverse constellation as well as of minor yet clinically relevant gender-specific differences in the manifestation of diabetic late complications.

The following experimental findings can be interpreted in favor of this rationale: (i) Certain GPI-AP with complete GPI anchor, such as CD73 and Gce1, were reported to be released from cultured and primary adipocytes in response to metabolically relevant stress factors, such as high levels of palmitate, $H_2O_2$, and anti-diabetic drugs (Aoki et al. 2007, Müller et al. 2008d-f, 2009a and b, 2010a, Ogawa et al. 2010). (ii) Plasma phospholipids analyzed by untargeted lipidomics have been shown to predict early neurodegeneration during presymptomatic Alzheimer's disease, that is presumably associated with metabolic disturbances (Chatterjee et al. 2015, Wurtman 2015, Casanova et al. 2016). (iii) GPI-AP, but only those harboring the complete GPI anchor, are known to be susceptible for transfer from donor to acceptor cells *in vitro* and *in vivo* in a functional state, thereby putatively transmitting biological materials within or between tissues (Rifkin and Landsberger 1990, Rooney et al. 1993, Anderson et al. 1996, Dunn et al. 1996, Ilangumaran et al. 1996, Kooyman et al. 1998, Miller 1998, Sloand et al. 1998). (iv) The level of the GPI-AP, CD73, in plasma was shown to be correlated with insulin sensitivity in diabetic mice and human probands (Lunkes et al. 2004, Guzman-Flores et al. 2015, Oyarzun et al. 2015). (v) Phospholipids in complex with membrane proteins have been reported to be released from vascular endothelial cells in culture and into plasma from mice following oxidative stress and high fat diet (Zhou et al. 2001, Mancek-Keber et al. 2015). (vi) Elevated plasma levels of GPI-AP displaying the complete GPI anchor have previously been measured for patients with ovarian cancer, glioblastoma brain tumors and lower grade colon adenocarcinomas, known to be exposed to metabolic stress as a consequence of their disease (Dolezal et al. 2014). (vii) Vesicle- and lipoprotein-like structures carrying GPI-AP have been identified in plasma (Väkevä et al.

1994, Rooney et al. 1993 and 1996, Neumann et al. 2007). Together, those data documenting elevated plasma levels of GPI-AP indicate that GPI modification is a potentially useful biomarker for the detection of certain human diseases. In fact, a GPI-specific lectin has been used previously as a diagnostic tool to detect GPI-AP in human plasma (Dolezal et al. 2014).

#### 1.2.4.2. The Study of GLEC-Associated GPI-AP

The presence of GPI-AP displaying the complete GPI anchor and GLEC in body fluids of T2D patients has not been studied so far. Typical conventional and commonly used methods for the analysis of the classical secretome in serum (e.g., ELISA, Western blotting, 2D-PAGE, mass spectrometry) are biased towards the detection of primarily high-abundance proteins. This necessitates complex and tedious fractionation procedures for enrichment of low-abundance components, such as GLEC. However, those methods most likely fail for the analysis of GLEC for the following reasons: (i) Loss in course of sample preparation by centrifugation or flotation; (ii) disrupture in course of sample solubilization for SDS-PAGE, (iii) inadequate sensitivity and quantitative or qualitative resolution; (iv) inadequate throughput due to complex procedures for enrichment.

To overcome these hurdles, a chip- and microfluidic channel-based sensor was introduced for the specific detection and biophysical characterization of large macromolecules, such as GLEC, even in the presence of serum and phospholipids that relies on their effects on SAW (Gronewold 2007 and 2009, Andrä et al. 2008, Saitakis and Gizeli 2012, Gruhl et al. 2013, Fogel et al. 2016) generated at the gold surface of chips (Gronewold et al. 2005, Voiculescu and Nordin 2012, Ding et al. 2013). Any interaction of GLEC with the gold surface will result in changes in the shape of the SAW, leading to both rightward-shifts in their phase (i.e., declines in frequency) and reductions in their amplitude. The major advantages of the SAW *vs*. the commonly used surface plasmon resonance (SPR) sensor (Biacore) rely on the possibility of measurement of large (lipid-containing) macromolecules even in the presence of turbid matrices, such as serum (Andrä et al. 2008, Voiculescu and Nordin 2012, Lee et al. 2013) as well as on the potential high sensitivity towards alterations in the composition (proteins, phospholipids) and structure (micelles, nanodiscs, vesicles, particles) of the GLEC. Albeit chip-based sensing *per se* does not enable the delineation of the type of GLEC contained in a given sample, the SAW signatures will be characteristic for the overall contents of all GLEC, either as summation signals or as 1D- / 2D-signatures of high informative value (see below).

Time-resolved phase shifts and amplitude reductions of the SAW (signatures) reflect alterations in mass loading of the GLEC to the chip surface and in their viscoelasticity, respectively, as validated with exosomes from primary rat adipocytes, which represent a subspecies of GLEC (Müller et al. 2018). Importantly, the presence of synthetic phospho-inositolglycans (PIG), that compete for the capturing of the GPI anchor of GPI-AP to α-toxin, enabled the differentiation between GLEC released from adipocytes of varying size. Pairwise comparison of SAW signatures produced by the GLEC in pooled or individual serum samples in the presence of PIG enabled differentiation according to either genotype (between normoinsulinemic-normoglycemic lean Wistar and lean ZF rats or hyperinsulinemic-normoglycemic obese Wistar and hyperinsulinemic-hyperglycemic obese ZDF rats or hyperinsulinemic-normoglycemic obese ZF and hyperinsulinemic-hyperglycemic obese ZDF rats) or body weight (between normoinsulinemic-normoglycemic lean ZF and hyperinsulinemic-normoglycemic obese ZF rats or hyperinsulinemic-normoglycemic lean ZDF and hyperinsulinemic-hyperglycemic obese ZDF rats or normoinsulinemic-normoglycemic lean Wistar and hyperinsulinemic-normoglycemic obese Wistar rats). Loss of the differentiating SAW signatures in course of elimination of GPI-AP, phospholipids or cholesterol and exposure toward physical stress is compatible with serum GLEC constituted by weak interactions between their components within a non-vesicular configuration. Thus, serum GLEC may be correlated to early stages of T2D and regarded as "phenomenological" biomarker for the differentiation and prediction of stress-related disorders.

In greater detail, the determination of changes in the size/mass (as reflected in phase shift) and viscoelasticity (as reflected in amplitude reduction) of the GLEC enables the differentiation according to both genotype (Wistar, ZF, ZDF) and phenotype (lean, obese) of the rats. However, the differences in phase shift and amplitude elicited by serum GLEC are more prominent for comparisons between rats (of similar body weight) differing in genotype than for those between lean and obese rats (of identical genotype). With regard to a putative association between the presence of GLEC in serum and the metabolic state in rats, the data are compatible with (i) hyperglycemia alone or in combination with genetic factors (obese ZDF *vs.* obese Wistar, obese ZDF *vs.* obese ZF), (ii) hyperinsulinemia alone (lean ZF *vs.* obese ZF, lean Wistar *vs.* obese Wistar) or in combination with genetic factors (obese ZF *vs.* obese Wistar) or (iii) both hyperglycemia and hyperinsulinemia alone (lean ZDF *vs.* obese ZDF) acting as driving forces, i.e., metabolic stress factors for the release of GLEC of the size/mass and viscoelasticity as detected by chip-based sensing of phase shift and amplitude. Remarkably, differences for GLEC were measured between rats which did not display significant deviations in fasting glucose and plasma insulin levels (lean Wistar *vs.* lean ZDF, lean Wistar *vs.* lean ZF). This raises the possibility that GLEC appear in serum prior to gross metabolic disturbances, which would make them particularly attractive as candidates for biomarkers for the prediction of T2D.

Importantly, the predictive power apparently relies on the "configuration" or "signature" formed by GPI-AP and phospholipids, since sensing for GPI-AP alone (α-toxin chip without "sandwich" detection of phospholipids) or sensing for phospholipids alone (annexin chip without prior capturing of GPI-AP) or the presence of detergent in the sample does not enable differentiation between normal and insulin-resistant/diabetic sera. Moreover, the protein moieties of GPI-AP, separated from the GPI anchor by lipolytic or proteolytic cleavage, *per se* are most likely not involved in eliciting the phase shift and amplitude differences since proteomic analysis of serum should already have identified those fragments and differences in their levels associated with the normal and insulin-resistant/diabetic state.

It seems unlikely that the maximal differences in phase shift and amplitude reduction observed between normal and insulin-resistant/diabetic rat sera are due to differences in the absolute composition of the GLEC, as would be reflected in differences in their mass and size, exclusively. Rather, differences in the relative "arrangement" of the components within the GLEC, in general, in the "configuration" or "neighborhood" formed by the GPI-AP together with "affine" serum components are presumably responsible for the production of the differential signatures measured with the chip-based sensor. As a consequence, those difference in the "arrangement" or "configuration" of the GLEC will not be detected by simple mass determination, e.g., using mass spectroscopy. Moreover, the structures, including mass and size, of the GLEC are probably not homogenous and uniform and may vary according to their cellular/tissue origin. This heterogeneity and the intrinsic instability of the GLEC would presumably prevent their (MALDI) mass spectroscopic analysis, which in principle can be performed upon coupling of the chip-based sensor (Treitz et al. 2008) and displacement of the GLEC from the chip.

Taken together, the findings obtained with GLEC so far (Müller et al. 2018) make the involvement of EV rather unlikely. Moreover, GLEC in rat serum exhibit considerably higher sensitivity toward physical stress compared to EV released from primary rat adipocytes and GLEC present in adipocyte incubation medium, despite the expression of GPI-AP and phospholipids in each of those entities as a prerequisite for their detection by chip-based sensing with α-toxin as capturing and annexin-V as detection molecules. Several explanations are possible for the apparent enrichment of GLEC compared to EV in course of chip-based sensing: (i) Serum from rats does not contain EV and the phase shifts and amplitude reductions monitored by the sensing are generated by "non-EV" GLEC, exclusively. (ii) Chip-based sensing of rat serum predominantly monitors "non-EV" GLEC due to methodological constraints (e.g., their small size may facilitate capturing by the chip compared to the large EV) or due to their overwhelming expression compared to EV. (iii) Serum from insulin-resistant rats contains EV from different cellular sources, among them those released from adipocytes, which exhibit relative resistance towards physical stress and others released from blood cells, which display exquisite sensitivity. Differential sensitivity of EV of different cellular origin towards physical stress has not been reported so far. However, it is well known that the major portion of EV typically contained in mammalian serum is released from macrophages, lymphocytes and other immune cells (Yanez-Mo et al. 2015). These findings strongly argue for tissues/cells different from adipose/adipocytes being responsible for the release of those GLEC which are responsible for the observed significant differences or trends in phase shift and amplitude between rats of similar body weight, but different genotype and *vice versa* between rats of identical genotype, but different body weight.

### 1.2.4.3. Working Plan, Scientific Concept and Methodology of the "GLEC Project"

The objectives and goals of the "GLEC project" rely on the identification and characterization of GLEC consisting of GPI-AP and phospholipids (and additional membrane proteins and lipids) in plasma samples, which differ between obese non-diabetic and obese diabetic rodents and humans for the prediction of T2D and vary within the population of diabetic subjects for the classification of T2D with regard to amount, composition and biophysical properties. Those differences will be reflected in the 1D- and 2D-signatures specific for the corresponding GLEC as revealed by chip-based sensing. Ideally, reliable and reproducible differences in the signatures, i.e., statistically significant correlations in the SAW signals between obese non-diabetics and obese diabetics, will be detectable using moderate sample volumes at early time points during the pathogenesis rather than at the onset of frank T2D, only. Introduction of novel T2D subtypes will depend on clustering of the individual 1D- and 2D-signatures deduced from the SAW signals of GLEC from obese diabetic rodents and humans into distinct well-separated patterns.

GPI-AP have been shown to exhibit high susceptibility for the release in GLEC from the surface of mammalian cells *in vitro* and *in vivo* in response to cellular and metabolic stress, such as high glucose, fatty acids and reactive oxygen species (Müller et al. 2009a and b). This is commonly thought to rely on their anchorage at the outer leaflet of the plasma membrane phospholipid bilayer by the GPI moiety, exclusively, that is covalently attached to the carboxyl-terminus of the polypeptide chain. However, the presence of GLEC in the plasma of T2D patients, that are faced with elevated blood glucose and free fatty acids levels, has not been studied so far. This is presumably due to conceptual restrictions (i.e., "reductionistic" and "holistic" thinking) (Figure 7) and technological challenges (see above).

For generation of the GLEC-specific SAW signal, GLEC are specifically bound to the chip surface *via* interaction of their GPI-AP and/or the phospholipid constituents with the capturing molecules, aerolysin (bacterial toxin which specifically binds to the core glycan of the GPI anchor) and annexin (peripheral membrane protein which specifically binds phosphatidylserine in the presence of $Ca^{2+}$), respectively. The capturing molecules are covalently coupled to 2D-SAM- or 3D-dextran-coated chips using routine chemical crosslinking techniques prior to measurement. Subsequent time-resolved measurement of phase shift and amplitude reduction as summation signals integrating the effects of all of the various GLEC at the various time points reflects the "overall" association kinetics of all GLEC contained in the sample and binding to the chip through either their GPI-AP or phospholipid constituents.

Dissociation of the GLEC bound to the chip is initiated by the addition of increasing concentrations of GPI glycan cores, i.e., synthetic PIG, which compete with the GPI-AP for binding to the aerolysin capturing molecules, or of EGTA, which disrupts the $Ca^{2+}$-dependent binding of the phospholipids to the annexin capturing molecules, respectively. The time-resolved data for phase shift and amplitude indicate the dissociation kinetics of GLEC for their GPI-AP or phospholipid constituents. This leads to one-dimensional (1D) signatures reflecting the sequential dissociation of the various species of GLEC in the sample depending on the relative number and accessibility for capturing of their GPI-AP and phospholipid constituents, respectively.

Cycles of sequential association with and dissociation from the chip of the GLEC through capturing *via* their GPI-AP and subsequently phospholipid constituents or *vice versa* will be run by using chip surfaces coated with both aerolysin and annexin. Alternative changes of the binding conditions enable the simultaneous or successive monitoring of the presence of GPI-AP and phospholipids in the various GLEC species along the following cycling protocol: (1) absence of PIG + absence of $Ca^{2+}$ for binding *via* GPI-AP, only; (2) absence of PIG + presence of $Ca^{2+}$ for binding *via* both GPI-AP and phospholipids; (3) presence of PIG + presence of $Ca^{2+}$ for displacement from GPI-AP and (re-)binding *via* phospholipids, only; (4) presence of PIG + presence of EGTA for release from phospholipids and resulting detachment from the chip. The time-resolved measurements of phase shift and amplitude reduction during each of the four cycles indicate the association and dissociation kinetics of all GLEC species contained in the sample *via* both their GPI-AP and phospholipid constituents. This two-dimensional (2D) analysis leads to signatures reflecting the sequential association and dissociation kinetics of the various GLEC species which depends on the relative number and accessibility of their GPI-AP and phospholipids for single (cycles 1, 3) and double (cycle 2) capturing and release (cycles 3, 4).

The presence of GLEC in plasma will be followed along the pathogenetic path from non-diabetic obesity to obese T2D *via* the various pre-diabetic states. Plasma will be collected from Zucker diabetic fatty (ZDF) rats (Chen and Wang 2005) and diet-induced obese (DIO) mice (Tschöp et al. 2011), which both have been fed with high fat diet and continuously monitored for metabolic parameters for several weeks. This procedure enables the discrimination between obese non-diabetic, obese pre-diabetic (for prediction) and obese diabetic (for stratification) animals. Over the years ZDF male rats have become a preferred animal model for obesity-linked T2D with predictable progression from the pre-diabetic to the diabetic state.

In addition to rodent models of T2D, plasma samples from overweight, obese and obese diabetic probands recruited as KORA cohort (Meisinger et al. 2016) and well-characterized with regard to metabolic parameters and diabetic late complications will be analyzed. Particular care will be taken to preserve the original structure of the GLEC during the procedures of blood collection and sample processing (e.g., GLEC enrichment, serum protein deprivement), if required. The information contents will be increasing from simple summation signals *via* 1D-signatures to 2D-signatures. For the quantitative evaluation of the summations signals, appropriate software has already been successfully introduced and made available by SAW Instruments Inc (Bonn, Munich, Germany). Corresponding tools for the analysis of 1D- and 2D- signatures will be developed in parallel with generation of the primary data. They have to reflect the shapes of the association and dissociation curves as well as the maximal changes in phase and amplitude during each cycle at each concentration of the displacing/releasing agents, such as PIG. The criteria for either summation signals or 1D/2D-signatures as the measure of GLEC will represent reproducibility, reliability and significance of their association with (pre-)diabetic phenotypes.

#### 1.2.4.4. Relevant Pilot Studies for the "GLEC Project"

Specific binding of a recombinant soluble variant of the human insulin receptor (commercial source) to human insulin was demonstrated using 2D-SAM-streptavidin-coated chip surface and biotinylated insulin as capturing molecule. The measured data ($K_D$, $k_{off}$, $k_{on}$) were in good agreement with literature data. Specific binding of detergent micelles harboring native rat insulin receptor and insulin receptor-containing rat liver membrane vesicles to insulin was demonstrated using 2D-SAM-streptavidin- or 3D-dextran-streptavidin-coated chip surface and biotinylated human insulin as capturing molecule. The association kinetics revealed specific detection of these unpurified detergent- and phospholipid-protein complexes by the chip-based sensor. So far, homogenous detection technologies for studying ligand-binding protein interactions (including optical principles) seemed to be restricted to purified samples and soluble binding partners rather than to be compatible with the (straightforward) analysis of lipid-protein complexes (Müller et al. 2014).

GLEC of the vesicle type harboring the GPI-AP CD73 and other extracellular nucleotidases were prepared from the supernatant of cultured rat adipocytes upon challenge with fatty acids and reactive oxygen species or by homogenization and subsequent differential centrifugation of total rat liver or by detergent solubilization of total rat liver plasma membranes. Subsequently, these GLEC were characterized as exosomes/microvesicles, mixed detergent-GPI-AP micelles or detergent-solubilized GPI-AP using established analytical procedures. These GLEC (adjusted to identical CD73 immune reactivity) were detected and differentiated from one another with high sensitivity and reliability by the chip-based sensor equipped with a 2D-SAM-streptavidin-coated chip surface and biotinylated AMP/ADP/ATP as capturing molecules for GPI-anchored extracellular purine-specific nucleotidases (Antonioli et al. 2013), among them CD73, and programmed for the measurement of the phase shifts and amplitude reductions as summation signal (Müller et al. 2014).

Subsequent experimentation with cultured primary rat adipocytes revealed that the established SAW sensing technology was able to resolve the differences in the releasing efficacy of CD73-harboring GLEC between the stress stimuli, palmitate, reactive oxygen species and the anti-diabetic sulfonylurea drug glimepiride, in a precision comparable to the previously used conventional and tedious preparative and analytical methods (Müller et al. 2009a and b). This was also true for the differences in the stress-induced release of GLEC between small and large adipocytes prepared from young and old rats, respectively, as well as between insulin-sensitive and insulin-resistant adipocytes generated by primary culture in the absence and presence of high concentrations of glucose and insulin, respectively (Müller et al. 2014). Quantitative deviations in relative changes between the phase shifts and amplitude reductions elicited by the various stimuli in the various types of adipocytes confirmed that the two SAW signals monitor different properties of the GLEC, mass/size and viscoelasticity.

For capturing of all GLEC harboring any GPI-AP and phospholipids, 3D-dextran chips were coated with either aerolysin or annexin, respectively, and then assayed for binding of CD73 (affinity-purified from detergent-solubilized rat liver membranes) and phosphatidylserine liposomes. Analysis of the SAW signals revealed the specific and sensitive detection of CD73 *via* capturing of its GPI anchor by the aerolysin coat and of the liposomes *via* capturing of their phosphatidylserine residues by the annexin-V coat. These successful capturing modes represent a prerequisite for the unbiased detection and of any GPI-AP-harboring GLEC. Preliminary data indicate that the detection is compatible with the presence of unfractionated mouse serum in the sample upon 10-fold dilution, at least (Müller et al. 2014).

Positive results, i.e., correlative data, would strongly argue for the principal usefulness of GLEC as biomarkers for the prediction of T2D and for its stratification into distinct pathophysiologically relevant subtypes (Häring 2016, Stefan et al. 2016). The elucidation of their nature (molecular composition, morphology, structure) will be addressed during subsequent studies using western blotting, mass spectrometry, dynamic light scattering, electron microscopy etc. upon their release from the chip. Moreover, a positive outcome will foster the systematic exploration of GLEC for T2D prediction in longitudinal studies with well-characterized patient cohorts available at the Helmholtz Center Munich as a member of the German Diabetes Center as well as for T2D stratification with well-characterized patients suffering from diabetic late complications.

Ethical, social and legal considerations will not be relevant due to the nature of the development of the bioanalytical method *in vitro* and the use of plasma samples from T2D animal models and human obese/diabetic subjects, which were generated primarily for other scientific reasons.

### 1.2.4.5. GLEC-Associated GPI-AP and the Differentiation and Prediction of (Metabolic) Stress States

The GLEC released into serum from cells/tissues of normal and insulin-resistant and diabetic rats most likely harbor the GPI-AP, the nucleotide-metabolizing enzymes Gce1 and CD73, since their removal by adsorption to agarose beads equipped with their corresponding substrates, cAMP and AMP, significantly diminished the phase shift and amplitude reduction of the SAW elicited by native GLEC. This finding prompts speculations about the nature of the cells/tissues releasing GLEC and their physiological role in the normal and disease state. The pronounced expression and metabolic stressor- (e.g., palmitate-, $H_2O_2$-) induced release of Gce1 and CD73 from primary rat adipocytes have previously been demonstrated (Müller et al. 2012b) raising the possibility of adipose tissue operating as the source for the GLEC. However, at present a contribution of other insulin target cells/tissues, such as liver and muscle, and blood cells, such as macrophages, lymphocytes and reticulocytes, cannot be excluded on basis of their relative abundance and proven capability of secreting EV (Lawson et al. 2016), which typically harbor GPI-AP (Bütikofer et al. 1989, Lopez-Cobo et al. 2016).

The enzymic functions of Gce1 and CD73 in (c)AMP metabolism on the surface of adipocytes leading in concert to the degradation of cAMP, which triggers lipolysis, and finally *via* AMP to adenosine, which blocks lipolysis (Cieslak and Roszek 2014, Dzeja and Terzic 2009, Antonioli et al. 2015), may be considered as relevant for a negative feedback regulation. Thereby excess incorporation into and release from lipids, respectively, of fatty acids operating in insulin-sensitive and resistant states, respectively, would be prevented. In this regard, the reduction of the coordinated release of Gce1 and CD73 into GLEC from the surface of adipocytes in response to metabolic stress factors would counteract the upregulation of lipolysis, which is typical for insulin-resistant adipose tissue, through elevated (c)AMP metabolism at the immediate adipose surface area by those plasma membrane-associated GPI-AP.

On the basis of the initial working hypothesis that the release of GLEC from metabolically relevant cells and tissues as a type of metabolic stress signal is associated with or even mediated by the development of the insulin resistant and diabetic state, its correlation with a decrease rather than an increase in phase shift was quite unexpected. Since right-warded shifts in the phase predominantly reflect mass loading to the (here α-toxin-coated) sensor chip surface (albeit overlapped by a minor component exerted by increases in the viscosity of the sample fluid passing through the channels), its decrease in lean ZDF compared to lean Wistar rats, lean ZDF compared to lean ZF rats, obese ZDF compared to obese Wistar rats and obese ZDF compared to obese ZF rats is compatible with a reduction in the mass and/or amount of the GLEC in the serum of the former compared to the latter rats. Serum levels of the GLEC will be critically determined by the rates of their release from cells/tissues and degradation in the circulation. Since neither the structure nor the mechanism of biogenesis of the GLEC are known at present, it can only be speculated about the molecular mechanisms of their turnover and the nature of their alterations in course of the development of insulin resistance and T2D.

#### 1.2.4.6. Model for the Release and Degradation of Serum GLEC in Normal and Insulin-Resistant/Diabetic Rats and Their Analysis by SAW Chip-Based Sensing

The presence of the GPI anchor *per se* appears to confer some important behavioral and functional attributes onto the protein moieties to which it is attached (Deeg and Davitz 1995, Ilangumaran et al. 1996, Nosjean and Roux 2003). In particular, the higher lateral mobility of GPI-AP within the plane of the phospholipid bilayer compared with typical transmembrane proteins may enable their association with specific membrane microdomains, such as lipid rafts (see above). In adipocytes, which are specifically devoted to glucose and lipid metabolism, GPI-AP and (glyco-)phospholipids, both equipped with saturated long-chain fatty acids, in concert with cholesterol have been reported to be enriched in lipid rafts at the extracellular leaflet of plasma membranes. Importantly, lipid rafts of higher and lower cholesterol content can be discriminated from one another in rat adipocyte plasma membranes with the GPI-AP typically residing at high cholesterol lipid rafts and moving to low cholesterol lipid rafts in signal-dependent fashion. Clustering in lipid rafts may be responsible for weaker association of GPI-AP with plasma membranes compared to transmembrane proteins and thereby facilitate release from the cell surface of GPI-AP with the complete GPI anchor remaining attached, such as CD73 and Gce1, in complex with cholesterol and phospholipids (i.e., as GLEC) and in positive correlation to the cholesterol content of the lipid rafts.

In the normoinsulinemic normoglycemic state of (wild-type) rats, the expression of high cholesterol lipid rafts favors the release of GLEC (Figure 9, thick arrow). Several configurations of the constituents within the GLEC are conceivable: Fragments from the extracellular plasma membrane leaflet encompassing a single or few copies of lipid raft(s) may form "open phospholipid monolayer sheets" initially. In course of release or shortly thereafter they may acquire a "closed" micellar structure spontaneously or upon integration of additional constituents or in course of fusion with other "sheets" in the blood. Alternatively, sequential bulbing, incision and fusion of the extracellular leaflet of plasma membranes at high cholesterol lipid rafts may result in the separation from the donor cell of "closed" micellar structures harboring the typical GLEC constituents. Importantly, for each of these mechanisms the membrane rigidity at high cholesterol lipid rafts may control the rate of the spontaneous as well as controlled release of GLEC in positive fashion, either through stabilization and extrusion of "open sheets" or through facilitating the shaping of "closed" micelles. The impact of the plasma membrane cholesterol content and thereby rigidity *vs.* fluidity of lipid rafts on their functionality has previously been underscored by demonstration of a correlation between cholesterol deprivation from lipid rafts and impairment of insulin signal transduction *via* the lipid raft-associated insulin receptor in rat adipocytes (Figure 9).

GLEC are thought to be remodeled and degraded in blood during a multi-step process initiated by lipolytic separation of the protein moiety-glycan core and phosphatidate residue of the GPI-AP by serum GPI-PLD (Figure 10). *In vitro* GPI-PLD action necessitates presentation of the GPI-AP within detergent micelles. In blood this requirement may be met by GLEC, since serum GLEC in the absence of exogenous detergent represent substrates for GPI-PLD *in vitro* as shown previously (Müller et al. 2018). Assuming low activity of GPI-PLD in the normoinsulinemic normoglycemic state, the GPI-AP coat of GLEC becomes degraded slowly only (thin arrow). Efficient release in concert with low degradation is responsible for the appearance of GLEC at high number and of high spec. mass and large size with extended, uniform and rigid GPI-AP coat of high elasticity in the serum from normal rats.

Detection of serum GLEC from normal rats (green line) by chip-based sensing in comparison to buffer alone (grey line) reveals (Figure 9, inset) (i) a pronounced SAW phase shift (red right-warded arrow) due to capturing of GLEC with a certain spec. mass and size at a certain number and (ii) a small reduction in the SAW amplitude (blue down-warded arrow), only, due to capturing of GLEC of high elasticity and low viscosity (for symbols see Figure 10).

In the hyperinsulinemic (hyperglycemic) state of obese Wistar and ZF (ZDF) rats, low cholesterol lipid rafts, originating from the high cholesterol lipid rafts, release GLEC of reduced spec. mass and size and/or at reduced number (Figure 10, thin arrow). It has been proposed that during intensive synthesis of triglycerides and accompanying biogenesis of lipid droplets in enlarging adipocytes, cholesterol is translocated from plasma membrane lipid rafts as its major storage site in small adipocytes to the emerging lipid droplets for insertion into their outer phospholipid monolayer, leading to transient depletion of cholesterol from lipid rafts. With regard to adipocytes, at least, this molecular mechanism driven by insulin stimulation of lipid synthesis could induce the transformation of high cholesterol lipid rafts in the plasma membranes of normoinsulinemic normoglycemic rats into low cholesterol lipid rafts of hyperinsulinemic (hyperglycemic) rats. Thereby, the release of GLEC into blood would be coupled to the metabolic state *via* the cholesterol content of the plasma membrane lipid rafts of the releasing cell with the higher the insulin plasma concentration, the lower/higher the amount of high/low cholesterol lipid rafts and the lower the efficacy of GLEC release. It remains to be clarified whether a similar mechanism could explain the apparent transfer of GPI-AP between cells *in vivo* which has been implicated a physiological role in intercellular material and/or information transfer.

Assuming high activity of GPI-PLD in the hyperinsulinemic (hyperglycemic) state, the GPI-AP coat of GLEC becomes intensely degraded (thick arrow). Extensive shaving of the GPI-AP coat from the surface of the GLEC would also provide an explanation for the seemingly contradictory decline in amplitude in parallel to the decrease in phase shift in course of development of the hyperinsulinemic (hyperglycemic) state. Typically, a portion of the amplitude reduction can be attributed to mere capturing of the GLEC, since mass loading *per se* will cause dampening of the maximal amplitude of the SAW. In addition, the viscoelasticity of the GLEC is presumably affected by the GPI-AP coat with the less extended and intact, the lower/higher the elasticity/viscosity of the GLEC and the higher the amplitude reduction provoked by energy absorption through the viscous GLEC. Consequently, hyperactivation of serum GPI-PLD in hyperinsulinemic (hyperglycemic) rats would lead to considerable increases in GLEC-induced amplitude reduction compared to the elastic GLEC with intact GPI-AP coat from normal serum.

Detection of serum GLEC from hyperinsulinemic rats (Figure 10, upper inset, green line) by chip-based sensing in comparison with buffer alone reveals (grey line) (i) a small SAW phase shift only (red right-warded arrow) due to capturing of GLEC with reduced specific mass and size and/or at diminished number and (ii) a pronounced reduction in the SAW amplitude (blue down-warded arrow) due to capturing of GLEC of low elasticity and high viscosity.

The direct comparison of serum between hyperinsulinemic and normoinsulinemic (normal) rats (Figure 10, lower inset) demonstrates for the former the considerably reduced phase shift (red left-warded arrow) due to downregulated specific mass and size and/or number in parallel to the diminished amplitude (down-warded blue arrow) due to elevated viscosity of the GLEC with incomplete GPI-AP coat.

Taken together, the parallel decline in amplitude (i.e., elasticity of GLEC) and phase shift (i.e., specific mass/size/amount of GLEC) when measuring serum from hyperinsulinemic (hyperglycemic) *vs*. normal rats can be reconciled best by a single molecular mechanism, the hyperactivation of serum GPI-PLD. In agreement, pretreatment of serum samples with GPI-PLD resulted in drastic and parallel declines in both phase shift and amplitude to up to concentrations which interfered with capturing due to complete elimination of the GPI-AP. It would be interesting to determine whether the single-nucleotide polymorphisms (SNP) in the GPI-PLD gene found associated with T2D patients in a recent genome-wide association study is correlated with an increase in the amount and/or activity of GPI-PLD.

#### 1.2.4.7. Value of Chip-Based Sensing of GLEC-Associated GPI-AP for Biomarker Research

Partially GPI-PLD-digested GLEC, as may be generated in serum from hyperinsulinemic (hyperglycemic) rats, are presumably of heterogenous nature differing in the relative portions of complete GPI-AP, cholesterol and phospholipids and, in consequence, in spec. mass, size and viscoelasticity. (Differential) Analysis of these parameters is not possible with classical biochemical methods (e.g., immunoblotting) or necessitates pre-fractionation steps (e.g., MS), which are difficult to perform due to the inherent instability of the GLEC. This reinforces the value of SAW chip-based sensing of macromolecular complexes, in general, and of GLEC, in particular, for the identification and application of disease biomarkers with higher information content than that provided by a single biomolecule species.

The following methodological points argue for the application of SAW chip-based sensing: The limited expenditure and costs for the preparation of adequate chips relying on the possibility for regeneration and multiple use of the chips, on the stability of the capturing molecule α-toxin, on the ease of its recombinant production and on efficient regeneration of the chips using mild conditions for the release of the captured GLEC (competitive displacement by phosphoinositolglycans rather than de-/renaturation of the capturing molecule using low pH and partial unfolding) are compatible with medium throughput and use in routine clinical practice (Müller and Tschöp 2015 and 2016). These advantages will facilitate longitudinal studies for clarification of the challenging question whether changes in the size/mass and/or viscoelasticity of serum GLEC represent (late or early) correlates for the (pre-)diabetic state and development of T2D, e.g., by SAW chip-based sensing of the GLEC in serum from ZDF rats along the development from normoglycemia to hyperglycemia in course of diet-induced (moderate) obesity.

Moreover, the clinical use of a biomarker necessitates the delineation of critical absolute thresholds values to avoid the pairwise comparison of the patient sample with an appropriate control as performed (Müller et al. 2018, in press) with the multiple comparative analyses of sera from diabetic obese ZDF rats *vs*. "healthy" lean ZDF, obese ZF, lean ZF, obese Wistar or lean Wistar rats. The data obtained raise the question about the relevance of the genetic background and body weight of those "controls" in the development of T2D, i.e., whether the patient samples have to be compared with those from "healthy" probands of matched genotype and/or body weight or from randomly collected populations.

The final experimental design for SAW chip-based sensing of serum GLEC enabled the elucidation of absolute values for phase shift and amplitude which differentiate between the genotype and/or body weight of rats without the need for parallel comparative analysis of "controls." It is based on the elimination of background signals caused by serum entities which are captured and detected by the sensor chip, but apparently are unrelated to the (pre-)diabetes-associated GLEC as revealed by their resistance toward physical stress. Future cross-sectional studies relying on the SAW chip-based sensing protocol already established (Müller et al. 2018) have to confirm that threshold values for phase shift and amplitude allow the differentiation according to (pre-)diabetic states for randomly collected rat cohorts and to demonstrate that GPI-AP assembled together with phospholipids in GLEC in serum are correlated to (pre-)diabetes in humans.

### 1.2.4.8. GLEC-Associated GPI-AP and the Problem of Causal Mechanisms

In theory novel biomarkers for the discrimination between the healthy and diseased states and some states in between can be discovered by exposing body fluids to any instrument and looking for differences in the resulting read-outs. By using a huge number and variety of instruments each characteristics of each component in the body fluid should be covered by one of those instruments, some or at least one of them will reveal a difference between the health and disease state. However, this apparent absence of bias cannot be achieved in the daily and experimental practice or in other words the complete lack of any bias for physiological as well as instrumental sensory systems is impossible. There is always selection of what can be recognized and of the conditions enabling that recognition, e.g., human eyes are biased for light of a certain wave length and for a certain period of stable appearance of the object in front of the lens.

On the basis of theoretical considerations and in analogy to the Heisenberg uncertainty principle of quantum mechanics, it cannot be excluded that simultaneous/ parallel determination of the biophysical characteristics, i.e., phase shift/amplitude, and the structure of the GLEC is impossible. The uncertainty principle implies that measurement of the position and impulse of objects requires the use of an apparatus equipped with stationary and mobile component(s), respectively, that cannot be realized within a single instrument. The determination of the structure of the GLEC would necessitate their stable formation independent of the instrument or the experimental conditions used. However, this might be incompatible with constitution/ expression of their biophysics, as becomes manifested in the effects on phase shift and amplitude of the SAW, that is heavily affected by the surrounding milieu of the GLEC at each time point of their life cycle. This milieu encompasses the blood stream *via* the handling and storage conditions to the microfluidic channels of the sensor chip and consequently will depend on both the instrument and the observer/scientist, who intends to unravel the signals produced by the instrument.

Nevertheless, the presented SAW chip-based sensing of labile GLEC, even without precise knowledge of the underlying molecular structures or biogenetic mechanisms, may be relevant for biomarker research. It fulfills the "correlation-is-enough" paradigm (Anderson 2008) and avoids the search for causality and causal mechanisms as primary goal, since it would impair the identification of novel predictive biomarkers by introducing a strong bias for the type of correlative partners. However, a completely unbiased approach is principally impossible. But nevertheless, on the basis of the awareness of a (limited and hopefully minimal) bias being inherent in each scientific approach, both the instruments and the researchers should remain open for the perception of unexpected phenomena.

In fact, the novel approach relies on a minimal bias, (i) the presence of GPI-AP harboring the complete anchor structure and (ii) their association with phospholipids (Müller et al. 2017a). Strikingly, the nature of the aggregates/complexes and the modes of their physiological or artificial generation as well as their physiological function or lack of function do not matter. Furthermore, the strength of the interaction between GPI-AP and phospholipids and other putative components may be very low and consequently the structure of the complex be of extraordinary instability. Presumably, GLEC do not survive conventional procedures of sample preparation or complex isolation, such as centrifugation, immune precipitation, filtration. Possibly, the components of the GLEC do not come into close/physical contact with each other, as is true for "real" typical macromolecular complexes. Rather GLEC components seem to reside in certain neighborhood/proximity, only, held together, i.e., kept in a certain configuration to one another over certain distances without direct physical contact by weak secondary (hydrophobic, Van der Waals) interactions. Those "configurations" or "Gestalten" (rather than typical complexes) and consequently the predictive information and power contained in them may have escaped detection so far by using conventional methods, by removing or destroying in course of sample preparation or analysis.

Moreover, the value of a biomarker for the prediction of (metabolic) diseases critically depends on (i) appearance and technical provability prior to disease onset, (ii) detection in easily accessible body fluids and (iii) threshold values for the differentiation of individuals according to (metabolic) disease genotype and phenotype. These criteria may be fulfilled better by GLEC rather than by individual molecular entities. Thus, GLEC seem to have the potential to fill the gap between (epi-)genetic and molecular/ phenotypic biomarkers for the prediction of common complex diseases.

According to the current working hypothesis, which is based on the reported experimental findings (Müller et al. 2018), GPI-AP equipped with complete anchor are released from donor cells in course of (metabolic) stress. Thereafter, the GLEC are formed at the boundary between cell and environment or even at a subsequent stage in the body fluid by the spontaneous interaction of the GPI anchor with certain serum components (phospholipids, cholesterol, transmembrane proteins). The GLEC are thought to build loosely associated micelle-like (or nanodisc-like) structures, which differ in composition, size and/or specific mass and/or configuration (conformation) between normal (healthy) and disease (diabetic) states, which may depend on the presence or absence of the respective GLEC-associated components in the body fluid. Importantly, the formation of GLEC may not be causally, i.e., mechanistically involved in the pathogenesis of T2D. Instead it may be secondary to the release of stress sensors, such as GPI-AP from the surface of relevant tissue cells, into the body fluid and their association with constituents expressed in the body fluid at relevant concentrations or not during (metabolic) stress (Müller et al. 2017b). The association may be weak and represent just a "configuration" driven by the immediate neighborhood of the released GPI-AP to certain body fluid components and weak secondary interactions. Those complexes may not be identified by focused approaches, as manifested in theoretical considerations about known molecular pathways or mechanisms of protein/vesicle/complex release from cells or in the measurement of released materials using established experimental procedures (e.g., centrifugation, OMICS).

Taken together, the chip-based sensing of GLEC-associated GPI-AP represents a minimally biased "hermeneutic phenomenological" approach (Figure 7). The time-resolved alterations in the amount and/or physicochemical characteristics of GLEC-associated GPI-AP operating at the boundary between the organism and its environment and representing components of both the organism and the environment, may be interpreted in a broader sense as "fields" or "Gestalten" originating from the (metabolic) state of that organism. "Fields" or "Gestalten" formed by GLEC-associated GPI-AP have to be recognized as being constructed by multiple human and non-human agents, among them the organism plus the environment plus the analyzing process constituted by the chip-based sensor instrument plus the interpreting scientists. This construction is neither random nor solely due to social forces but dependent on a multitude of human and non-human agents as has been adequately discussed in the "acteur"-network-theory (Latour 2010). Consequently, the SAW chip-based sensing of GLEC-associated GPI-AP will eliminate the need for the epistemological separation or distinction between the observed phenomenon and the agencies of observation.

This "back to the things," i.e., here life science phenomenons, may enable the recognition of correlates for transcendental processes, exceeding the traditional read-outs based on the (environmental) stimulus/cause – ("organismal") response/effect scheme, which is commonly used, for instance for the prediction of common complex diseases. The concept of the recognition of "fields" and "Gestalten," which initially have been introduced by Anthony Jack and Tim Shallice (Jack and Shallice 2001) as correlates for intentionality and directedness, could be transformed to (metabolic) stress situations and then lead to a novel level of the prediction of common complex (metabolic) diseases in course of merging realistic and "constructivistic" views about their development into the so-called "agential realism".

#### 1.2.4.9. GLEC-Associated GPI-AP under Hermeneutic Phenomenological View

The analysis of GLEC-associated GPI-AP using chip-based sensing on the basis of SAW technology can be regarded as typical example for an underlying type of theory of knowledge or philosophy of science which may be characterized best as "hermeneutic phenomenology." According to a simplified view, this approach represents a synthesis of three philosophical paths for thinking, arguing and acting as a researcher, the phenomenology, hermeneutics and analytical or language philosophy introduced by Edmund Husserl (Husserl 1996), Martin Heidegger (Heidegger 2009), Maurice Merleau-Ponty (Merleau-Ponty 1966) and Ludwig Wittgenstein (Wittgenstein 1984) during the first half of the 20$^{th}$ century. Unter the credo "back to the things" the objects of research should be investigated under maximal reduction of the influences of preformed opinions, common sense thinking and traditional ways of scientific arguing and acting about all relevant research activities with accompanying maximal maintenance of open-mindedness and sensitivity of our perception for phenomena being so far outside the range of our research activities. The preformed opinions and unconscious phenomena, respectively, which are of greatest relevance include, but are not restricted to the choice of the research topics, the methodology used, the type of data generated, the interpretation of the results obtained, encompassing the initial aim and intention of the study as well as the consequences derived from its outcome for future research programs and, last but not least, the kind of presentation of the study in the scientific community and public as scientific paper, editorial, lecture or press release. However, it has to be recognized and accepted, that complete elimination of preformed opinions about research activities and those related phenomenons of relevance is impossible. Human beings, in general, and scientists, in particular, are always embedded in specific forms of living, culture and tradition (i.e., "Lebensform" or "Lebenswelt"), thinking and speaking which are intrinsically and deeply interconnected and cannot be separated from one another, such as mind and language. Consequently, it is essential that this limitation is kept in mind and that the impact of society, culture, tradition, language and individual life history, i.e., of the "Lebensform" or "lifeworld" is made transparent and concomitantly kept at a minimal level. This will greatly facilitate creation of the starting point for the following, in an ideal sense, mere descriptive investigation of the phenomenon envisioned thereby omitting any explanation, argumentation or theory. At best, pure data bases will be generated exclusively, such as Phenomenon (Sample) X1 > Measurement (Instrument) Y1 > Dataset Z1; Phenomenon (Sample) X2 > Measurement (Instrument) Y2 > Dataset Z2; and so on. The resulting data bases will "automatically" lead to the elucidation of correlations or associations, such as between Phenomenon X1 and Phenomenon X2. However and this is of critical importance, the data bases must not be used for the construction of seemingly causal (mechanistic) relationships, such as between Phenomenon X1 and Phenomenon X2, according to the phenomenological view "correlation is enough" (Anderson 2008). In fact, the hope is still alive that the embedding of correlative thinking in the context of the preformed opinion, tradition, lifeworld and thus "Lebensform" of the individual scientist and the scientific community may create an additional level of watching participative ("schauender"), but not theoretical argumentative, interpretation, which represents the central demand of the "hermeneutic phenomenology" (Müller 2016).

With regard to GLEC-associated GPI-AP, the "scientific ground," created and accepted by the scientific community, relied on the exclusive anchorage of GPI-AP at the outer leaflet of the plasma membrane bilayer at the boundary between organism and environment. The minimal assumption or bias derived from this "phenomenon" was the idea that this elusive topology of GPI-AP may account for their sensitivity towards changes in the (i) (patho)physiology of the cells/tissues, including the viscoelasticity of their plasma membranes, and (ii) composition and characteristics of the surroundings (e.g., body fluid, culture medium), including the mechanical stress exerted by the ECM or neighboring cells.

The use of α-toxin and SAW chip-based sensing of GPI-AP equipped with the complete GPI anchor is based on (i) the aim to identify proteins released from cell surfaces into blood, which so far had escaped detection by traditional proteomics techniques (Lim et al. 2014) and (ii) the intention to generate complexes between GPI-AP, phospholipids and cholesterol with higher information content compared to the (sum of the) individual constituents. In fact, this represents the only bias, and it is a (minimal) working hypothesis, which, nevertheless has to be recognized by the scientists involved.

Importantly, the measurement of SAW signatures was aimed to yield (material) mere correlates for certain (patho)physiological states. Initially, there was no intention to decipher the underlying molecular mechanisms and structures as well as causal relationships between the studied phenomenons, i.e., the release of complete GPI-AP and the emergence of certain metabolic genotypes or phenotypes. The overall process of the "phenomenon of the release of GPI-AP with complete anchor and its (pathophysiological meaning" can be divided into three parts: (i) Become aware of the (limited) bias and preformed opinion (i.e., working hypothesis based on tradition, state-of-the-art knowledge) underlying this phenomenon (i.e., working hypothesis of the reasons for choosing GPI-AP as indicators of the boundary between cell and environment, (ii) describe the phenomenon under avoidance of any hypothetical explanation or causal argumentation (i.e., serum samples were measured for their SAW signatures which were compiled using data bases, exclusively ("descriptive phenomenology") and (iii) convert the data base into a narrative (scientific publication, "paper") with limited interpretation under strict consideration of the initial bias (see i) and future intentions ("hermeneutic phenomenology"). Nevertheless, considering the life world ("Lebenswelt") of the author and of the relevant scientific community, the mere description of the phenomenons may be extended to a limited degree towards some introspections with regard to the mode of release of the GPI-AP and putative applications in the future. This does neither exceed the "hermeneutic phenomenological" approach nor correspond to a mere description or intentional narrative, which in scientific publications are often masked by the neutral, objective, non-personalized and untouchable style of writing (Müller 2017).

Importantly, from a "transcendental" point of view the proposed transformation of life-scientific thinking, acting and writing from the traditional explanatory, causal and narrative style, as typically used by an apparent external neutral observer, into the novel merely descriptive, correlative and hermeneutic "Schauen" of individual scientific actors as well as the scientific network (community), in total, which does not discriminate between internal and external, subject and object, does resemble the so-called interpretative phenomenological analysis (IPA). IPA has been introduced about two decades ago by the psychologist Jonathan A. Smith (Smith 2007) for studying (patho)psychological phenomena, in general, and the behavior of patients suffering from chronic complex diseases, such as T2D, in particular (Müller 2016).

In greater detail, IPA represents an approach of qualitative research initially specifically designed for psychology and social sciences with an idiographic focus. Thus it tries to provide insights into how a proband with a certain background makes sense or meaning of a given phenomenon (Reid et al. 2005, Gill 2014). Typically, those phenomena are related to personal experiences of (major) significance, such as a serious life event. IPA has been founded on the basis of the philosophical theories of phenomenology and hermeneutics (Larkin et al. 2006). Consequently, IPA has to be considered as one of several approaches of qualitative social sciences combining phenomenological, psychological, interpretative and idiographic components (Brocki and Wearden 2006).

Data collection and analysis preformed by IPA is not designed to test (working) hypotheses, which is typically the case for experimental approaches in natural sciences with the underlying paradigm of hypothesis-driven research and the aim to verify or falsify the working hypothesis. In contrast, the IPA scientist reflects upon the proband's preconceptions about the data and tries to eliminate those in order to emphasize the importance of the lifeworld of the proband. Transcripts of the interviews are analyzed in considerable detail with the main focus shifting forth and back from the major claims of the proband to the scientist's interpretation of the sense, meaning and aims of those claims. IPA's hermeneutic process is one of analysis in combination of creation of meaning. Thus, the IPA researcher tries to create meaning of the proband's efforts to create meaning of their self-experiences, resulting in a double-hermeneutic process. According to textbook knowledge, IPA could be applied for research questions aimed to elucidate what a certain experience made by the proband was like (phenomenology) and how the proband made sense of this experience (interpretation)(Smith 1996).

In fact, the analytical procedure of IPA is considered as being "bottom-up," with the scientist producing "codes" from the transcribed interview data or identifying "codes," which might fit to the data, rather than being founded on a pre-existing working hypothesis or theory. Strikingly, IPA studies do not test theories. However, they may be used for the further development or modification of existing theories. A sound IPA study is designed to balance the mere phenomenological description with moderate, yet thoughtful interpretation and to connect this interpretation in the probands' accounts. Moreover, IPA has to consider an idiographic focus to avoid any loss of specific variations and to maintain a strong emphasis on meaning rather than to stick to causal or mechanistic relations. A considerable extent of transparency, which is predominantly reflected in contextual details about the selected probands, a clear understanding of the process, meaningful commentaries on the interview data, major points indicated by verbatim quotes, is also critical for the evaluation of the transferability and plausibility of an IPA study. The inclusion of solid criteria for credibility, such as triangulation, cross-validation, independent audit or cooperative inquiry, will probably contribute to an improvement of the reader's confidence (Smith 2011).

According to Martin Heidegger, natural sciences have meanwhile reached a completely distinct meaning which does not favor wisdom as a kind of draft for the foundation of the existence of (material and non-material) entities. They even try to avoid the "opening" of those areas or frames for research which lack clear-cut boundaries with regard to the topics, objects and methods employed. In fact, modern western sciences do focus on the explanations and calculations to accomplish control over those material entities which seem to be of relevance, meaning and usefulness for their practical (often technical) application dependent on the period of time and the many acteurs involved, such as society, scientific community, prosperity (Heidegger 2014, p. 249). Consequently, for Heidegger the modern natural sciences do not create wisdom about the being of entities but, at best, they produce knowledge about the characteristics of some of their (selected) properties or functions. Thereby natural sciences are limited to the fixation and confirmation of those areas of putative knowledge which have already been accepted as being firm in a dogmatic fashion and thus be regarded as "closed" and sealed *vs*. distant as well as neighboring perspectives.

In Heidegger's sense sciences are in strict conflict with genuine philosophy with its original request of "formally" pointing to the solely creatively acting and pure transcendental nature of life (Heidegger 2014, p. 255), i.e., to the questioning and draft-like character of the being (Heidegger 1967, p. 336). It is this fragile and creative questioning, designing and pleading beyond the current understanding of selected and individual life phenomena towards novel areas, topics and methods, thereby leaving behind causal explanations and exact calculations for the phenomenon as well as for the complete control over it, which is missing completely in modern sciences. With regard to the terminology of the latter, the study object, i.e., being and life, becomes reduced to its dismembering into biological, psychological, typological, molecular (etc.) constituents, i.e., to the type of (experimental) experience as well as demands and intentions, in a simple and cheap fashion (Heidegger 2014, p. 258). In other words, thinking beyond the own boundaries has been replaced by attributing to something or, in the best case, by putting something down to its cause.

> "Die Farbe leuchtet auf und will nur leuchten. Wenn wir sie verständig messend in Schwingungszahlen zerlegen, ist sie fort. Sie zeigt sich nur, wenn sie unentborgen und unerklärt bleibt. Die Erde läßt so jedes Eindringen in sie an ihr selbst zerschnellen." ["Color is bright and wants to be bright only. If we decompose it into frequencies by explanatory measurement, it vanishes. It only shows up, if it remains veiled and unexplained. Thereby the earth itself manages to prevent anything from penetrating into it."] (Heidegger 1977, p. 33)

Heidegger used the example of psychoanalysis to illustrate how scientism tries to attribute current psychological attitudes or the individual behavior to inborn impulses, the evolutionarily "old" natural instinct or life experiences in the past (e.g., childhood, education). In analogy to the natural sciences, psychoanalytical explanations may be successful in merely making current events plausible on the basis of experiences in the past, with the critical difference that natural sciences, but not psychoanalysis, offer the advantage of providing predictive power and prognosis based on calculations. Nevertheless, in any case each explanation ultimately represents a belated reduction of a new hitherto unknown entity (e.g., experience, perception) to an already known entity. As such, the scientific explanation may be plausible, convincing and useful. However, at variance with genuine philosophy, natural sciences will never operate creatively, albeit certain explanatory theories may reflect (a pronounced extent of) the creativity of their creators, the researchers. Natural sciences have to stick strictly to reductionism with all its science-specific requirements.

This reductionistic non-transcendental thinking makes the critical difference between science and genuine philosophy for Ludwig Wittgenstein, too. He stated that it will not and must never be the matter or task of philosophers to attribute something to something else or to explain something with something else (Wittgenstein 2013, p. 39). This restriction of philosophy to generate mere descriptions and questioning drafts and to strictly avoid any explanations and hypotheses distinguishes philosophy from science in a radical and unambiguous sense.

> "Es darf nichts Hypothetisches in unsern Betrachtungen sein. Alle Erklärung muß fort, und nur Beschreibung an ihre Stelle treten." ["There must not be anything hypothetical in our considerations. All explanations must be left aside and have to be replaced by description only."] (Wittgenstein 1984a, §109)

> "Hat die Verachtung Goethes für das Experiment im Laboratorium und die Aufforderung in die freie Natur zu gehen und dort zu lernen, hat dies mit dem Gedanken zu tun, dass die Hypothese (unrichtig aufgefasst) schon eine Fälschung der Wahrheit ist?" ["Does Goethe's contempt for experiments in the laboratory and his request to go into free nature for learning there result from the idea that the hypothesis (incorrectly understood) is already a forgery of the truth?"] (Wittgenstein 1984b, p. 465)

In this regard Ludwig Wittgenstein argued like a typical phenomenologist.

> "Was ich brauche, ist eine psychologische oder vielmehr phänomenologische Farbenlehre, keine physikalische und ebenso wenig eine physiologische. Und zwar muss es eine rein phänomenologische Farbenlehre sein, in der nur von wirklich Wahrnehmbarem die Rede ist und keine hypothetischen Gegenstände - Wellen, Zellen etc. - vorkommen." ["What I need is a psychological or rather a phenomenological theory of color, neither a physical nor a physiological one. And, it must be a pure

> phenomenological theory of color, in which only really perceptible matters are mentioned rather than hypothetical objects - waves, cells etc."] (Wittgenstein 1970, p. 273)

And the same holds true for the founder of the actor-network-theory Bruno Latour.

> "Kein Forscher sollte die Aufgabe erniedrigend finden, beim Beschreiben zu bleiben. Sie ist, im Gegenteil, die höchste und seltenste Leistung." ["No researcher should find the task degrading to stick to the act of description. It is, on the contrary, the highest and rarest achievement."] (Latour 2010, p. 237)

For these philosophers or "anti-philosophers" it is most important for genuine philosophy, how questions are asked, and prior to this operational problem that questions about the being are asked at all (Heidegger 2014, p. 274). In contrast, the scientific questioning is aimed at the testing, i.e., verification or falsification, of a pre-formed working hypothesis, which typically is based on previous perceptions, experiences, knowledge and tradition. This falling back of natural sciences on pre-formed assumptions or opinions makes the very reason, that they do not put "open" or "transcendental" questions. In face of the apparently distinct or even exclusionary intentions and operational ways of proceeding of philosophy and (pure) natural sciences, in general, and the dominating scientistic understanding of being and life in the (applied) technosciences as generated by the current (western) societies, in particular, it remains to be clarified whether a broadening of scientific research to "open" and "transcendental" questioning could represent a desirable and fruitful aim and if so, whether and how this could be managed in practice.

In fact, the determination of GPI-AP assembled into GLEC in body fluids with the help of chip-based sensing for the description of certain (patho)physiological states could be regarded as pilot project for this "opening" of natural sciences, in general, and life sciences, in particular. As mentioned above, this approach of "hermeneutic phenomenology" offers the advantage of a minimal bias (i.e., release of constituents from cellular membranes in response to endogenous cellular and/or exogenous environmental stimuli) in combination with the mere description of the appearance of GLEC in relevant body fluids (on the basis of monitoring of SAW signatures) and their correlation with (patho)physiological (here metabolic disease) states, only. Thereby, the GPI-AP equipped with the complete anchor could "point" to a life phenomenon exceeding the boundaries between cell and extracellular space, organism and environment, internal and external. Most important, the chip-based sensing of GLEC-associated GPI-AP is not aimed at the generation of explanations for the delineation of causal relationships between and acquisition of control over the described phenomena *via* elucidation of the underlying molecular structures (of the GLEC) and mechanisms (of the GLEC release). This self limitation represents a prerequisite for the subsequent interpretative analysis (performed in a moderate fashion) and, in combination with it, will lead to "opening" of the formerly narrow area of scientific questioning and concomitantly to dampening or even to the elimination of the reductionistic scientistic process as a result of "hermeneutic phenomenology."

## 2. GPI-AP for Cell Surface Engineering

The transfer of GPI-AP *in vitro* was described independently by two groups even before the initial identification of GPI anchors. During the study of phospholipid exchange between liposomes and cells, it was found that certain membrane proteins become transferred selectively from erythrocytes to liposomes in a reversible manner (Bouma et al. 1977). Importantly, the transferred proteins were found to be localized to the extracellular leaflet of the erythrocyte plasma membrane bilayer, such as AChE. Subsequently, incubation of the detergent-solubilized and purified human erythrocyte membrane protein DAF or CD55 (see above) with sheep erythrocytes was demonstrated to result in their association with the erythrocytes, presumably mediated by insertion of their GPI anchor into the outer plasma membrane leaflet (Medof et al. 1984). Importantly, erythrocyte-associated DAF maintained the biological activity of its detergent-solubilized version with preventing human complement attack and the resulting erythrocyte lysis. Thereafter, the detection of modification of membrane proteins with GPI anchors and their structural characterization led to additional experimental documentations of the transfer of GPI-AP from liposomal or cellular donor membranes to target cell membranes as well as the insertion of enriched or purified detergent-solubilized GPI-AP into liposomal or target cell membranes.

### 2.1. Functionality of Inserted GPI-AP

Since in principal any protein of interest can be expressed as GPI-anchored variant, which upon incubation with cells of interest will insert into their plasma membranes *in vitro* under putative preservation of the authentic function of interest, cell surface "engineering" or "painting" was assumed to represent an attractive alternative to classical gene transfer for the manipulation of the characteristics of cell surfaces and their composition for specific biotechnological and pharmaceutical applications. The principal validity of this approach was investigated with biotinylated human DAF55 and AChE as well as murine GPI-anchored B7-1 and B7-2 as reporters for typical GPI-AP for the determination of their plasma membrane distribution and signaling function upon transfer to HeLa or Chinese hamster ovary cell plasma membranes and insertion into the outer bilayer leaflet (Premkumar et al. 2001). Three modes of association with the cell surface were observed for each of those reporter GPI-AP, (i) unspecific adsorption and/or peripheral embedding, (ii) insertion into non-raft membrane areas and (iii) insertion into lipid rafts according to the criterion of insolubility in cold non-ionic detergent (see above). Movement from localization (i) or (ii) to localization (iii) occurred, however at slow rate only, and led to the acquisition of the native function, i.e., inhibition of complement-induced formation of the MAC and surface expression of enzymic activity, respectively. No differences were reported between the four reporter GPI-AP with regard to insertion efficacy and the type of detergent used as well as their final localization in lipid rafts and colocalization with caveolin proteins and endogenous cellular GPI-AP (Premkumar et al. 2001). These data argue for maintenance of the native function of exogenously inserted GPI-AP and thus for the principal usefulness of cell surface engineering with GPI-AP, for instance, to create cells with modified cancer cell antigens for administration to tumor patients in order to elicit specific anti-tumor response.

A number of studies about GPI-AP transfer originated from the investigation of the acquired disease paroxysmal nocturnal hemoglobinuria (PNH). The molecular cause of PNH relies on defective synthesis of precursors for GPI anchors in stem cells for the blood cell lineage. In consequence, the erythrocytes of PNH patients are devoid of GPI-AP, in general, which thus can be used as biomarker for PNH. As a diagnostic tool for monitoring the lack of GPI-AP expressing the complete GPI anchor the GPI-binding lectin alpha-toxin can be used (Shin et al. 2004, Wu and Guo 2010). The complement regulatory/inhibitory components CD55, CD59 and the homologous restriction factor (HRF) are among the missing GPI-AP, which results in extreme sensitivity towards attack and lysis provoked by the autologous complement (Rosse 1993). In greater detail, CD55 interferes with the classical and alternative pathways of complement activation in course of blockade of the formation of the C3-C5 convertase complex, which is responsible for the generation of C5b. C5b triggers the formation of MAC, which induces the generation of pores in the target cell membranes. Thus, CD59 and HRF interfere with the formation of MAC. Strikingly, isolated and washed erythrocytes from PNH patients (re-)gained considerable resistance towards complement-mediated destruction in isotonic buffer by mere treatment with detergent-solubilized and purified CD55 (Medof et al. 1985), CD59 (Wilcox et al. 1991) or HRF (Zalman et al. 1987). Apparently, the transferred or inserted GPI-AP manage to exert their biological activity of MAC inhibition and therefore seem to be inserted into the outer plasma membrane leaflet of the target cells in correct fashion.

Since many mammalian GPI-AP are involved in signal transmission across T-cell plasma membranes upon their crosslinking by antibodies (Robinson et al. 1991b and 1997) or in response to ligand binding to peptide receptor Qa-2 (Robinson et al. 1989), the question emerged whether transferred or inserted GPI-AP manage to transduce signals across the plasma membranes to downstream effector systems. Strikingly, immediately after insertion into mouse lymphocytes or human neutrophils, neither rat brain Thy-1 (Zhang et al. 1992) nor sheep CD59 (Morgan et al. 1993), respectively, succeeded in efficient signal transduction across plasma membranes to downstream effectors. However, upon prolonged incubation with the target cells and following redistribution in plasma membrane lipid rafts, human CD59, inserted into U937 monocytes, elicited intracellular $Ca^{2+}$-fluxes, which was taken as strong indication for the acquisition of signaling competence (van den Berg et al. 1995). Consequently, it was concluded that upon exogenous insertion into target cell plasma membranes, GPI-AP gain functionality in course of their redistribution to those sites or membrane subdomains, which represent the preferred localization for the majority of GPI-AP, the lipid rafts.

## 2.2. Mechanisms of GPI-AP Insertion

Ample experimental evidence is available for mediation of the direct insertion of exogenous GPI-AP into the outer plasma membrane leaflets by their saturated long-chain fatty acyl chains as initially had been demonstrated for glycophospholipids (Saqr et al. 1993): (i) The insertion strictly depends on the presence of the diacylglycerol or phosphatidate moiety of the GPI anchor as shown by its abrogation as a result of their enzymatic or chemical removal (Walter et al. 1992). (ii) Membrane insertion of GPI-AP with GPI anchors harboring two fatty acyl chains is more stable compared to that of GPI anchors with only a single acyl residue left, as revealed by efficient extraction using liposomes or serum lipid-binding proteins (Walter et al. 1992). (iii) The efficient insertion of CD55 and trypanosomal VSG into erythrocyte membranes which had been pretreated with proteinase K and sialidase, respectively (Rifkin and Landsberger 1990) as well as the failure of high-salt treatment to extract inserted CD55 (Medof et al. 1984) argue against mere peripheral association of GPI-AP with proteinaceous or lipidic cell surface components *via* electrostatic interactions. (iv) Hydrophobic interactions of the GPI fatty acyl chains with plasma membranes of high lipophilicity, such as lipid rafts, can be excluded on basis of the finding that insertion of the mycobacterial GPI-AP lipoarabinomannans into lymphocytes is specifically abrogated by addition of excess of their own GPI anchor, but not by excess of phospholipids (Ilangumaran et al. 1995). (v) The insertion of detergent-solubilized and purified GPI-AP *in vitro* is temperature-dependent with 37°C being considerably more efficient than at 4°C (Medof et al. 1984, Ilangumaran et al. 1995, Nagarajan et al. 1995), similar to that of gangliosides (Saqr et al. 1993). This temperature dependence may rely on several cellular processes affecting the exogenous insertion of GPI-AP, such as the fluidity of plasma membranes as well as lateral mobility of their components, the endocytosis of plasma membrane constituents and the inclusion/exclusion of GPI-AP into/from plasma membrane domains, such as the lipid rafts (Johannes and Mayor 2010). (vi) The insertion of detergent-solubilized and purified GPI-AP is blocked by the simultaneous presence of lipid- and fatty acid-binding proteins, such as albumin or lipoproteins (HDL and LDL), whereas GPI-AP, once inserted, cannot be extracted by those agents (Medof et al. 1984, Nagarajan et al. 1995).

Importantly, the experimental protocols of the above studies enabled the insertion of the purified GPI-AP into the acceptor cell plasma membranes from aqueous milieu lacking any detergent at all (Zhang et al. 1992, Ilangumaran et al. 1995) or containing only very low concentrations of (non-ionic) detergent, which usually do not affect the intactness of plasma membranes (Medof et al. 1984, Walter et al. 1992). On basis of the above findings, a number of different molecular mechanisms have been suggested so far for the insertion of exogenous GPI-AP into biological acceptor membranes: (i) The fatty acyl chains of monomeric GPI-AP may insert directly from stabilizing detergent micelles or upon their dissociation from lipid- or fatty acid-binding proteins, without gaining access to the aqueous milieu of the incubation medium. (ii) GPI-AP, assembled into oligomers upon removal of detergent molecules during their purification (Kuchel et al. 1978), may insert in course of dissociation of the oligomer in response to contact with hydrophobic membrane domains or lipid- or fatty acid-binding (serum) proteins. The resulting local appearance of monomers at intimate neighborhood to the plasma membrane (lipid rafts) could drive the direct insertion, as described for (i). (iii) GPI-AP may insert as oligomers into the luminal leaflet of the endosomal membranes in course of internalization by constitutive endocytosis and subsequent dissociation into monomers in the endosomal compartment, as described for (ii). After vectorial vesicular transport to the cell surface, the recycled GPI-AP would be expressed at the extracellular leaflet of the plasma membranes. These hypothetical mechanisms may share the final insertion step with direct insertion of the GPI anchor fatty acyl chains into lipid rafts containing high concentrations of saturated long-chain glyco(sphingo)lipids and cholesterol as has been demonstrated previously for lipoarabinomannans (Ilangumaran et al. 1995). Alternatively, GPI-AP, once inserted and distributed over the cell surface in random fashion, may diffuse laterally into lipid rafts and thereby acquire the typical structural (i.e., resistance towards extraction by cold non-ionic detergent) and functional characteristics (e.g., signal transducing capability)(van den Berg et al. 1995).

Since it is generally assumed that GPI-AP are targeted to their typical (sub)cellular localization in course of their biosynthesis, leading to their characteristic distribution between distinct (e.g., apical *vs.* basolateral) plasma membrane domains and between plasma membranes and intracellular (e.g., Golgi apparatus, ER) membranes and are expressed in cell- and tissue-specific fashion, it remains unclear whether promiscuous transfer of certain GPI-AP between the membranes of different cells occurs *in vivo* under certain (patho)physiological conditions. Two decades ago the first reports about the operation of GPI-AP transfer *in vivo* were made available (Pearce et al. 1990, Kooyman et al. 1995, McCurry et al. 1995). Those dealt with complement regulatory serum proteins, exclusively and suggested that these GPI-AP do not reside at plasma membrane areas of donor and acceptor cells which come into close neighborhood or even contact to one another. This unequal distribution of GPI-AP between the donor and acceptor plasma membranes would enable the transfer of GPI-AP between two neighboring cells within the same tissue depot.

As an alternative, it has been suggested that fatty acid- or lipid-binding serum proteins may support the transfer of GPI-AP (in the oligomeric state) through shielding of the hydrophobic portion of the GPI anchor from access to the aqueous milieu of the blood and simultaneous possibly receptor-mediated catalysis of its insertion into the acceptor plasma membranes (Ilangumaran et al. 1996). HDL was regarded as candidate for a lipid-binding serum protein exerting this role, since it was found that it interacts with CD59 in human serum and facilitates the transfer of GPI-AP between plasma membranes under serum free conditions (Väveka et al. 1994). However, at present it cannot be excluded that serum lipoproteins with bound GPI-AP become internalized upon interaction with their cognate receptors. In this case it is conceivable that the GPI-AP dissociate from their carrier protein during endocytosis and subsequently insert into the luminal leaflet of the endosomal membranes. Subsequent vesicular transport and recycling to the cell surface would result in expression of the GPI-AP at the extracellular leaflet of the plasma membranes.

At variance, the demonstrated *in vivo* transfer of GPI-AP from erythrocytes as donor membranes could involve EV, which harbor GPI-AP at a level enriched *vs.* total plasma membranes and are known to be released from erythrocytes in course of mechanical stress, age or ATP depletion (Bütikofer et al. 1989). After attachment to the acceptor plasma membranes, those EV could (i) mediate the direct insertion of their passenger GPI-AP, (ii) undergo membrane fusion directly or (iii) become internalized and recycled back to the cell surface. The operation of these putative mechanisms of EV-mediated transfer and final exposure at the outer plasma membrane leaflet of GPI-AP *in vivo* under (patho)physiological conditions remains to be demonstrated.

The possibility of differential insertion of GPI-AP into distinct areas of plasma membranes, which may be of considerable interest for biotechnological or pharmaceutical applications was addressed with two recombinant GPI-anchored green fluorescent proteins (GFP) equipped with the GPI signal sequence (see above) of either DAF (GFP-GPI/DAF) or TRAIL-R3 (GFP-GPI/TRAIL). Interestingly, GFP-GPI/DAF was recovered with lipid rafts, exclusively, whereas GFP-GPI/TRAIL was found to partition predominantly into non-raft areas of plasma membranes in both cultured HEK293 cells and primary lymphocytes (Legler et al. 2005). Engineering of the target cell surface using these chimeric GPI-AP occurred in rapid and stable fashion under preservation of the function of the passenger (GFP) protein. Interestingly, no difference with regard to the differential localization at plasma membrane raft and non-raft areas was detected between engineering (i.e., incubation of the cells with detergent-solubilized and purified GFP-GPI) and transfection (i.e., recombinant expression of the GFP-GPI in the cells). As expected, cell surface engineering was strictly dependent on the presence of the GPI anchor and stimulated by co-incubation with lipids enriched in lipid rafts, such as sphingomyelin, cholesterol and dipalmitoyl-phosphatidylethanolamine (Legler et al. 2005). These data suggest that the use of different GPI signal sequences for cell surface engineering causes differential targeting of the chimeric passenger protein-GPI to distinct plasma membrane microdomains, which may substantially broaden the range of potential pharmaceutical and biotechnological applications.

Studies with monolayers of PC at the air-water interface as a model membrane system and the GPI-AP, bovine intestinal mucosa AP, indicate that the degree of unsaturation of the phospholipids of the target plasma membranes, as reflected in similar exclusion pressures and different Langmuir isotherms for dipalmitoyl-PC in comparison to palmitoyloleoyl-PC, considerably affects the interface interaction induced by insertion of the GPI-AP (Kouzayha and Besson 2005). This characteristics of the GPI-AP-phospholipid interaction may explain the differential targeting of GPI-AP to lipid raft and non-raft domains of target plasma membranes, guided by both the GPI signal sequence and the phospholipid composition of the target plasma membranes.

Finally, in an erythrocyte-liposome system the question was addressed whether the insertion of GPI-AP into membranes can occur spontaneously without the aid of a catalyst (Suzuki and Okumura 2000a). Analysis of the buoyant density distribution of the membranes present in the total incubation mixture revealed that the GPI-AP AChE does not appear at the liposomes, but instead is recovered with small light as well as large heavy vesicles, which both were shed from the erythrocytes in the presence of liposomes. Apparently, the liposomes initially induced the release of AChE from the erythrocytes into large heavy vesicles as well as the subsequent transfer from the large heavy into the small light vesicles, which thereby become enriched in this GPI-AP (Suzuki and Okumura 2000a). These results suggest that GPI-AP fail to spontaneously transfer between the plasma membranes of mammalian cells with the important consequence that the cell surface engineering with exogenously added (recombinant chimeric) GPI-AP relies on the activity of a so far unknown catalyst. This catalyst has to be expressed at the acceptor cell surface and apparently is present at those cells, which have been engineered with GPI-AP and reported so far.

## 2.3. Pharmaceutical Cell Surface Engineering

Valid experimental documentation has been presented over the past two decades that certain infectious microorganisms (*Mycobacterium tuberculosis*) and parasitic protozoans (*Plasmodium falciparum, Trypanosoma brucei, Schistosoma mansoni*) use the transfer of pathogen-derived GPI-AP, which often represent virulence factors and (extremely densely packed) coat proteins of the microorganisms (e.g., trypanosomal VSG, mycobacterial lipoarabinomannans, *Plasmodium* merozoite surface proteins) to escape the host immune defense and complement system (McConville and Ferguson 1993, Schofield and Hackett 1993, Tachado and Schofield 1994, Brennan and Nikaido 1995). This raised the question whether certain strategies of those microorganisms in taking advantage of GPI-AP transfer could be useful for novel medical and pharmaceutical applications. An example may be provided by transplantation medicine. The acute rejection of xenografts derived from phylogenetically distant donors by human patients as acceptors is caused by the host complement factors and is currently tried to be prevented by inhibition of complement activation, e.g., by administration of cobra venom factor or soluble versions of the complement factor. In the future, the transfer of host complement regulatory proteins mediated by GPI anchors may turn out to represent an alternative approach for the prevention of acute transplant rejection with improved compliance and prolonged xenograft survival (McCurry et al. 1995).

There are additional examples for the application of the technology of GPI-AP transfer to manipulate host immune responses. Cytolytic T lymphocytes (CTL) manage to eliminate tumor and virus-infected cells only after their appropriate stimulation by antigenic peptides, which have to be presented on the target cells in combination with major histocompatibility (MHC) class I molecules. In fact, the purified GPI-modified version of the MHC class I molecule HLA-A2.1 was shown to be transferred to MHC-class-I-negative cells together with an antigenic peptide, such as hepatitis B peptide, bound to the antigen-binding cleft, which was dependent on the intact GPI anchor (Huang et al. 1994). Those cells exogenously equipped with antigen-presenting MHC class I molecules were able to present the surface-engineered antigen to a T-cell in an MHC-restricted fashion and thereby succeeded in the stimulation of CTL upon specific response to the antigenic peptide. Moreover, the immunogenicity of tumor cells was demonstrated to be increased upon GPI-dependent transfer of GPI-modified versions of the B7-1 and B7-2 molecules, which are able to induce co-stimulatory signals responsible for CTL activation (Brunschwig et al. 1995, McHugh et al. 1995). Taken together, the transfer of GPI-AP may be useful for the engineering of cell surfaces by the aimed insertion of GPI-modified versions of (patho)physiologically relevant proteins with the potential for antiviral and anticancer immunotherapy.

In fact, much subsequent work and expenditure has been devoted for the application of the methodology of GPI-AP surface engineering for antigen-presenting cells (APC). (i) As primary cells APC suffer from poor transfectability by recombinant means, in general, which can be overcome by surface engineering with GPI-AP for therapeutic use. In particular, transfer of GPI-AP was demonstrated to adapt the antigenic repertoire of APC for therapeutic needs by engineering of their surface with prebuilt heterotrimeric complexes, which are equipped with GPI-anchored class I MHC heavy chain, $ß_2M$ light chain, and antigenic peptides (Huang et al. 1994). Thus, cell surface engineering guarantees the fine tuning of the density of antigens exposed at the APC surface. Following engineering of APC in adequate fashion, they may be used for *ex vivo* T-cell amplification. (ii) The transfer of GPI-AP apparently offers a path for the transformation of APC from antigen-specific T-cell activating into inhibitory cells. Such a functional transformation can be achieved by engineering the surface of certain APC with GPI-anchored CD8 (Hambor et al. 1990). CD8 is known to mediate its interfering potency by transduction of negative signaling *via* MHC class I molecules on the T-cell surface (Chen et al. 1992 and 1994). (iii) Transfer of GPI-AP may be able to alter the adhesive characteristics of APC as is exemplified by the successful artificial construction of intercellular adhesion sites on the basis of the insertion of GPI-anchored adhesin (Weber et al. 1994). (iv) Previous data argue for successful upregulation of the APC stimulating function by GPI-AP transfer (Brunschwig et al. 1995, McHugh et al. 1995 and 1999), since the GPI-anchored versions of the cell surface costimulators B7-1 and B7-2 apparently are able to maintain the activity of the corresponding unmodified versions.

Taken together these findings may be of relevance for the area of immunization against cancer. In fact, it was already reported that ectopic expression of these costimulators with the help of classical recombinant gene transfer leads to considerable upregulation of the immunogenicity of tumor APC (Chen et al. 1994, Tykocinski et al. 1996). Consequently, it is conceivable that syngeneic tumor cells engineered accordingly can serve as cellular cancer vaccines with the aim to induce adequate tumor-specific systemic T-cell responses. The strategy of cell surface insertion of GPI-anchored costimulators would also circumvent the very low rate of recombinant transfection observed with the majority of primary tumor cells and simultaneously represent a relatively convenient and straightforward method for the reliable and controlled engineering of tumor cells with immunogenic potential for safe clinical use.

In addition to surface engineering of APC, the insertion of exogenous GPI-AP into target plasma membranes could fulfill therapeutic aims, which however necessitate the GPI-AP transfer between cells *in situ* within tissues rather than *ex vivo* between an artificial GPI-AP source and the isolated target/acceptor cells. Such a therapeutic aim represents the engineering *in situ* of vascular endothelial cells of grafts with GPI-AP for the blockade of thrombotic events and inhibition of rejection processes.

The technological possibility to produce microgram to even milligram amounts of pure GPI-AP using modern methods for recombinant expression and biochemical purification has paved the path for their insertion into the outer leaflet of plasma membranes, thereby creating cell surfaces with novel and desired biophysical properties and physiological functions. It has been proposed that this process, also been termed cell surface painting or engineering, offers a number of advantages in comparison to classical transfection of cells with genes coding for transmembrane proteins for their subsequent expression at the cell surface (Medof et al. 1996): (i) Cell surface engineering with GPI-AP is efficient also in cells which require high expenditure for successful gene transfection, such as primary cells, immune cells or bone marrow progenitors. (ii) Cell surface engineering is efficient with low numbers of cells, in particular, if the cells are difficult to propagate. (iii) Cell surface engineering does not depend on the cell type. (iv) Cell surface engineering enables the accurate control of the amount of protein which is ultimately exposed at the cell surface. (v) The efficacy of cell surface engineering depends solely on the amount of purified GPI-AP used for the insertion rather than on the capacity of the cell for biosynthesis and vesicular trafficking of the (transmembrane or GPI-anchored) cell surface protein. (vi) Cell surface engineering can be performed for a number of GPI-AP in parallel, which are inserted into the plasma membranes of the same target cells in sequential or simultaneous fashion.

Meanwhile a number of studies have been designed for the evaluation of (the most) critical parameters for the experimental insertion process (Medof et al. 1984, Davitz et al. 1987, Cardoso de Almeida et al. 1988, Low and Prasad 1988, Walter et al. 1992), such as (i) detergent concentration of less than 1/10th of the critical micellar concentration, (ii) absence of serum proteins including fatty acid- or lipid-binding proteins as well as of GPI-PLD, (iii) incubation temperature between 30° and 37°C and (iv) the amount of GPI-AP added to the incubation mixture with 1-3% insertion rate per hour. Furthermore, experimental criteria have been defined for the discrimination of exogenously added GPI-AP that achieved complete and functional plasma membrane insertion *vs.* those that remained non-specifically attached at the cell surface, similar to peripheral membrane proteins (Medof et al. 1996, Tykocinski et al. 1996), such as (i) temperature dependence of the insertion, (ii) inhibition of the insertion by albumin or lipoproteins, (iii) failure to extract the GPI-AP from membranes by incubation with high salt or liposomes, (iv) failure to cleave the GPI-AP by GPI-PLD, (v) functionality of the GPI-AP at a level comparable to that of the endogenous (transmembrane or GPI-anchored) version and (vi) missing effect of dilution of the incubation mixture on the functionality of the GPI-AP.

## 3. GPI-AP (AND THEIR INTERCELLULAR TRANSFER) AS PUTATIVE DRUG TARGETS

A series of experimental studies has been performed during the past two decades which strongly argue for the principal applicability of the method of GPI-AP transfer for a variety of therapeutic and pharmaceutical aims. The successful transfer of GPI-AP to GPI-deficient cells, i.e., to cells with defective GPI biogenesis and lacking GPI-AP, was shown for cultured cells (Kooyman et al. 1995, Anderson et al. 1996, Liu et al. 2002) animals (McCurry et al. 1995) and even humans, particular for patients with PNH (see above). This is a chronic hemolytic disease caused by defective biosynthesis of GPI anchors (Rosse 2001, Almeida et al. 2006). The general lack of expression of GPI-AP leads to extreme sensitivity towards complement-mediated lysis, which is primarily due to the absence of the GPI-AP CD55 and CD59, two inhibitors of the formation of the

MAC of complement. It was demonstrated that in course of transfer of CD55 and CD59 from EV, which expose the two GPI-AP at the outer leaflet of their bilayer membranes, to erythrocytes from PNH patients at the outer leaflet of their plasma membrane bilayer led to significant reduction of the susceptibility of the PNH erythrocytes towards complement-mediated lysis (Sloand et al. 1998). Furthermore, recombinant GPI-anchored CD59 was found to be transferred from erythrocytes to vascular endothelial cells as well as from seminal fluid to prostasomes (Kooyman et al. 1995, Rooney et al. 1996). Moreover, the intercellular transfer of the GPI-AP Thy-1 was reported for chimeric murine embryoid bodies, which were constituted by a mixture of normal and PNH cells (Dunn et al. 1996). In patients infected with trypanosomal parasites, the transfer of VSG from the membranes of the trypanosomes to human erythrocytes was detected, which seems to be causative for the infection-associated anemia through immune cell-triggered destruction of the erythrocytes (Rifkin and Landsberger 1990). Interestingly, the acquisition of CD55 from the infected host by certain parasites seems to protect them against complement-mediated lysis (Pearce et al. 1990).

### 3.1. Reproduction Biology

A number of GPI-AP have been found to be transferred from epididymal luminal fluids (ELF) onto the surface of spermatozoa during the path of sperm maturation (Martin-DeLeon 2015). Those GPI-AP play multiple roles in immunoprotection, as is true for CD52, CD59 and CD73 (Kirchhoff and Hale 1996, Kirchhoff et al. 1997, Yeung et al. 1997), as well as in key processes of fertilization, among them sperm-egg interactions, acrosomal exocytosis and capacitation, as is true for P34H (Sullivan et al. 2006, Moskovtsev et al. 2007) and sperm adhesion molecule 1 (SPAM1)(Chen et al. 2006). The mechanism of their transfer to sperm cells seems to involve both EV, termed epididymosomes, which were recently demonstrated to dock to lipid rafts on the sperm membrane *via* hydrophobic interactions (Griffiths et al. 2008a), and lipid- or fatty acid-binding proteins, termed lipid carriers, which support the transport of GPI-AP through the aqueous and membrane-free milieu of the ELF (Zhang and Martin-DeLeon 2003, Zhang et al. 2004). The former vesicle-independent mechanism turned out to be more efficient than the latter vesicular one. However, both of them apparently rely on the direct insertion of the GPI anchor into the outer (extracellular) plasma membrane leaflet of the sperm, which is presumably driven by hydrophobic interactions (Griffiths et al. 2008b and 2009). Recent proteomic studies have revealed that epididymosomes from several mammalian species are equipped with a number of transmembrane proteins, among them plasma membrane $Ca^{2+}$-ATPase 4 (PMCA4)(Girouard et al. 2011), which is produced in the testis and fulfills an essential function as the predominant efflux pump for $Ca^{2+}$ in spermatozoa of mice. In fact, the transfer of PMCA4 from bovine and mouse epididymosomes to spermatozoa in course of epididymal maturation *in vivo* and *in vitro* argues for binding of the epididymosomes to the sperm surface prior to their fusion with the sperm (Patel et al. 2013), which seems to be supported by CD9 *via* the generation of membrane sites with competence for membrane fusion.

In addition to a fusogenic mechanism for transmembrane proteins, such as PMCA4, a non-vesicular mechanism seems to predominantly operate during the transfer of GPI-AP, such as SPAM1. Previous data already argued for a transfer of GPI-AP from ELF to sperm cells in the absence of vesicles (Rooney et al. 1996, Griffiths et al. 2008a). In fact, SPAM1 was detected in murine ELF in both a soluble (60%) and a vesicular (40%) fraction (Zhang and Martin-DeLeon 2003). Upon incubation of each of these fractions with spermatozoa, SPAM1 was transferred onto the sperm surface in both cases with the soluble fraction being more efficient (Griffiths et al. 2008b). Furthermore, SPAM1 of the soluble fraction was demonstrated to consist of monomers as well as oligomers (Griffiths et al. 2009). Importantly, the oligomers turned out to be inactive in transferring SPAM1 onto the sperm surface, but presumably serve as reservoir for monomers after their dissociation. The transfer competence of the monomers in the aqueous milieu of the ELF is most likely preserved by their association with fatty acid- or lipid-binding proteins, such as apolipoprotein J or clusterin (Griffiths et al. 2009, Martin-DeLeon 2011). A lipid-exchange model relying on lipid carriers such as clusterin may explain the transfer of GPI-AP to the sperm surface (Martin-DeLeon 2015).

### 3.2. Neurogenerative Prion Diseases

At variance with other amyloid-associated protein misfolding diseases, such as Alzheimer's, prion diseases are well-known to be transmitted between individuals which relies on spreading of protein aggregates between tissue cells. Moreover, only the pathogenetic mechanism of prion diseases rather than that of all the other protein misfolding diseases is based on the improper folding of a GPI-AP. Recently, the impact of GPI modification on the propagation and spreading of protein aggregates was investigated in cultured neuronal cells using ectopic expression of a GPI-modified version of the amyloidogenic yeast protein Sup35NM (Speare et al. 2010). Incubation of the cells with fibrils constituted by GPI-anchored Sup35NM, but not with an unmodified version, resulted in the generation of aggregates of Sup35NM, which displayed great similarity to typical prion-like protein aggregates on the basis of the criteria of protease resistance, detergent insolubility and self propagation. Importantly, analysis of the intercellular transfer of GPI-modified Sup35NM aggregates using live-cell imaging revealed the involvement of contact sites between cells which contain and lack those aggregates, respectively, as well as the transfer of GPI-modified Sup35NM from the former to the latter cells (Speare et al. 2010). Consequently, it was concluded that membrane anchorage of prion protein by GPI fosters the formation and intercellular transfer of protein aggregates and thereby may increase the transmission and pathogenetic potential of prion diseases compared to other protein misfolding diseases.

A number of studies have shown unequivocally that the replacement of the GPI anchor for a transmembrane domain causes the trafficking of the cellular form of prion protein from lipid rafts to non-raft areas of the plasma membranes and that this non-raft version of prion protein is protected against conversion into the misfolded scrapie form upon transfection into scrapie-infected (cultured) neuronal cells. A very recent investigation was aimed to clarify whether this apparent resistance is caused by the segregation of the transmembrane cellular and GPI-anchored scrapie prion protein forms into separate membrane areas (Marshall et al. 2017). Interestingly, infection of cultured neuronal cells lacking endogenous prion protein with wild-type GPI-anchored prion proteins from brain microsomes or with purified amyloid fibrils formed by anchor-less prion protein demonstrated that only the former but not the latter prion form supports the propagation of misfolded prion protein. These data are compatible with the GPI anchor being responsible for targeting of the cellular form of prion protein to plasma membrane lipid rafts where it undergoes conversion into the scrapie form with accompanying formation of amyloid fibrils (Marshall et al. 2017). Thus GPI anchorage and the resulting lipid raft localization of cellular prion protein would represent prerequisites for its ability to spread as a prion.

However, the conclusion that GPI anchoring of the prion protein favors misfolding of the cellular form into the disease-linked scrapie form as well as the propagation and infectivity of prions seems to be in conflict with recent findings that prion proteins lacking the GPI anchor lead to increased production and spreading of prions *in vivo* and to a higher risk for the transmission of prion diseases. Compatible with these observations are the thermodynamic characterization and kinetic analysis of the conversion into amyloid fibrils of carboxy-terminally truncated anchor-less forms of prion protein (Kovac et al. 2016). Remarkably, the shortest anchor-less form turned out to be the most labile one in course of temperature increase and to be converted into amyloid fibrils most rapidly upon exposure to acidic pH compared to longer anchor-less forms or full-length GPI-anchored prion protein. Moreover, a difference in length as small as a single amino acid only was reflected in an altered lag time of fibrillization and stability *vs.* low pH (Kovac et al. 2016). Further detailed experimentation will be required to clarify the contribution of the GPI anchor for the replication and propagation of the scrapie form of prion protein.

The likely dependence of the physiological role of the cellular form of prion protein on its subcellular localization received an additional aspect in a recent study, which analyzed the destination of the endogenous and ectopically expressed cellular forms of prion protein in the brain of old mice lacking any signs of neurodegenerative prion disease (Faris et al. 2017). Surprisingly, the cellular form of prion protein equipped with fully processed mature N-linked glycans was detected in mitochondria, seemingly as a transmembrane variant with the amino-terminus protruding into the intermembrane space and the carboxy-terminus facing the matrix. Importantly, in mice transgenic for an anchor-less form of prion protein, a considerable portion of total prion protein was recovered with mitochondria (Faris et al. 2017). Together these data argue for the expression of the cellular form of prion protein in brain mitochondria which does not rely on its modification with a GPI anchor. Considering the lack of an obvious disease phenotype despite old age and the expression of mutant variants or of high amounts of the cellular form of prion protein, its role for the (patho)physiology of the brain, in general, and the function of brain mitochondria, in particular, remains to be determined.

### 3.3. Development

Cripto-1 (CR-1) operates as a co-receptor for the ligand Nodal, which is a member of the transforming growth factor-ß (TGF-ß) family. Thereby CR-1 controls downstream Nodal signaling, which is essential for vertebrate development (Shen 2007). Interestingly, upon expression as membrane-anchored version, which can be achieved by either a typical transmembrane domain or a GPI anchor, CR-1 was found to exert its co-receptor function in assisting Nodal signaling in both cell-autonomous and non cell-autonomous fashion (Watanabe et al. 2007). In contrast, a soluble version of CR-1, generated by either carboxy-terminal truncation or mutation, completely failed in Nodal downstream signaling (Bamford et al. 2000). These data suggested that membrane attachment of CR-1, but not the presence of a transmembrane domain *per se*, is prerequisite for its operation as a Nodal co-receptor and concomitantly raised the question how CR-1 manages to elicit its paracrine activity. Subsequent flow cytometric analysis demonstrated the intercellular transfer of the GPI-anchored version of CR-1 (Watanabe and Salomon 2010). This was accompanied by the increased formation of unique membrane extensions in cells expressing the GPI-anchored compared to the soluble version, which resulted in upregulation of their adhesion properties. It was concluded that expression of GPI-anchored CR-1 causes alterations in the physicochemical characteristics of the plasma membranes which facilitate the intercellular transfer of CR-1, in comparison to its transmembrane version, and thereby enable or enhance its paracrine activity.

### 3.4. Coagulation

The protein C (PC)-system was identified as an anticoagulant and cytoprotective signaling pathway which seems to be effective in multiple diseases according to results with *in vivo* disease models. In endothelial cells PC is known to bind to a receptor

(EPCR), which operates as a stimulatory cofactor for the PC and activated PC (APC)-mediated protease-activated receptor (PAR) downstream signaling pathways. Importantly, inflammatory diseases are often associated with diminished cell surface expression of EPCR and thus with reduced PC signaling and APC activity in relevant tissues. Interestingly, a recent approach was introduced for the restoration of the functionality of EPCR with the help of cell surface engineering using GPI-anchored EPCR (EPCR-GPI). The efficacy of the insertion of EPCR-GPI into plasma membranes of EPCR-deficient endothelial cells was strongly dependent on the incubation time and concentration (Bouwens et al. 2015). Engineering of the endothelial cell surface with EPCR-GPI was found to be long-lasting and to lead to restoration of PC signaling in EPCR-deficient cells, as manifested in APC-induced cleavage of PAR1 and PAR3 as well as PAR1-dependent AK phosphorylation. These findings strongly argue for the acquisition of functionality of the transferred EPCR-GPI in the engineered cells. Strikingly, EPCR was recovered with plasma membrane lipid rafts, characterized by the expression of caveolin-1. It was concluded that EPCR becomes targeted to and inserted into caveolae of the target cells which apparently is mediated by its GPI anchor (Bouwens et al. 2015). Taken together, this novel approach of cell surface engineering led to physiologically effective levels of the corresponding target protein (EPCR) at the surface of relevant (EPCR-deficient endothelial) cells and simultaneously to restoration of APC downstream signaling, which induces APC binding, PC stimulation, upregulation of APC-induced PAR cleavage and finally enhancement of cytoprotective activity. Thus, cell surface engineering with natural or artificial GPI-AP may be regarded as useful technology for the restoration of the expression and activity of those (cell surface) proteins which are missing in relevant cells and tissues in course of the development of certain diseases.

### 3.5. Cancer

Tissue inhibitors of metalloproteinases (TIMP) were recognized as modulators for a variety of (patho)physiological processes encompassing the downregulation of the proteolytic cleavage of a number of growth factors as well as of the proteolytic degradation of ECM proteins (Chirco et al. 2006, Nagase et al. 2006). Currently it is thought that, at least in part, these biological functions are determined by the ratio of the concentrations of TIMP, including their cell surface binding-proteins, *vs*. their substrates, the matrix metalloproteinases (MMP), as well as by the strength of their 1:1 stoichiometric interaction (Notohamiprodjo et al. 2012). TIMP-1 was identified as a secreted protein that can be recovered from cell surfaces through its interaction with specific binding-proteins, only. In order to increase the ratio of cell surface-bound TIMP-1 *vs*. MMP using the technique of cell surface engineering, a recombinant GPI-anchored version of TIMP-1 (TIMP-1-GPI) was incubated with primary dermal fibroblasts following its purification and solubilization with detergent (Djafarzadeh et al. 2013). It was observed that the insertion of TIMP-1-GPI into the outer leaflet of plasma membranes correlated well to a marked decline of the secretion of MMP and simultaneously to a considerable lowering of the ratio of cell surface associated MMP *vs.* constitutive cell surface components. Moreover, incubation of the fibroblasts with TIMP-1-GPI led to concentration-dependent abrogation of their proliferation and transient lowering of the expression level of fibrosis-related genes, which was accompanied by upregulation of the proapoptotic phenotype as well as the expression of apoptosis-relevant proteins (Djafarzadeh et al. 2013). Consequently, it was concluded that GPI-anchored TIMP-1 may be used for the tight and efficient control of extracellular MMP-mediated proteolysis, which presumably represents an efficient target for the therapy of fibrosis and scar formation. Moreover, peritumoral administration of GPI-anchored TIMP-1 was reported to inhibit the growth of colon carcinoma in Rag-2 gamma chain-deficient mice (Raggi et al. 2009).

In analogy, interference with TGF-ß1-induced signaling and gene expression in renal carcinoma cells was accomplished by cell surface engineering with TIMP-1, resulting in elevated apoptosis and diminished proliferation (Notohamiprodjo et al. 2012). In fact, a variety of putative anti-cancer effects were detected in course of incubation of tumor cells with TIMP-1-GPI *in vitro*, which encompass the clustering of MMP at the cell surface, the decrease in the release of MMP from tumor cells, the concentration-dependent interference with the growth, invasion and migration of renal carcinoma cells and the elevation of the susceptibility of the cancer cells for FAS-induced apoptosis (Djafarzadeh et al. 2006 and 2009). Together these findings considerably expand the range of putative applications of the cell engineering technology to cancer therapy.

## 4. GPI-AP as Therapeutic Proteins

By nature, the therapeutic proteins, in marked contrast to small-molecule drugs, are characterized by exquisite potency and selectivity, as exemplified best by typical receptor–ligand or enzyme–substrate interactions. As a consequence, protein therapeutics in general have the huge advantage of operating “intrinsically” with high safety as well as of being discovered on a rational basis and developed using a straightforward strategy. Their wild-type versions represent “naturally” occurring cellular components. Therefore, upon their identification therapeutic proteins may be used for the treatment either directly by substitution of defective or missing gene products, such as insulin in case of diabetes, or indirectly by blockade of hyperactive or overproduced gene products using neutralizing antibodies or binding-proteins, such as monoclonal antibodies against constitutively activated receptor tyrosine kinases in case of certain tumors. However, therapeutic proteins and antibodies have to overcome a number of critical hurdles prior to beneficial systemic action, which are caused predominantly by their unfavorable pharmacokinetic profile, in particular low oral adsorption and rapid metabolism (by hepatic clearance and proteolytic degradation). Oral adsorption necessitates the passage along the gastrointestinal tract and the transport across intestinal epithelial cells into the circulation and, possibly, subsequently across vascular endothelial cells into the interstitial tissue spaces. For therapeutic proteins with the major site of action in the serum (e.g., plasma enzymes) or interstitial spaces between tissue cells for reaching the surface (e.g., receptors) of target cells (in case of polypeptide hormones), this so-called oral bioavailability may represent the only barrier for gaining access to their primary target cell and triggering the corresponding molecular mechanism of drug action. Gene products with the primary target and molecular mechanism of action operating in the cytoplasm (e.g., transcription factors, signaling proteins, enzymes) have to be transported from the interstitial spaces *via* the ECM across the plasma membranes of the target cells, in addition. This is typically not managed by therapeutic proteins and antibodies, and therefore they have to be injected according to their site of action (e.g., subcutaneously, intravenously, intramuscularly, intraperitoneally). Thus, disease prevention and therapy based on the injection of therapeutic proteins and antibodies into the circulation are not feasible in case of intracellular targets due to their inability of being transported into the target cells. Furthermore, even in case of an extracellular target of the therapeutic protein, its administration by injection is meanwhile known to be considerably hampered by the low compliance of the patients towards their non-oral administration. This limitation is of particular relevance for common complex diseases, such as T2D and obesity, which during the initial stages of the pathogenesis are often characterized by only low to moderate impact for the individual's life quality, however, with negative consequences for the acceptance of regular drug injections.

In fact, the long path from the mouth to the drug target within the body across the intestinal and cellular barriers has remained the major hurdle for the use of protein therapeutics in humans so far. The intestine is lined by a monolayer of epithelial cells, so-called enterocytes, which are interconnected to each other by specific protein complexes, so-called tight junctions. Macromolecules, such as polypeptide hormones, antibodies, and binding-proteins, typically cannot cross the tight junctions for the so-called paracellular passage from the intestinal lumen directly into the circulation. Rather they are first broken down *via* peptides to their constituent free amino acids by various proteolytic enzymes located throughout the gastrointestinal tract (Bai and Chang 1995). These constituents are then taken up by the gastrointestinal epithelium. Of course, this route of absorption leads to almost complete elimination of the physiological activity of the original protein and provides an explanation why typical oral bioavailabilities of proteins (of medium size) are usually less than 1-2% [Iyer et al. 2010]. Therefore, despite the fact that recombinant DNA technology has allowed rapid and straightforward production of therapeutic proteins and antibodies for steadily increasing numbers, the delivery of these drugs still depends on the subcutaneous or intravenous injection. For the past eight decades, a great deal of work has focused on attempts to develop non-invasive methods of delivering protein drugs, with the oral route clearly being the most studied and desired one.

In fact, bypassing of the intestinal epithelial transport barrier has been tackled by several distinct approaches (Carino and Mathiowitz 1999). The most common strategies followed so far rely on the use of protease inhibitors (Yamamoto et al. 1994) and permeation enhancers (Palin et al. 1986, Mesiha et al. 1994) that has been demonstrated to considerably increase the efficacy of oral protein delivery. A general limitation of the former approach is that the long-term effect of chronic administration of protease inhibitors remains unknown so far and may result in protein malabsorption and other adverse side effects. The use of permeation enhancers could potentially damage or even dissolve the gastrointestinal barrier and thereby pave the path for undesired small and large molecules, eventually derived from the food, from the intestinal lumen into the circulation, which will be accompanied with increased risks for local inflammations and gastrointestinal infections.

## 4.1. Oral Delivery of Therapeutic Proteins

According to the definition from the Royal Society of London in July 2004, nanotechnologies are the design, characterization, production, and application of structures, devices, and systems by controlling shape and size at the nanometer-scale (Dapergolas and Gregoriadis 1976, Schwendeman et al. 1996, Takka and Acarturk 1999, Rowsen et al. 2000, Ebbesen and Jensen 2006). In recent years an in-depth understanding of nanostructured substances has been provided by the use of new sophisticated tools, such as atomic force microscopy. The recent advances in nanoparticles (NP) and nanotechnologies make them very promising tools in the delivery of biomolecules including proteins for disease prediction and prevention (Bhavsar and Amiji 2007, Davis 2009, Agueros et al. 2011, Bimbo et al. 2011). NP with a hydrophilic surface are especially desirable to transport protein therapeutics to the target tissues or cells because they can escape the uptake by mononuclear phagocytes, macrophages, and the reticuloendothelial system (RES) in the blood and organs (Woitiski et al. 2010). In addition, a biological signal can be chemically conjugated onto the surface of NP to recognize specific tissues or cells. Although NP as a drug carrier have their own drawbacks, such as low loading capacity and wide size distribution, they have attracted increasing attention from chemists, biologists, engineers, and pharmacologists since they provide the possibility of transporting bioactive compounds to specific tissues, cells, and cell compartments (Yokoyama 2005).

NP offer numerous advantages over conventional dosage forms, including the ability to protect biomolecules from biodegradation (Lasch and Schönbrunner 1936, Woitiski et al. 2011) and to target drugs to the site of action. They are made by forming drug-polymer complexes in which the drug is uniformly dispersed or by creating nanoscale-vesicles (such as liposomes and micelles) to entrap the drug molecules. The surface characteristics of NP are important in determining their susceptibility to uptake by the RES. Modification of surface properties has been reported to affect both the circulation time and ultimate fate of the NP. It has been found that NP with more hydrophobic surfaces tend to taken up rapidly by the liver, spleen, and lungs (Schwendeman et al. 1996, Mathiowitz et al. 1997), whereas liposomes coated with hydrophilic surfaces exhibit significantly prolonged circulation times *in vivo* (Dapergolas and Gregoriadis 1976). It is believed that the high surface concentration of hydrated polymer chains sterically hinders protein absorption and opsonization of the liposomes. The synthetic hydrophilic polymer polyethyleneglycol (PEG) is the most common polymer used for NP recognition by the RES. The surface charge as well as the presence of targeting signals at the NP surface are other important parameters in determining how NP interact with cells, the membranes of which are usually negatively charged.

In recent years paracellular transport of protein drugs by encapsulation into NP assembled from various biomaterials, such as chitosan, alginate, mannans, and γ-polyglutamic acid at specific ratios, have been demonstrated (Damge et al. 1988, Schwendeman et al. 1996, Mathiowitz et al. 1997, Ebbesen and Jensen 2006, Soussan et al. 2009). These NP apparently adhere to the negatively charged mucus and glycocalyx layers of the small intestine, which is mediated by their positive net surface charge. Following infiltration of these layers, the NP trigger opening of the tight junctions, presumably by causing the redistribution of some of their constituent proteins, for example, F-actin, occludin, and ZO-1. This allows subsequent passage of the NP between neighboring enterocytes along the concentration gradient from the intestinal lumen into the interstitial spaces of the underlying intestinal tissue, which is formed by vascular endothelial and smooth muscle cells. However, immediately after the passage the tight junctions reseal again to maintain the intestinal permeability barrier.

Polymeric NP not only improve the water solubility and bioavailability of drugs but also allow for drug targeting by conjugating biological signals onto their surface or by using stimuli-sensitive polymers. Despite the accumulation of drug molecules at the desired target sites, such as tumors, at higher concentrations, provoked by the introduction of targeting signals to the NP, considerable portions of drug molecules can still be found entrapped in the liver. This is possibly due to leakage of the drug molecules from the NP or the retention of the NP within the sophisticated liver structure rich in the RES. Therefore future research in the field of drug delivery of protein therapeutics should be focused on enhancing drug stability in the NP before reaching the target site and before accumulation of the drug at the target sites by using stable polymer core-shell NP and specific targeting signals.

*4.1.1. Structures*

Many different macromolecular NP structures, such as drug polymer conjugates, micelles, liposomes, and dendrimers, have been designed to transport drugs to their intended target tissues and cells. Micelles can be made from amphiphilic block copolymers that self-assemble into small, spherical structures (Dapergolas and Gregoriadis 1976, Damge et al. 1998, Takka and Acarturk 1999, Uhrich et al. 1999, Rowsen et al. 2000, Matsumara 2008). Liposomes are vesicles made of phospholipid bilayers that can encapsulate drugs in their luminal cores or interfaces between the bilayers (Soussan et al. 2009). NP are generally polymeric matrices in the form of nanosized colloids that can encapsulate a drug through physical entrapment (association between the drug and polymer) or chemical conjugation (formation of a chemical bond between the drug and polymer)(Soussan et al. 2009).

Typically, NP with a size smaller than 100 nm in diameter accumulate in solid tumors, which are characterized by extensive angiogenesis, defective vascular architecture, impaired lymphatic drainage, and an increased production of permeability factors (Gullotti and Yeo 2009). This phenomenon is known as the enhanced permeability and retention (EPR) effect. It can be exploited to improve drug delivery by NP to tumors and other sites, including areas of inflammation and infarct, which possess similar pathological characteristics. The EPR effect is considered as a passive targeting mechanism, but more specific drug targeting can be achieved by binding targeting ligands to the surface of the NP, such as monoclonal antibodies, peptides or sugar moieties. Drug targeting can also be achieved by introducing stimuli-sensitive materials into the NP (Mathiowitz et al. 1997). Temperature- (Lin et al. 2007) and pH-sensitive (Yang et al. 2007, Mathiowitz 2008) polymers have been used in drug targeting since many pathological processes are known to be accompanied by local temperature increases and/or acidosis. Poly(*N*-isopropylacrylamide) and its copolymers are the most extensively studied thermosensitive polymers. One of the drawbacks of using thermosensitive NP is that heating is inaccessible to the depth of larger tissues and organs. Recently polymer core-shell NP with both temperature and pH sensitivity have been developed with pH changes inducing the temperature sensitivity (Kirpotin et al. 2006). These NP were stable in the normal physiological environment but become deformed and precipitated in acidic environments, such as the tumor interstitium and intracellular compartments. They are superior to carriers that are either thermosensitive or pH-sensitive due to their potential to target deep tissues, where it may not be possible to trigger drug release by hyperthermia. In addition, their phase alternation triggered by external pH changes is much more abrupt than that of pH-sensitive NP (Chaw et al. 2004, Yang et al. 2006).

*4.1.2. Materials*

Polymers used for the fabrication of NP include polylactide (Liu et al. 2005), poly(lactide-co-glycolide) (Liu et al. 2005), PLA/PLGA-*b* PEG (Lee et al. 2005), PEG-poly(ß-benzyl-L-aspartate) block copolymers (Kataoka et al. 2000), PEG-poly(€-caprolactone)-PEG (Zhang and Zhuo 2005a), and poly(trimethylene carbonate)-PEG-poly(trimethylene carbonate) triblock copolymers (Zhang and Zhuo 2005b). A drug can be loaded into the NP by various methods, such as by an oil-in-water emulsion method if the drug is water insoluble (Dong and Feng 2005) or a water-in-oil-in-water double-emulsion process if the drug is water soluble (Zambaux et al. 1998). Membrane dialysis is a widely used process for the production of drug-loaded micellar NP (Kataoka et al. 2000, Wei et al. 2005). The drug-loading capacity of the NP depends on the properties of the drug, the production parameters, and the structural compatibility between the drug and the hydrophobic segments of the amphiphilic polymers (Allen et al. 1999). It has always been a challenge to incorporate water-soluble drugs into polymeric NP. Recently, ionic complexation was used to load drugs, such as doxorubicin and verapamil, into dextran sulfate, an anionic polymer with relatively high loading levels (Wong et al. 2004).

NP formulations using muco-adhesive polymers, such as chitosan, poly(lactic-)co-glycolic acid, poly-γ-glutamic acid and alginate, poly(alkyl-)cyanoacrylate micro-particles, and ß-cyclodextrin liposomes, have been studied extensively over the last decade (Wong et al. 2004, Pavlou and Reichert 2004) and found to encapsulate the protein drugs apparently physically protected from proteolytic degradation. Furthermore, it has been shown clearly that such NP cross the epithelial layer through Peyer's patches (Yang et al. 2008, Singh and Lillard 2009, McLendon et al. 2010). While some of the encapsulated protein drugs have been demonstrated to exert the desired physiological effect in appropriate animal models, such as lowering of blood glucose in *Streptozotocin*-diabetic rats in the case of insulin, further development has not been reported so far. Presumably, these nanomaterials "actively" induce the transient opening of the tight junctions in course of their direct binding to protein components of the zona occludens and thereby enable the paracellular passage of the encapsulated therapeutic proteins into the circulation. One critical disadvantage of this "site specific" delivery and "colonic absorption" could be the inability to adequately correct and compensate for all physiological aspects of the missing or defective endogenous susceptibility gene product.

## 4.2. Oral Delivery of GPI-AP

Very recent experimental results hint to the exciting possibility of a novel strategy for both oral delivery and tissue targeting in parallel. It is based on a "natural" co-translational modification of the therapeutic protein with a glycolipid structure, the covalent coupling to GPI (Müller 2010a and b). In analogy to adipocytes for which the uptake and transcellular transport of GPI-AP *via* plasma membrane lipid rafts, LD and EV have been amply documented (Müller et al. 2009a and b), the multistep oral delivery of therapeutic proteins and antibodies in the course of transcellular transport and/or translocation onto the surface of cytoplasmic LD and subsequent release *via* EV and translocation into the cytoplasm of enterocytes has been proposed (Müller 2010a and b). It is based on the unique property of GPI-AP to associate through their GPI anchor with detergent micelles, plasma membrane lipid rafts, LD and EV in a fashion compatible with inter- and intracellular trafficking as well as physiological function (Figure 13).

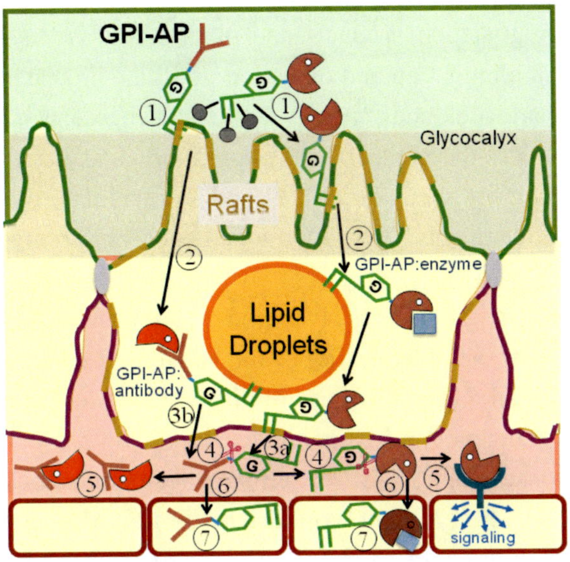


Figure 13. Hypothetical model for the oral delivery of GPI-AP. The epithelial cell layer lining the gastrointestinal tract is a tightly bound assembly of enterocytes with minimal leakage across (in transcellular fashion) or between (in paracellular fashion) the cells which forms a physical barrier to absorption. It is made up of a single layer of columnar epithelial cells supported by the lamina propria and muscularis mucosa. The epithelium folds to form the so-called brush border membrane and each intestinal epithelial cell also possesses microvilli, uniform one-micrometer long finger-like projections, on its apical surface. Together, these structures increase the absorptive area of the intestinal tract by approximately two orders of magnitude. However, these structures hinder the absorption of proteins as the microvilli also contain digestive enzymes. In addition to these formidable barriers, on top of the epithelial layer lies the glycocalyx, a layer of sulphated mucopolysaccharides, and a layer of mucus consisting of glycoproteins, enzymes, electrolytes and water. Glycocalyx and mucus present yet another physical barrier to the transport of proteins. After ingestion and arrival at the intestinal lumen the therapeutic protein or antibody has to pass the glycocalyx and mucus layers and then be transported into blood vessels and underlying tissue cells. Typically, proteins do not follow the transcellular route across the microvilli of the apical plasma membranes or the paracellular route across the protein complexes, which form the tight junctions (grey ovals). These connect the neighboring enterocytes to yield the intact and sealed intestinal epithelial cell layer which prevents macromolecules contained in the food from gaining access to the circulation. Upon oral administration of GPI-AP, such as receptors or enzymes, in the absence or presence of detergent, they become spontaneously inserted into lipid rafts (brown lines) expressed at the outer leaflet of the brush border plasma membranes (step 1, green lines). Subsequently, the GPI-AP are translocated from the apical plasma membranes of the enterocyte onto the phospholipid monolayer at the surface of the enterocyte LD (step 2). Thereafter, the GPI-AP are liberated from the LD and migrate to the cytoplasmic leaflet of the basolateral plasma membranes (pink lines) triggering their translocation across the phospholipid bilayer to the enterocyte outer surface, which may either involve lipid rafts (step 3a) or be independent of raft function (step 3b). Upon arrival of the GPI-AP at the vessel lumen, their anchor may be cleaved off by action of specific proteases or the GPI-PLC/D in the plasma relieving the soluble protein moiety (step 4), which may exert its enzymic or signaling function in the circulation (step 5). Alternatively, the circulating GPI-AP harboring the complete anchor due to missing protease and GPI-PLC/D action are translocated across the apical and basolateral plasma membranes of vascular endothelial cells into interstitial spaces of the underlying tissues as well as across the plasma membranes of neighboring tissue cells into their cytoplasm (step 6), where the GPI-AP fulfill their physiological function (step 7).

Initial experiments with mixed micelles consisting of phosphatidic acid, cholesterol, negatively charged glycosphingolipids, and therapeutic GPI-AP, such as human insulin, and encapsulated together with some of the above biomaterials could represent a major advantage in oral protein delivery. When assayed *in vitro* for paracellular transport, these NP were shown to trigger the opening and closure of the tight junctions in a reversible and efficient fashion, presumably due to the amphiphilic overall nature exerted by the GPI anchor (Figure 13). Moreover, when enwrapped by an acid-stable film, these NP have demonstrated their potential to increase plasma insulin and lower blood glucose in insulin-free diabetic rats. Importantly, during these preliminary studies, accumulation of these NP or of their degradation products and other safety issues were not observed, possibly due to efficient elimination and excretion of all the biomaterials used, including the GPI anchor. The pharmacokinetic and -dynamic challenges, such as the adjustment of the time-action profile and the control of the interindividual variability of the anchor removal, will be similar for the paracellular and transcellular routes of oral protein delivery on the basis of therapeutic GPI-AP, but do not seem to represent insurmountable hurdles for the future.

The amphiphilic nature of the GPI glycolipid structure, which typically embeds the attached protein moiety in the outer leaflet of the eukaryotic plasma membrane bilayer, imposes novel biophysical characteristics to the carboxy-terminally coupled protein moiety with regard to adherence to hydrophobic and amphiphilic surfaces and materials (see above). As a consequence, purified GPI-AP can be reconstituted into mixed micelles, liposomes (Milhiet et al. 2002), model lipid rafts (Milhiet et al. 2002), natural membranes or LD harboring a core of neutral triglycerides (Gong et al. 2009) as well as be adsorbed to natural and artificial polymers. In addition, GPI-AP can be cross-linked to each other by chemical means or covalently linked to biomaterials, which constitute the shell of the NP. These assemblies, aggregates, and "bubbles" with grossly differing properties are currently being evaluated for a number of different purposes, among them the transport across intestinal enterocytes.

## 4.3. Tissue Targeting of GPI-AP

The size of the NP and the thickness and type of the nanomaterials used apparently determine their systemic trafficking in the circulation following their transit across the intestinal barrier. This could open the possibility for specific targeting of therapeutic proteins and antibodies to the affected tissues, such as muscle, fat, liver. The strategy of "passive targeting" relies on the relatively selective extravasation and retention of long-circulating NP on the basis of specific structural features of the diseased tissue. The strategy of "active targeting" is based on the modification of the surface of NP with ligands that can specifically recognize the diseased cells. These two clinically relevant and NP-based targeting strategies rely on the specific interaction between the ligands (peptide hormones, binding proteins, antibodies, peptide mimetics) on the carrier surface and the corresponding receptors expressed on the diseased cells. For example, human epidermal growth factor receptor-2 (HER2)(Mamot et al. 2005, Yang et al. 2007), folic acid receptor and vasoactive intestinal peptide receptor (VIP-R)(Dagar et al. 2003) have been investigated as biomarkers for NP targeted to breast tumors during the past two decades. Additional new targeting strategies have emerged as a way of improving the targeting efficiency of NP.

The recently reported observations with adipocytes (Nishino et al. 2008, Parra et al. 2010) raised the possibility for the operation of similar mechanisms of inter- and intracellular trafficking of GPI-AP between as well as in other tissue cells (Lin et al. 2009), such as pancreatic ß-, muscle or liver cells, with their transfer from EV to plasma membrane lipid rafts and subsequent translocation to the surface of cytoplasmic LD. Both lipid rafts and LD are present in virtually all mammalian cell types studied so far, albeit at considerably varying amounts (Danielsen and Hansen 2008). The subcellular location at the LD surface enables the therapeutic protein or antibody to substitute for missing or subnormal activity or to dampen hyperactivity, respectively, of a susceptibility gene product expressed in the cytoplasm or nucleus. The apparent role of LD in the specific retention and release of regulatory and structural proteins suggests new therapeutic options for a variety of common diseases, including obesity and T2D.

# Conclusion

Many cellular proteins undergo a variety of post-translational modifications, including proteolytic processing, glycosylation, and phosphorylation. The covalent modification of proteins by lipid structures is a relatively novel area of research. Although the modification of bacterial cell-wall proteins with *N*-acyl diglycerides, resulting in the proteolipid class of proteins, has been established for almost five decades, it has only become evident for the last two to three decades that a vast array of cellular (and viral) proteins can be modified by fatty acids, polyisoprenoid moieties or glycolipids. Research on these modifications is bringing together workers from a range of diverse disciplines: membrane and protein biochemists, immunologists, geneticists, cell biologists, bacteriologists, parasitologists, virologists, pharmacologists, and bioinformaticians. Such modifications have a pronounced effect on the structure and function of the protein, and play critical roles in many biological systems. The processes involved in the attachment as well as removal of the lipid moieties are the focus of a wide range of basic and applied research, and are becoming targets for therapeutic intervention.

Lipid modification of proteins include the covalent attachment of long-chain acyl groups, particularly myristate and palmitate, coupling of long-chain prenyl groups and the carboxy-terminal anchorage of proteins *via* GPI moieties. Such modifications introduce the amphipatic nature or, at least, increase dramatically the hydrophobic character of those proteins, thereby facilitating their interaction with cellular membranes, and hence can play a role in cellular signaling events, cellular targeting of proteins as well as growth and differentiation. Indeed, various key growth control proteins can be modified in this way. During the past three decades some of the specific signals that direct lipid modifications have been deciphered and the enzymology of these processes is being unraveled. Such enzymes provide potential targets for novel therapeutic agents, particularly in the control of cell growth and division. With the rapidly increasing interest in lipid modification processes and the development of technologies for studying such proteins at the cellular and structural levels, it is now timely to investigate putative applications for important biomedical and pharmaceutical applications.

In the case of GPI anchors, a variety of (patho)physiological functions has been suggested in addition to mere membrane anchorage. These include the mediation of cell adhesion and other cell-cell interactions, allowing increased lateral membrane mobility of proteins and providing more efficient and dense packing of cell surface proteins (compared to transmembrane proteins). The susceptibility of GPI-AP for lipolytic cleavage of their anchor structure and the tissue-specific expression of (G)PI-PLC/D suggest that the release of GPI-AP from cellular membranes may be a (patho)physiologically regulated and controlled event. Several functional aspects of GPI anchors deserve particular attention. In polarized (epithelial or endothelial) cells, GPI anchoring correlates with apical localization. Furthermore, recombinant ectopic expression of a GPI anchor for basolateral or regulated secretory proteins confers apical cell surface localization. GPI attachment therefore appears to act as a dominant apical transport signal implicating this post-translational modification in the intracellular sorting and trafficking of membrane proteins. A potential role for GPI in cellular signaling has proved more controversial. GPI anchors and fragments derived from the GPI glycan core have been implicated as mediators of insulin action, but the potency, the precise molecular nature of such mediators and their origin have remained elusive. Since the initial observation that the differentiation antigen Thy-1 was mitogenic for T-cells, a multitude of distinct GPI-AP proteins has now been implicated in signaling processes in hematopoietic cells and particularly in T-cell activation but, again, the precise mechanism by which the GPI-AP transduce the activation signals across the plasma membranes is still unclear.

During the past two decades another putative function of GPI-AP, which could also explain the reason why during evolution the expression of certain cell surface proteins as GPI-AP had been introduced rather than as typical bi- or polytopic transmembrane protein, has been the object of intense research efforts. This refers to the release of GPI-AP from the extracellular face of the plasma membranes without (proteolytic or lipolytic) cleavage of the anchor moiety, which instead remains attached to the protein moiety in completion, and in fact represents the main focus of this book. Albeit a number of GPI-AP have meanwhile been identified which can be recovered from extracellular aqueous compartments, such as body fluids, and are equipped with the complete GPI anchor, the underlying structures (lipoprotein-like particles, vesicles, multimers, binding proteins, micelle-like phospholipid complexes GLEC), the responsible molecular mechanisms (shedding, exocytosis, spontaneous liberation, "extraction") and the (patho)physiological processes mediated thereby remain to be elucidated. However, the release of GPI-AP together with the complete GPI anchor could be used in a pilot study to demonstrate the successful transformation of a phenomenon into biomedical applications, such as for biomarkers for the prediction, diagnosis, prognosis and stratification of common complex diseases, without the need for this molecular, mechanistic and causal "wisdom" of canonic hypothesis-driven "reductionistic" research. At variance, a "hermeneutic phenomenological approach" relying on (i) mere description and correlative analysis on the basis of a minimal and well-recognized (and accepted) bias and (scientific) tradition, (ii) presentation of the findings as database rather than as narrative in thc typical neutral and irrefutable style of a scientific publication and (iii) very moderate interpretation leading to understanding rather than to explanation and lacking theories and models according to the credo of "the end of theory" (Anderson 2008): "Correlation is enough"... and ... "supersedes causation, and science can advance even without coherent models, unified theories, or really any mechanistic explanation at all."

# REFERENCES

Abbasi, F., Chu, J. W., Lamendola, C., McLaughlin, T., Hayden, J., Reaven, G. M., and Reaven, P. D. 2004. "Discrimination between obesity and insulin resistance in the relationship with adiponectin." *Diabetes* 53:585-90.

Acosta, J., Hettinga, J., Fluckiger, R., Krumrei, N., Goldfine, A., Angarita, L., and Halperin, J. 2000. "Molecular basis for a link between complement and the vascular complications of diabetes." *Proc. Natl. Acad. Sci. USA* 97:5450-55.

Adeyo, O., Goulbourne, C. N., Bensadoun, A., Beigneux, A. P., Fong, L. G., and Young, S. G. 2012. "Glycosylphosphatidylinositol-anchored high-density lipoprotein-binding protein 1 and the intravascular processing of triglyceride-rich lipoproteins." *J. Intern. Med.* 272:528-40.

Agouni, A., Lagrue-Lak-Hal, A. H., Ducluzeau, P. H., Mostefai, H. A., Draunet-Busson, C., Leftheriotis, G., Heymes, C., Martinez, M. C., and Andriantsitohaina, R. 2008. "Endothelial dysfunction caused by circulating microparticles from patients with metabolic syndrome." *Am. J. Pathol.* 173:1210-19.

Agueros, M., Espuelas, S., Esparza, I., Calleja, P., Peñuelas, I., Ponchel, G., and Irache, J. M. 2011. "Cyclodextrin-poly(anhydride) nanoparticles as new vehicles for oral drug delivery." *Expert Opin. Drug Deliv.* 8:721-34.

Ahmed, K. A., and Xiang, J. 2011. "Mechanisms of cellular communication through intercellular protein transfer." *J. Cell. Mol. Med.* 15:1458-73.

Ahmed, S. N., Brown, D. A., and London, E. 1997. "On the origin of sphingolipid/cholesterol-rich detergent-insoluble cell membranes: Physiological concentrations of cholesterol and sphingolipid induce formation of a detergent-insoluble, liquid-ordered lipid phase in model membranes." *Biochemistry* 36:10944-53.

Airas, L., Niemela, J., Salmi, M., Puurunen, T., Smith, D. J., and Jalkanen, S. 1997. "Differential regulation and function of CD73, a glycosyl-phosphatidylinositol-linked

70-kD adhesion molecule, on lymphocytes and endothelial cells." *J. Cell. Biol.* 136:421-31.

Aitken, A., and Cohen, P. 1984. "Identification of N-terminal myristyl blocking groups in proteins." *Meth. Enzymol.* 106:205-10.

Almeida, A., Murakami, Y., Layton, D. M., Hillmen, P., Sellick, G. S., Maeda, Y., Richards, S., Patterson, S., Kotsianidis, I., Mollica, L., Crawford, D. H., Baker, A., Ferguson, M., Roberts, I., Houlston, R., Kinoshita, T., and Karadimitris, A. 2006. "Hypomorphic promoter mutation in PIGM causes inherited glycosylphosphatidylinositol deficiency." *Nat. Med.* 12:846-51.

Al-Nedawi, K., Meehan, B., and Rak, J. 2009b. "Microvesicles: messengers and mediators of tumor porgression." *Cell Cycle* 8:2014-8.

Al-Nedawi, K., Meehan, B., Kerbel, R. S., Allison, A. C., and Rak, J. 2009a. "Endothelial expression of autocrine VEGF upon the uptake of tumor-derived microvesicles containing oncogenic EGFR." *Proc. Natl. Acad. Sci. USA* 106:3794-9.

Al-Nedawi, K., Meehan, B., Micallef, J., Lhotak, V., May, L., Guha, A., and Rak, J. 2008. "Intercellular transfer of the oncogenic receptor EGFRvIII by microvesicles derived from tumour cells." *Nat. Cell. Biol.* 10:619-24.

Alpers, D. H., Eliakim, R., and DeSchryver-Kecskemeti, K. 1989. "Secretion of hepatic and intestinal alkaline phosphatases: similarities and differences." *Clin. Chim. Acta* 186:211-24.

Allan, D., and Thomas, P. 1981. "$Ca^{2+}$-induced biochemical changes in human erythrocytes and their relation to microvesiculation." *Biochem. J.* 198:433-40.

Allan, D., Thomas, P., and Limbrick, A. R. 1980. "The isolation and characterization of 60 nm vesicles ('nanovesicles') produced during ionophore A23187-induced budding of human erythrocytes." *Biochem. J.* 188:881-7.

Allen, C., Maysinger, D., and Eisenberg, A. 1999. „Nano-engineering block copolymer aggregates for drug delivery." *Colloids Surf. B Biointerf.* 16:3-27.

Allen, J. P., Feher, G., Yeates, T. O., Komiya, H., and Rees, D. C. 1987. "Structure of the reaction center from *Rhodobacter sphaeroides* R-26: the cofactors." *Proc. Natl. Acad. Sci. USA* 84:5730-4.

Anderson, C. 2008. "The end of theory: The data deluge makes the scientific method obsolete." *Wired* 06.23.08.

Anderson, S. M., Yu, G., Giatina, M., and Miller, J. L. 1996. "Intercellular transfer of a glycosylphosphatidylinositol (GPI)-linked protein: release and uptake of CD4-GPI from recombinant adeno-associated virus-transduced HeLa cells." *Proc. Natl. Acad. Sci. USA* 93:5894-8.

Andrä, J., Böhling, A., Gronewold, T. M. A., Schlecht, U., Perpeet, M., and Gutsmann, T. 2008. "Surface acoustic wave biosensor as a tool to study the interaction of antimicrobial peptides with phospholipid and lipopolysaccharide model membranes." *Langmuir* 24:9148-53.

Antonioli, L., Blandizzi, C., Csóka, B., Pacher, P., and Haskó, G. 2015. "Adenosine signalling in diabetes mellitus - pathophysiology and therapeutic considerations." *Nat. Rev. Endocrinol.* 11:228-41.

Antonioli, L., Pacher, P., Vizi, E. S., and Haskó, G. 2013. "CD39 and CD73 in immunity and inflammation." *Trends Mol. Med.* 19:355-67.

Aoki, N., Jin-no, S., Nakagawa, Y., Asai, N., Arakawa, E., Tamura, N., Tamura, T., and Matsuda, T. 2007. "Identification and characterization of microvesicles secreted by 3T3-L1 adipocytes: redox- and hormone-dependent induction of milk fat globule-epidermal growth factor 8-associated microvesicles." *Endocrinology* 148:3850-62.

Aoki, N., Yokoyama, R., Asai, N., Ohki, M., Ohki, Y., Kusubata K., Heissig, B., Hattori, K., Nakagawa, Y., and Matsuda, T. 2010. "Adipocyte-derived microvesicles are associated with multiple angiogenic factors and induce angiogenesis *in vivo* and *in vitro*." *Endocrinology* 151:2567-76.

Ashiru, O., Boutet, P., Fernandez-Messina, L., Agüera-González, S., Skepper, J. N., Valés-Gómez, M., and Reyburn, H. T. 2010. "Natural killer cell cytotoxicity is suppressed by exposure to the human NKG2D ligand MICA*008 that is shed by tumor cells in exosomes." *Cancer Res.* 70:481-9.

Ashiru, O., Lopez-Cobo, S., Fernandez-Messina, L., Pontes-Quero, S., Pandolfi, R., Reyburn, H. T., and Valés-Gómez, M. 2013. "A GPI anchor explains the unique biological features of the common NKG2D-ligand allele MICA*0008." *Biochem. J.* 454:295-302.

Ast, T., Cohen, G., and Schuldiner, M. 2013. "A network of cytosolic factors targets SRP-independent proteins to the endoplasmic reticulum." *Cell* 152:1134-45.

Attwood, T. K., Eliopoulos E. E., and Findlay, J. B. C. 1991. "Multiple sequence alignment of protein families showing low sequence homology: a methodological approach using database pattern-matching discriminators for G-protein-linked receptors." *Gene* 98:153-9.

Aupeix, K., Hugel, B., Martin, T., Bischoff, P., Lill, H., Pasquali, J. L., and Freyssinet, J. M. 1997. "The significance of shed membrane particles during programmed cell death *in vitro* and *in vivo* in HIV-1 infection." *J. Clin. Invest.* 99:1546-54.

Ayala-Sarmiento, A. E., Estudillo, E., Perez-Sanchez, G., Sierra-Sánchez, A., González-Mariscal, L., Martínez-Fong, D., and Segovia, J. 2016. "GAS1 is present in the cerebrospinal fluid and is expressed in the choroid plexus of the adult rat." *Histochem. Cell Biol.* 146:325-36.

Ayers, L., Nieuwland, R., Kohler, M., Kraenkel, N., Ferry, B., and Leeson, P. 2015. "Dynamic microvesicle release and clearance within the cardiovascular system: triggers and mechanisms." *Clin. Sci.* 129:915-31.

Bahceci, M., Gokalp, D., and Bahceci, S. 2007. "The correlation between adiposity and adiponectin, tumor necrosis factor alpha, interleukin-6, and high sensitivity

C-reactive protein levels. Is adipocyte size associated with inflammation in adults?" *J. Endocrinol. Invest.* 30:210-4.

Dähr, M., von Holtey, M., Müller, G., and Eckel, J. 1995. "Direct stimulation of myocardial glucose transport and glucose transporter-1 (GLUT1) and GLUT4 protein expression by the sulfonylurea glimepiride." *Endocrinology* 136:2547-53.

Bahram, S., Bresnahan, M., Geraghty, D. E., and Spies, T. 1994. "A second lineage of mammalian major histocompatibility complex class I genes." *Proc. Natl. Acad. Sci. USA* 91:6259-63.

Bai, J. P. F., and Chang, L. L. 1995. "Transepithelial transport of insulin: Insulin degradation by insulin-degrading enzyme in small intestinal epithelium." *Pharm. Res.* 12:1171-5.

Baj-Krzyworzeka, M., Szatanek, R., Weglarczyk, K., Baran, J., and Zembala, M. 2007. "Tumour-derived microvesicles modulate biological activity of human monocytes." *Immunol. Lett.* 113:76-82.

Baj-Krzyworzeka, M., Szatanek, R., Weglarczyk, K., Baran, J., Urbanowicz, B., Brański, P., Ratajczak, M. Z., and Zembala, M. 2006. "Tumour-derived microvesicles carry several surface determinants and mRNA of tumour cells and transfer some of these determinants to monocytes." *Cancer Immunol. Immunother.* 55:808-18.

Bamford, R. N., Roessler, E., Burdine, R. D., Saplakoğlu, U., dela Cruz, J., Splitt, M., Goodship, J. A., Towbin, J., Bowers, P., Ferrero, G. B., Marino, B., Schier, A. F., Shen, M. M., Muenke, M., and Casey, B. 2000. "Loss-of-function mutations in the EGF-CFC gene CFC1 are associated with human left-right laterality defects." *Nat. Genet.* 26:365-9.

Bandlow, W., Wied, S., and Müller, G. 1995. "Glucose induces amphiphilic to hydrophilic conversion of a subset of glycosyl-phosphatidylinositol-anchored ectoproteins in yeast." *Arch. Biochem. Biophys.* 324:300-16.

Baragano-Raneros, A., Suarez-Alvarez, B., and Lopez-Larrea, C. 2014. "Secretoy pathways generating immunosuppressive NKG2D ligands: New targets for therapeutic intervention." *Oncoimmunology* 3:e28497.

Barber, S. R., Davies, M. J., Khunti, K., and Gray, L. J. 2014. "Risk assessment tools for detecting those with pre-diabetes: a systematic review." *Diabetes Res. Clin. Pract.* 105:1-13.

Barboni, E., Rivero, B. P., George, A. J., Martin, S. R., Renoup, D. V., Hounsell, E. F., Barber, P. C., and Morris, R. J. 1995. "The glycophosphatidylinositol anchor affects the conformation of Thy-1 protein." *J. Cell Sci.* 108:487-97.

Bastarache, J. A., Fremont, R. D., Kropski, J. A., Bossert, F. R., and Ware, L. B. 2009. "Proagulant alveolar microparticles in the lungs of patients with acute respiratory distress syndrome." *Am. J. Physiol. Lung Cell. Mol. Physiol.* 297:L1035-41.

Bauer, B., Davidson, M., and Orwar, O. 2009. "Proteomic analysis of plasma membrane vesicles." *Angew. Chem. Int. Ed.* 48:1656-9.

Bauer, S., Groh, V., Wu, J., Steinle, A., Phillips, J. H., Lanier, L. L., and Spies, T. 1999. "Activation of NK cells and T cells by NKG2D, a receptor for stress-inducible MICA." *Science* 285:727-9.

Beigneux, A. P., Davies, B. S., Bensadoun, A., Fong, L. G., and Young, S. G. 2009. "GPIHBP1, a GPI-anchored protein required for the lipolytic processing of triglyceride-rich lipoproteins." *J. Lipid Res.* 50(Suppl.):S57-62.

Beigneux, A. P., Davies, B. S., Gin, P., Weinstein, M. M., Farber, E., Qiao, X., Peale, F., Bunting, S., Walzem, R. L., Wong, J. S., Blaner, W. S., Ding, Z. M., Melford, K., Wongsiriroj, N., Shu, X., de Sauvage, F., Ryan, R. O., Fong, L. G., Bensadoun, A., and Young, S. G. 2007. "Glycosylphosphatidylinositol-anchored high-density lipoprotein-binding protein 1 plays a critical role in the lipolytic processing of chylomicrons." *Cell Metab.* 5:279-91.

Benting, J. H., Rietveld, A. G., and Simons, K. 1999. "N-glycans mediate the apical sorting of a GPI-anchored, rat-associated protein in Madin-Darby canine kidney cells." *J. Cell Biol.* 146:313-20.

Berckmans, R. J., Nieuwland, R., Tak, P. P., Böing, A. N., Romijn, F. P., Kraan, M. C., Breedveld, F. C., Hack, C. E., and Sturk, A. 2002. "Cell-derived microparticles in synovial fluid from inflamed arthritic joints support coagulation exclusively *via* a factor VII-dependent mechanism." *Arthritis Rheum.* 46:2857-66.

Berckmans, R. J., Sturk, A., van Tienen, L. M., Schaap, M. C., and Nieuwland, R. 2011. "Cell-derived vesicles exposing coagulant tissue factor in saliva." *Blood* 117:3172-80.

Bergman, A. S., and Carlsson, S. R. 1994. "Saponin-induced release of cell-surface-anchored Thy-1 by serum glycosylphosphatidylinositol-specific phospholipase D." *Biochem. J.* 298:661-8.

Bernal-Mizrachi, L., Jy, W., Jimenez, J. J., Pastor, J., Mauro, L. M., Horstman, L. L., de Marchena, E., and Ahn, Y. S. 2003. "High levels of circulating endothelial microparticles in patients with acute coronary syndromes." *Am. Heart J.* 145:962-70.

Berryman, D. E., and Bensadoun, A. 1995. "Heparan sulfate proteoglycans are primarily responsible for the maintenance of enzyme activity, binding, and degradation of lipoprotein lipase in Chinese hamster ovary cells." *J. Biol. Chem.* 270:24525-31.

Berteloot, A., Chabot, J. G., and Hugon, J. S. 1981. "Turnover of mouse intestinal brush border membrane proteins and enzymes in organ culture. A direct evaluation from studies on the evolution of enzyme activities during the culture." *Biochim. Biophys. Acta* 678:423-36.

Bessler, M., Schaefer, A., and Keller, P. 2001. "Paroxysmal nocturnal hemoglobinuria: insights from recent advances in molecular biology." *Transfus. Med. Rev.* 15:255-67.

Bhakdi, S., Weller, U., Walev, I., Martin, E., Jonas, D., and Palmer M. 1993. "A guide to the use of pore-forming toxins for controlled permeabilization of cell membranes." *Med. Microbiol. Immunol.* 182:167-75.

Bhavsar, M. D., and Amiji, M. M. 2007. "Polymeric nano- and microparticle technologies for oral gene delivery." *Expert Opin. Drug Deliv.* 4:197-213.

Bimbo, L. M., Mäkilä, E., Laaksonen, T., Lehto, V. P., Salonen, J., Hirvonen, J., and Santos, H. A. 2011. "Drug permeation across intestinal epithelial cells using porous silicon nanoparticles." *Biomaterials* 32:2625-33.

Björge, L., Jensen, T. S., Kristoffersen, E. K., Ulstein, M., and Matre, R. 1996. "Identification of the complement regulatory protein CD59 in human colostrum and milk." *Am. J. Reprod. Immunol.* 35:43-50.

Björge, L., Jensen, T. S., Vedeler, C. A., Ulvestad, E., Kristoffersen, E. K., and Matre, R. 1993. "Soluble CD59 in pregnancy and infancy." *Immunol. Lett.* 36:233.

Black, B. L., Yoneyama, Y., and Moog, F. 1980. "Microvillus memrbane vesicle accumulation in media during culture of intestine of chick embryo." *Biochim. Biophys. Acta* 601:343-8.

Blanchard, P. J., Lankar, D., Faure, F., Regnault, A., Dumont, C., Raposo, G., and Hivroz, C. 2002. "TCR activation of human T cells induces the production of exosomes bearing the TCR/CD3/zeta complex." *J. Immunol.* 168:3235-41.

Blouin, C. M., Le Lay, S. L., Lasnier, F., Dugail, I., and Hajduch, E. 2008. "Regulated association of caveolins to lipid droplets during differentiation of 3T3-L1 adipocytes." *Biochem. Biophys. Res. Commun.* 376:331-5.

Bonauer, A., Boon, R. A., and Dimmeler, S. 2010. "Vascular microRNAs." *Curr. Drug Targets* 11:943-9.

Bond, J. S., and Butler, P. E. 1987. "Intracellular proteases." *Annu. Rev. Biochem.* 56:333-64.

Bordier, C. 1981. "Phase separation of integral membrane proteins in Triton X-114 solution." *J. Biol. Chem.* 256:1604-7.

Bosson, R., Jaquenoud, M., and Conzelmann, A. 2006. "GUP1 of *Saccharomyces cerevisiae* encodes an O-acyltransferase involved in remodeling of the GPI anchor." *Mol. Biol. Cell* 17:2636-45.

Boulanger, C. M., Scoazec, A., Ebrahimian, T., Henry, P., Mathieu, E., Tedgui, A., and Mallat, Z. 2001. "Circulating microparticles from patients with myocardial infarction cause endothelial dysfunction." *Circulation* 104:2649-52.

Bouma, S. R., Drislane, F. W., and Huestis, W. H. 1977. "Selective extraction of membrane-bound proteins by phospholipid vesicles." *J. Biol. Chem.* 252:6759-63.

Boursier, J., Chaigneau, J., Roullier, V., Lainé, F., Sandrini, J., Michalak, S., Hubert, I., Dib, N., Oberti, F., Bertrais, S., Hunault, G., Deugnier, Y., Rousselet, M. C., Cavaro-Ménard, C., Gallois, Y., Aubé, C., and Calès, P.; Study group UNAM-METABOL. 2011. "Steatosis degree, measured by morphometry, is linked to other liver lesions and metabolic syndrome components in patients with NAFLD." *Eur. J. Hepatol.* 54:1011-9.

Bouwens, E. A., Stavenuiter, F., and Mosnier, L. O. 2015. "Cell painting with an engineered EPCR to augment the protein C system." *Thromb. Haemost.* 114:1144-55.

Bowen, R. F., Raikwar, N. S., Olson, L. K., and Deeg, M. A. 2001. "Glucose and insulin regulate glycosylphosphatidylinositol-specific phospholipase D expression in islet beta cells." *Metabolism* 50:1489-92.

Brasitus, T. A., and Schachter, D. 1980. "Lipid dynamics and lipid-protein interactions in rat enterocyte basolateral and microvillus membranes." *Biochemistry* 19:2763-9.

Brasoveanu, L. I., Fonsatti, E., Visintin, A., Pavlovic, M., Cattarossi, I., Colizzi, F., Gasparollo, A., Coral, S., Horejsi, V., Altomonte, M., and Maio, M. 1997. "Melanoma cells constitutively release an anchor-positive soluble form of protectin (sCD59) that retains functional activities in homologous complement-mediated cytotoxicity." *J. Clin. Invest.* 100:1248-55.

Braun-Breton, C., Rosenberry, T. L., and da Silva, L. P. 1988. "Induction of the proteolytic activity of a membrane protein in Plasmodium falciparum by phosphatidyl inositol-specific phospholipase C." *Nature* 332:457-9.

Brennan, P. J., and Nikaido, H. 1995. "The envelope of mycobacteria." *Annu. Rev. Biochem*. 64:29-63.

Brett, M., and Findlay, J. B. C. 1979. "Investigation of the organization of rhodopsin in the sheep photoreceptor membrane by using cross-linking reagents." *Biochem. J.* 177:215-23.

Brewis, L. A., Turner, A. J., and Hooper, N. M. 1994. "Activation of the glycosyl-phosphatidylinositol-anchored membrane dipeptidase upon release from pig kidney membranes by phospholipase C." *Biochem. J.* 303:633-8.

Briscoe, J., and Therond, P. 2005. "Hedgehog signaling: from the Drosophila cuticle to anti-cancer drugs." *Dev. Cell* 8:143-51.

Brocki, J. J. M., and Wearden A. J. 2006. "A critical evaluation of the use of interpretative phenomenological analysis (IPA) in health psychology." *Psychol. Health* 21:87-108.

Brodbeck, U. 1998. "Signalling properties of glycosylphosphatidylinositols and their regulated release from membranes in the turnover of glycosylphosphatidylinositol-anchored proteins." *Biol. Chem.* 379:1041-4.

Brodsky, S. V., Zhang, F., Nasjietti, A., Goligorsky, M. S. 2004. "Endothelium-derived microparticles impair endothelial function *in vitro*." *Am. J. Physiol. Heart Circ. Physiol*. 286:H1910-5.

Brooimans, R. A., van Wieringen P. A., van Es L. A., and Daha, M. R. 1992. "Relative roles of decay-accelerating factor, membrane cofactor protein, and CD59 in the protection of human endothelial cells against complement-mediated lysis." *Eur. J. Immunol.* 22:3135-40.

Broomfield, S. J., and Hooper, N. M. 1993. "Characterization of an antibody to the cross-reacting determinant of the glycosyl-phosphatidylinositol anchor of human membrane dipeptidase." *Biochim. Biophys. Acta.* 1145:212-8.

Brown, D. A., and London, L. 1998. "Functions of lipid rafts in biological membranes." *Annu. Rev. Cell. Dev. Biol.* 14:111-36.

Brown, D. A., and Rose, J. K. 1992. "Sorting of GPI-anchored proteins to glycolipid-enriched membrane subdomains during transport to the apical surface." *Cell* 68:533-44.

Brunner, G., Gabrilove, J., Rifkin, D. B., and Wilson, E. L. 1991. "Phospholipase C release of basic fibroblast growth factor from human bone marrow cultures as a biologically active complex with a phosphatidylinositol-anchored heparan sulfate proteoglycan." *J. Cell Biol.* 114:1275-83.

Brunner, G., Metz, C. N., Nguyen, H., Gabrilove, J., Patel, S. R., Davitz, M. A., Rifkin, D. B., and Wilson, E. L. 1994. "An endogenous glycosylphosphatidylinositol-specific phospholipase D releases basic fibroblast growth factor-heparan sulfate proteoglycan complexes from human bone marrow cultures." *Blood* 83:2115-25.

Bruno, S., Grange, C., Deregibus, M. C., Calogero, R. A., Saviozzi, S., Collino, F., Morando, L., Busca, A., Falda, M., Bussolati, B., Tetta, C., and Camussi, G. 2009. "Mesenchymal stem cell-derived microvesicles protect against acute tubular injury." *J. Am. Soc. Nephrol.* 20:1053-67.

Brunschwig, E. B., Levine, E., Trefzer, U., and Tykocinski, M. L. 1995. "Glycosyl phosphatidylinositol modified murine B7-1 and B7-2 retain costimulator function." *J. Immunol.* 155:5498-5505.

Buchheim, W., Welsch, U., and Patton, S. 1988. "Electron microscopy and carbohydrate histochemistry of the human milk fat globule membrane." In *Biology of the Human Milk*, edited by L. A. Hanson, 27. New York: Raven Press.

Bülow, R., Nonnengässer, C., and Overath, P. 1989. "Release of the variant surface glycoprotein during differentiation of bloodstream to procyclic forms of *Trypanosoma brucei*." *Mol. Biochem. Parasitol.* 32:85-92.

Bulut, D., Tuns, H., and Mugge, A. 2009. "CD31$^{+}$/Annexin V$^{+}$ microparticles in healthy offsprings of patients with coronary artery disease." *Eur. J. Clin. Invest.* 39:17-22.

Buscaglia, C. A., Campo, V. A., Di Noia, J. M., Torrecilhas, A. C., De Marchi, C. R., Ferguson, M. A., Frasch, A. C., and Almeida, I. C. 2004. "The surface coat of the mammal-dwelling infective trypomastigote stage of *Trypanosoma cruzi* is formed by highly diverse immunogenic mucins." *J. Biol. Chem.* 279:15860-9.

Busija, A. R., Patel, H. H., and Insel, P. A. 2017. "Caveolins and cavins in the trafficking, maturation, and degradation of caveolae: implications for cell physiology." *Am. J. Physiol. Cell Physiol.* 312:C459-77.

Bütikofer, P., Kuypers, F. A., Xu, C. M., Chiu, D. T., and Lubin, B. 1989. "Enrichment of two glycosyl-phosphatidylinositol-anchored proteins, acetylcholinesterase and decay accelerating factor, in vesicles released from human red blood cells." *Blood* 74:1481-5.

Caiolfa, V. R., Zamai, M., Malengo, G., Andolfo, A., Madsen, C. D., Sutin, J., Digman, M. A., Gratton, E., Blasi, F., and Sidenius. N. 2007. "Monomer-dimer dynamics and distribution of GPI-anchored uPAR are determined by cell surface protein assemblies." *J. Cell Biol.* 179:1067-1082.

Caird, J. D., and Wiseman, G. M. 1970. "Purification of the delta toxin of *Staphylococcus aureus*." *Can. J. Microbiol.* 16:703-8.

Campion, J., Milagro, F. I., and Martinez, J. A. 2009. "Individuality and epigenetics in obesity." *Obes. Rev.* 10:383-92.

Camussi, G., Deregibus, M. C., Bruno, S., Cantaluppi, V., and Biancone, L. 2010. "Exosomes/microvesicles as a mechanism of cell-to-cell communication." *Kidney Int.* 78:838-48.

Cardoso de Almeida, M. L., Turner, M. J., Stambuk, B. B., and Schenkman, S. 1988. "Identification of an acid-lipase in human serum which is capable of solubilising glycophosphatidylinositol-anchored proteins." *Biochem. Biophys. Res. Commun.* 150:476-82.

Carino, P. C., and Mathiowitz, E. 1999. "Oral insulin delivery." *Adv. Drug Deliv. Rev.* 35:249-57.

Carr, J. M., Dvorak, A. M., and Dvorak, H. F. 1984. "Circulating membrane vesicles in leukemic blood." *Cancer Res.* 45:5944-51.

Casanova, R., Varma, S., Simpson, B., Kim, M., An, Y., Saldana, S., Riveros, C., Moscato, P., Griswold, M., Sonntag, D., Wahrheit, J., Klavins, K., Jonsson, P. V., Eiriksdottir, G., Aspelund, T., Launer, L. J., Gudnason, V., Legido Quigley, C., and Thambisetty, M. 2016. "Blood metabolite markers of preclinical Alzheimer's disease in two longitudinally followed cohorts of older individuals." *Alzheimers Dement.* 12:815-22.

Castellana, D., Toti, F., and Freyssinet, J. M. 2010. "Membrane microvesicles: Macromessengers in cancer disease and progression." *Thrombosis Res.* 125:S84-8.

Castillion, G. A., Aguilera-Romero, A., Manzano-Loppez, J., Epstein, S., Kajiwara, K., Funato, K., Watanabe, R., Riezman, H., and Muñiz, M. 2011. "The yeast p24 complex regulates GPI-anchored protein transport and quality control by monitoring anchor remodeling." *Mol. Biol. Cell* 22:2924-36.

Caughey, B., Baron, G. S., Chesebro, B., and Jeffrey, M. 2009. "Getting a grip on prions: oligomers, amyloids, and pathological membrane interactions." *Annu. Rev. Biochem.* 78:177-204.

Cavallone, D., Malagolini, N., and Serafini-Cessi, F. 2001. "Mechanism of release of urinary Tamm-Horsfall glycoprotein from the kidney GPI-anchored counterpart." *Biochim. Biophys. Res. Commun.* 280:110-4.

Ceriani, R. L., and Blank, E. W. 1988. "Experimental therapy of human breast tumors with $^{131}$I-labeled monoclonal antibodies prepared against the human milk fat globule." *Cancer Res*. 48:4664-72.

Cermelli, S., Guo, Y., Gross, S., and Welte, M. A. 2006. "The lipid-droplet proteome reveals that droplets are a protein-storage depot." *Curr. Biol*. 16:1783-95.

Ciesla, M., Skrzypek, K., Kozakowska, M., Loboda, A., Jozkowicz, A., and Dulak, J. 2011. "MicroRNAs as biomarkers of disease onset." *Anal. Bioanal. Chem.* 401:2051-61.

Cieślak, M., and Roszek, K. 2014. "Purinergic signaling in the pancreas and the therapeutic potential of ecto-nucleotidases in diabetes." *Acta Biochim. Pol.* 61:655-62.

Clayton, A., Court, J., Navabi, H., Adams, M., Mason, M. D., Hobot, J. A., Newman, G. R., and Jasani, B. 2001. "Analysis of antigen presenting cell derived exosomes, based on immuno-magnetic isolation and flow cytometry." *J. Immunol. Methods* 247:163-74.

Clemente, R., Jones, D. R., Ochoa, P., Romero, G., Mato, J. M., and Varela-Nieto, I. 1995. "Role of glycosyl-phosphatidylinositol hydrolysis as a mitogenic signal for epidermal growth factor." *Cell. Signal*. 7:411-21.

Coca-Prieto, I., Kroupa, O., Gonzales-Santos, P., Magne, J., Olivecrona, G., Ehrenborg, E., and Valdivielso, P. 2011. "Childhood-onset chylomicronaemia with reduced plasma lipoprotein lipase activity and mass: identification of a novel GPIHBP1 mutation." *J. Intern. Med.* 270:224-8.

Cocucci, E., Racchetti, G., and Meldolesi, J. 2009. "Shedding microvesicles: artefacts no more." *Trends Cell Biol*. 19:43-51.

Cocuzzi, E., Szczotka, L. B., Brodbeck, W. G., Bardenstein, D. S., Wei, T., and Medof, M. E. 2001. "Tears contain the complement regulator CD59 as well as decay-accelerating factor (DAF)." *Clin. Exp. Immunol.* 123:188-95.

Colburn, W. A. 2003. "Biomarkers in drug discovery and development. From target identification through drug marketing." *J. Clin. Pharmacol*. 43:329-41.

Cole, P. 2008. "Insights into inflammatory mechanisms and new therapeutic strategies." *Drug News Perspect.* 21:573-9.

Comfurius, P., Senden, J. M., Tilly, R. H., Schroit, A. J., Bevers, E. M., and Zwaal, R. F. 1990. "Loss of membrane phospholipid asymmetry in platelets and red cells may be associated with calcium-induced shedding of plasma membrane and inhibition of aminophospholipid translocase." *Biochim. Biophys. Acta* 1026:153-60.

Conde-Vancells, J., Rodriguez-Suarez, E., Embade, N., Gil, D., Matthiesen, R., Valle, M., Elortza, F., Lu, S. C., Mato, J. M., and Falcon-Perez, J. M. 2008. "Characterization and comprehensive proteome profiling of exosomes secreted by hepatocytes." *J. Proteom. Res.* 7:5157-66.

Conde-Vancells, J., Rodriguez-Suarez, E., Gonzalez, E., Gonzalez, E., Berisa, A., Gil, D., Embade, N., Valle, M., Luka, Z., Elortza, F., Wagner, C., Lu, S. C., Mato, J. M., and Falcon-Perez, M. 2010. "Candidate biomarkers in exosome-like vesicles purified from rat and mouse urine samples." *Proteomics Clin. Appl.* 4:416-25.

Conen, D., Rexrode, K. M., Craeger, M. A., Ridker, P. M., and Pradhan, A. D. 2009. "Metabolic syndrome, inflammation, and risk of symptomatic peripheral artery disease in women: a prospective study." *Circulation* 120:1041-7.

Cook, S. L., Bouma, S. R., and Huestis, W. H. 1980. "Cell to vesicle transfer of intrinsic membrane proteins." *Biochemistry* 19:4601-7.

Copic, A., Latham, C. F., Horlbeck, M. A., D'Arcangelo, J. G., and Miller, E. A. 2012. "ER cargo properties specify a requirement for COPII coat rigidity mediated by Sec13p." *Science* 335:1359-62.

Coppinger, J. A., Cagney G., Toomey, S., Kislinger, T., Belton, O., McRedmond, J. P., Cahill, D. J., Emili, A., Fitzgerald, D. J., and Maguire, P. B. 2004. "Characterization of the proteins released from activated platelets leads to localization of novel platelet proteins in human atherosclerotic lesions." *Blood* 103:2096-104.

Cross, G. A. M. 1990. "Cellular and genetic aspects of antigenic variation in trypanosomes." *Ann. Rev. Immunol.* 8:83-110.

Cui, J., Panse, S., and Falkner, B. 2011. "The role of adiponectin in metabolic and vascular disease: a review." *Clin. Nephrol.* 75:26-33.

Chan, B. L., Lisanti, M. P., Rodriguez-Boulan, E., and Saltiel, A. R. 1988. "Insulin-stimulated release of lipoprotein lipase by metabolism of its phosphatidylinositol anchor." *Science* 241:1670-2.

Chargaff, E., and West, R. 1946. "The biological significance of the thromboplastic protein of blood." *J. Biol. Chem.* 166:189-97.

Charriere, S., Peretti, N., Bernard, S., Di Filippo, M., Sassolas, A., Merlin, M., Delay, M., Debard, C., Lefai, E., Lachaux, A., Moulin, P., and Marçais, C. 2011. "GPIHBP1 C89F neomutation and hydrophobic C-terminal domain G175R mutation in two pedigrees with severe hyperchylomicronemia." *J. Clin. Endocrinol. Metab.* 96:E1675-9.

Chatterjee, P., Lim, W. L., Shui, G., Gupta, V. B., James, I., Fagan, A. M., Xiong, C., Sohrabi, H. R., Taddei, K., Brown, B. M., Benzinger, T., Masters, C., Snowden, S. G., Wenk, M. R., Bateman, R. J., Morris, J. C., and Martins, R. N. 2015. "Plasma phospholipid and sphingolipid alterations in presenilin1 mutation carriers: A pilot study." *J. Alzheimers Dis.* 50:887-94.

Chaw, C. S., Chooi, K. W., Liu, X. M., Tan, C. W., Wang, L., and Yang, Y. Y. 2004. "Thermally responsive core-shell nanoparticles self-assembled from cholesteryl end-capped and grafted polyacrylamides: Drug incorporation and *in vitro* release." *Biomaterials* 26:5064-74.

Chen, D., and Wang, M. W. 2005. "Development and application of rodent models for type 2 diabetes." *Diabetes, Obesity, Metabolism* 7:307-17.

Chen, H., Griffiths, G., Galileo, D. S., and Martin-DeLeon, P. A. 2006. "Epididymal SPAM1 is a marker for sperm maturation in the mouse." *Biol. Reprod.* 74:923-30.

Chen, J. K., Taipale, J., Cooper, M. K., and Beachy, P. A. 2002. "Inhibition of Hedgehog signalling by direct binding of cyclopamine to Smoothened." *Genes Dev.* 16:2743-8.

Chen, L., Ashe, S., Brady, W. A., Hellström, I., Hellström, K. E., Ledbetter, J. A., McGowan, P., and Linsley, P. S. 1992. "Costimulation of antitumor immunity by the B7 counterreceptor for the T lymphocyte molecules CD28 and CTLA-4." *Cell* 71:1093-102.

Chen, L., McGowan, P., Ashe, S., Johnston, J., Li, Y., Hellström, I., and Hellström, K. E. 1994. "Tumor immunogenicity determines the effect of B7 costimulation on T cell-mediated tumor immunity." *J. Exp. Med.* 179:523-32.

Chen, M. H., Li, Y. J., Kawakami, T., Xu, S. M., and Chuang, P. T. 2004. "Palmitoylation is required for the production of a soluble multimeric Hedgehog protein complex and long-range signaling in vertebrates." *Genes Dev.* 18:641-59.

Chen, Y., Li, G., Liu, Y., Werth, V. P., Williams, K. J., and Liu, M. L. 2016. "Translocation of endogenous danger signal HMGB1 from nucleus to membrane microvesicles in macrophages." *J. Cell Physiol.* 231:2319-26.

Chen, Y., Veracini, L., Benistant, C., and Jacobson, K. 2009. "The transmembrane protein CBP plays a role in transiently anchoring small clusters of Thy-1, a GPI-anchored protein, to the cytoskeleton." *J. Cell Sci.* 122:3966-72.

Cheng, J. P. X., and Nichols, B. J. 2016. "Caveolae: One Function or Many?" *Trends Cell Biol.* 26:177-89.

Cheruvanky, A., Zhou, H., Pisitkun, T., Kopp, J. B., Knepper, M. A., Yuen, P. S., and Star, R. A. 2007. "Rapid isolation of urinary exosomal biomarkers using a nanomembrane ultrafiltration concentrator." *Am. J. Physiol. Renal Physiol.* 292:F1657-61.

Chirco, R., Liu, X. W., Jung, K. K., and Kim, H. R. 2006. "Novel functions of TIMPs in cell signaling." *Cancer Metastasis Rev.* 25:99-113.

Chistiakov, D. A., Orekhov, A. N., and Bobryshev, Y. V. 2015. "Extracellular vesicles and atherosclerotic diseases." *Cell. Mol. Life Sci.* 72:2697-708.

Cho, L. W., Kilpatrick, E. S., Keevil, B. G., Coady, A. M., and Atkin, S. L. 2009. "Effect of metformin, orlistat and pioglitazone treatment on mean insulin resistance and its biological variability in polycystic ovary syndrome." *Clin. Endocrinol.* 70:233-7.

Chou, W. L., Chuang, L. M., Chou, C. C., Wang, A. H., Lawson, J. A., FitzGerald, G. A., and Chang, Z. F. 2007. "Identification of a novel prostaglandin reductase reveals the involvement of prostaglandin E2 catabolism in regulation of peroxisome proliferator-activated receptor gamma activation." *J. Biol. Chem.* 282:18162-72.

Dagar, S., Krishnadas, A., Rubinstein, I., Blend, M. J., and Onyüksel, H. 2003. "VIP grafted sterically stabilized liposomes for targeted imaging of breast cancer: *in vivo* studies." *J. Control. Rel.* 91:123-33.

Dagher, G., Donne, N., Klein, C., Ferre, P., and Dugail, I. 2003. "HDL-mediated cholesterol uptake and targeting to lipid droplets in adipocytes." *J. Lipid Res.* 44:1811-20.

Dalley, J. A., and Bulleid, N. J. 2003. "How does the translocon differentiate between hydrophobic sequences that form part of either a GPI (glycosylphosphatidylinositol)-anchor signal or a stop transfer sequence?" *Biochem. Soc. Trans.* 31:1257-9.

Damge, C., Michael, C., Aprahamian, M., and Couvreur, P. 1988. "New approach for oral administration of insulin with polyalkylcyanoacrylatenanocapsules as oral carrier." *Diabetes* 37:247-51.

Dangaj, D., Abbott, K. L., Mookerjee, A., Zhao, A., Kirby, P. S., Sandaltzopoulos, R., Powell, D. J. Jr., Lamazière, A., Siegel, D. L., Wolf, C., and Scholler, N. 2011. "Mannose receptor (MR) engagement by mesothelin GPI anchor polarizes tumor-associated macrophages and is blocked by anti-MR human recombinant antibody." *PLoSOne* 6(12):e28386.

Danielsen, E. M., and Hansen, G. H. 2008. "Lipid raft organization and function in the small intestinal brush border." *J. Physiol. Biochem.* 64:377-82.

Danielli, J. F., and Davson, H. 1935. "A contribution to the theory of permeability of thin films." *J. Cell Comp. Physiol.* 5:495-508.

Dapergolas, G., and Gregoriadis, G. 1976. "Hypoglycaemic effect of liposome-entrapped insulin administered intragastrically into rats." *Lancet* 2:824-7.

Davies, B. S., Beigneux, A. P., Barnes, R. H., Tu, Y., Gin, P., Weinstein, M. M., Nobumori, C., Nyrén, R., Goldberg, I., Olivecrona, G., Bensadoun, A., Young, S. G., and Fong, L. G. 2010. "GPIHBP1 is responsible for the entry of lipoproein lipase into capillaries." *Cell Metab.* 12:42-52.

Davies, C. S., Harris, C. L., and Morgan, B. P. 2005. "Glycation of CD59 impairs complement regulation on erythrocytes from diabetic subjects." *Immunology* 114:280-6.

Davis, A., and Lachmann, P. J. 1993. "Membrane defence against complement lysis: the structure and biological properties of CD59." *Immunol. Res.* 12:258-75.

Davis, A., Simmons, D. L., Hale, G., Harrison, R. A., Tighe, H., Lachmann, P. J., and Waldmann, H. 1989. "CD59, an LY-6-like protein expressed in human lymphoid cells, regulates the action of the complement membrane attack complex on homologous cells." *J. Exp. Med.* 170:637-54.

Davis, E. M., Kim, J., Menasche, B. L., Sheppard, J., Liu, X., Tan, A. C., and Shen, J. 2015. "Comparative haploid genetic screens reveal divergent pathways in the biogenesis and trafficking of glycophosphatidylinositol-anchored proteins." *Cell Reports* 11:1727-36.

Davis, M. E. 2009. "The first targeted delivery of siRNA in humans via a self-assembling, cyclodextrin polymer-based nanoparticle: From concept to clinic." *Mol. Pharmaceut.* 6:659-68.

Davison, M. D., and Findlay, J. B. C. 1986a. "Modification of ovine opsin with the photosensitive hydrophobic probe 1-azido-4(125I)iodobenzene. Labelling of the chromophore-attachment domain." *Biochem. J.* 234:413-20.

Davison, M. D., and Findlay, J. B. C. 1986b. "Identification of the sites in opsin modified by photoactivated azido(125)iodobenzene." *Biochem. J.* 236:359-95.

Davitz, M. A., Hereld, D., Shak, S., Krakow, J., Englund, P. T., and Nussenzweig, V. 1987. "A glycan-phosphatidylinositol-specific phospholipase D in human serum." *Science* 238:81-4.

Davy, H. 1812. "Elements of Chemical Philosophy." In *The Collected Works of Sir Humphry Davy (1839-1840) Vol. 4*, edited by J. Davy, 37. Leipzig.

De Broe, M. E., Borgers, M., and Wieme, R. J. 1975. "The separation and characterization of liver plasma membrane fragments circulating in the blood of patients with cholestasis." *Clin. Chim. Acta* 59:369-72.

De Broe, M. E., Wieme, R. J., Logghe, G. N., and Roels, F. 1977. "Spontaneous shedding of plasma membrane fragments by human cells *in vivo* and *in vitro*." *Clin. Chim. Acta* 81:237-45.

De Cat, B., and David, G. 2001. "Developmental roles of the glypicans." *Semin. Cell Dev. Biol.* 12:117-25.

De Ferranti, S., and Mozaffarian, D. 2008. "The perfect storm: obesity, adipocyte dysfunction, and metabolic consequences." *Clin. Chem.* 54:945-55.

de Hoop, M. J., and Dotti, C. G. 1993. "Membrane traffic in polarized neurons in culture." *J. Cell Sci. Suppl.* 17:85-92.

De Luca, C., and Olefsky, J. M. 2008. "Inflammation and insulin resistance." *FEBS Lett.* 582:97-105.

De Nardo, C., Fonsatti, E., Sigalotti, L., Calabrò, L., Colizzi, F., Cortini, E., Coral, S., Altomonte, M., and Maio, M. 2002. "Recombinant transmembrane CD59 (CD59-TM) confers complement resistance to GPI-anchored defective melanoma cells." *J. Cell Physiol.* 190:200-6.

De Wever, O. 2015. "Biological properties of extracellular vesicles and their physiological functions." *J. Extracell. Vesicles* 4:27066.

Deeg, M. A., and Davitz, M. A. 1995. "Glycosylphosphatidylinositol-phospholipase D: a tool for glycosylphosphatidylinositol structural analysis." *Methods Enzymol.* 250:630-40.

Deeg, M. A., Brass, E. P., and Rosenberry, T. L. 1993. "Inositol glycan phosphate derived from human erythrocyte acetylcholinesterase glycolipid anchor and inositol cyclic 1,2-phosphate antagonize glucagon activation of glycogen phosphorylase." *Diabetes* 42:1318-23.

Deisenhofer, J., Epp, O., Miki, K., Huber, R., and Michel, H. 1985. "Structure of the protein subunits in the photosynthetic reaction centre of *Rhodopseudomonas viridis* at 3A resolution." *Nature* 318:618-24.

Del Conde, I., Shrimpton, C. N., and Thiagarajan, P. 2005. "Tissue-factor-bearing microvesicles arise from lipid rafts and fuse with activated platelets to intiate coagulation." *Blood* 106:1604-11.

Delacour, D., Cramm-Behrens, C. I., Drobecq, H., Le Bivic, A., Naim, H. Y., and Jacob, R. 2006. "Requirement for galectin-3 in apical protein sorting." *Traffic* 8:251-8.

Delacour, D., Greb, C., Koch, A., Salomonsson, E., Leffler, H., Le Bivic, A., and Jacob, R. 2007. "Apical sorting by galectin-3-dependent glycoprotein clustering." *Traffic* 8:379-88.

Delahunty, M. D., Stafford, F. J., Yuan, L. C., Shaz, D., and Bonifacino, J. S. 1993. "Uncleaved signals for glycosylphosphatidylinositol anchoring cause retention of precursor proteins in the endoplasmic reticulum." *J. Biol. Chem.* 268:12017-27.

Denzer, K., Kleijmeer, M. J., Heijnen, H. F., Stoorvogel, W., and Geuze, H. J. 2000. "Exosome: from internal vesicle of the multivesicular body to intercellular signaling device." *J. Cell Science* 113:3365-74.

Deregibus, M. C., Cantaluppi, V., and Calogero, R. 2007. "Endothelial progenitor of cell derived microvesicles activate an angiogenic program in endothelial cells by a horizontal transfer of mRNA." *Blood* 110:2440-8.

Derganc, J., Antonny, B., and Copic, A. 2013. "Membrane bending: the power of protein imbalance." *Trends Biochem. Sci.* 38:576-84.

DeSchryver-Kecskemeti, K., Eliakim, R., Carroll, S., Stenson, W. F., Moxley, M. A., and Alpers, D. H. 1989. "Intestinal surfactant-like material: a novel secretory product of the enterocyte." *J. Clin. Invest.* 84:1355-61.

DeSchryver-Kecskemeti, K., Eliakim, R., Green, K., and Alpers, D. H. 1991. "A novel intracellular pathway for rat intestinal digestive enzymes (alkaline phosphatase and sucrase) *via* a lamellar particle." *Lab. Invest.* 65:365-73.

Diamant, M., Nieuwland, R., Pablo, R. F., Sturk, A., Smit, J. W., and Radder, J. K. 2002. "Elevated numbers of tissue-factor exposing microparticles correlate with components of the metabolic syndrome in uncomplicated type 2 diabetes mellitus." *Circulation* 106:2442-7.

Diep, D. B., Nelson, K. L., Raja, S. M., Pleshak, E. N, and Buckley, J. T. 1998. "Glycosylphosphatidylinositol anchors of membrane glycoproteins are binding determinants for the channel-forming toxin aerolysin." *J. Biol. Chem.* 273:2355-60.

Dignat-George, F., and Boulanger, C. M. 2011. "The many faces of endothelial microparticles." *Arterioscler. Thromb. Vasc. Biol.* 31:27-33.

Ding, X., Li, P., Lin, S. C., Stratton, Z. S., Nama, N., Guo, F., Slotcavage, D., Mao, X., Shi, J., Costanzo, F., and Huang, T. J. 2013. "Surface acoustic wave microfluidics." *Lab Chip* 13:3626-49.

Distler, J. H., Pisetsky, D. S., Huber, L. C., Kalden, J. R., Gay, S., and Distler, O. 2005. ”Microparticles as regulators of inflammation: novel players of cellular crosstalk in the rheumatic diseases.” *Arthritis Rheum.* 52:3337-48.

Djafarzadeh, R., Milani, V., Rieth, N., von Luettichau, I., Skrablin, P. S., Hofstetter, M., Noessner, E., and Nelson, P. J. 2009. “TIMP-1-GPI in combination with hyperthermic treatment of melanoma increases sensitivity to FAS-mediated apoptosis.” *Cancer Immunol. Immunother.* 58:361-71.

Djafarzadeh, R., Noessner, E., Engelmann, H., Schendel, D. J., Notohamiprodjo, M., von Luettichau, I., and Nelson, P. J. 2006. “GPI-anchored TIMP-1 treatment renders renal cell carcinoma sensitive to FAS-mediated killing.” *Oncogene* 25:1496-508.

Djafarzadeh, R., Notohamiprodio, S., Rieth, N., Hofstetter, M., Noessner, E., and Nelson, P. J. 2013. “Treatment of dermal fibroblasts with GPI-anchored human TIMP-1 protein moderates processes linked to scar formation.” *J. Invest. Dermatol.* 133:803-11.

Doering, T. L., and Schekman, R. 1996. “GPI anchor attachment is required for Gas1p transport from the endoplasmic reticulum in COPII vesicles.” *EMBO J.* 15:182-91.

Doery, H. M., Magnusson, B. J., Gulasekharam, J., and Pearson, J. E. 1965. “The properties of phospholipase enzymes in staphylococcal toxins.” *Nature* 329:846-8.

Dolezal, S., Hester, S., Kirby, P. S., Nairn, A., Pierce, M., and Abbott, K. L. 2014. “Elevated levels of glycosylphosphatidylinositol (GPI) anchored proteins in plasma from human cancers detected by *C. septicum* alpha toxin.” *Cancer Biomarkers* 14:55-62.

Dong, Y., and Feng, S. S. 2005. “Poly(D,L-lactide-co-glycolide)/montmorillonite nanoparticles for oral delivery of anticancer drugs.” *Biomaterials* 26:6068-76.

Doria, A., Patti, M. E., and Kahn, C. R. 2008. “The emerging genetic architecture of type 2 diabetes.” *Cell Metab.* 8:186-200.

Ducharme, N. A., and Bickel, P. E. 2008. “Lipid droplets in lipogenesis and lipolysis.” *Endocrinology* 149:942-9.

Dugail, I., and Hajduch, E. 2007. “A new look at adipocyte lipid droplets: towards a role in the sensing of triacylglycerol stores?” *Cell. Mol. Life Sci.* 64:2452-8.

Dugail, I., Le Lay, S., Varret, M., Le Liepvre, X., Dagher, G., and Ferré, P. 2003. “New insights into how adipocytes sense their triglyceride stores. Is cholesterol a signal?” *Horm. Metab. Res.* 35:204-10.

Duncan, B. B., Schmidt, M. I., Pankow, J. S., Ballantyne, C. M., Couper, D., Vigo, A., Hoogeveen, R., Folsom, A. R., and Heiss, G. Atherosclerosis Risk in Communities Study. 2003. “Low-grade systemic inflammation and the development of type 2 diabetes: the atherosclerosis risk in communities study.” *Diabetes* 52:1799-805.

Duncan, T., Kutty, G., Chader, G. J., and Wiggert, B. 1994. A glycoprotein binding retinoids and fatty acids is present in *Drosophila*. *Arch. Biochem. Biophys.* 312:158-66.

Dunn, D. E., Yu, J., Nagarajan, S., Devetten, M., Weichold, F. F., Medof, M. E., Young, N. S., and Liu, J. M. 1996. "A knock-out model of paroxysmal nocturnal hemoglobinuria: Pig-a(-) hematopoiesis is reconstituted following intercellular transfer of GPI-anchored proteins." *Proc. Natl. Acad. Sci. USA* 93:7938-43.

Dunnick, J. K., Rooke, J. D., Aragon, S., and Kriss, J. P. 1976. "Alteration of mammalian cells by interaction with artificial lipid vesicles." *Cancer Res*. 36:2385-9.

Durrant, L. G., and Spendlove I. 2001. "Immunization against tumor cell surface complement-regulatory proteins." *Curr. Opin. Investig. Drugs* 2:959-66.

Dyck, W. P., Hall, F. F., and Ratliff, C. F. 1973. "Hormonal control of intestinal alkaline phosphatase secretion in the dog." *Gastroenterology* 65:445-50.

Dzeja, P., and Terzic, A. 2009. "Adenylate kinase and AMP signaling networks: metabolic monitoring, signal communication and body energy sensing." *Int. J. Mol. Sci*. 10:1729-72.

Eardley, D. D., and Koshland, M. E. 1991. "Glycosylphosphatidylinositol: a candidate system for IL-2 signal transduction." *Science* 251:78-81.

Eaton, S. 2006. "Release and trafficking of lipid-linked morphogens." *Curr. Opin. Genet. Dev*. 16:17-22.

Eaton, S. 2008. "Multiple roles for lipids in the Hedgehog signalling pathway." *Nat. Rev. Mol. Cell Biol.* 9:437-45.

Ebbesen, M., and Jensen, T. G. 2006. "Nanomedicine: Techniques, potentials and ethical implications." *J. Biomed. Biotechnol*. 5:51516.

Edgar, J. R., Manna, P. T., Nishimura, S., Banting, G., and Robinson, M. S. 2016. "Tetherin is an exosomal tether." *eLife* 5:e17180.

Eggeling, C., Ringemann, C., Medda, R., Schwarzmann, G., Sandhoff, K., Polyakova, S., Belov, V. N., Hein, B., von Middendorff, C., Schönle, A., and Hell, S. W. 2009. "Direct observation of the nanoscale dynamics of membrane lipids in a living cell." *Nature* 457:1159-62.

Eilertsen, K. E., and Osterud, B. 2005. "The role of blood cells and their microparticles in blood coagulation." *Biochem. Soc. Trans*. 33:418-22.

Eisenberg-Bord, M., Shai, N., Schuldiner, M., and Bohnert, M. 2016. "A tether is a tether is a tether: Tethering at membrane contact sites." *Dev. Cell* 39:395-409.

Eisenhaber, B., Bork, P., and Eisenhaber, B. 2001. "Post-translational GPI lipid anchor modification of proteins in kingdoms of life: analysis of protein sequence data from complete genomes." *Protein Eng*. 14:17-25.

Eisenhaber, B., Bork, P., and Eisenhaber, F. 1999. "Prediction of potential GPI-modification sites in proprotein sequences." *J. Mol. Biol*. 292:741-58.

Eisenhaber, B., Maurer-Stroh, S., Novatchkova, M., Schneider, G., and Eisenhaber, F. 2003. "Enzymes and auxiliary factors for GPI lipid anchor biosynthesis and post-translational transfer to proteins." *BioEssays* 25:367-85.

Eliakim, R., Alpers, D. H., Oren, R., Fich, A., and DeSchryver-Kecskemeti, K. 1996. "Abundance of surfactant-like particles reflects mucosal integrity in patients with peptic ulcer disease." *Gut* 39:353-9.

Eliakim, R., Becich, M. J., Green, K., and Alpers, D. H. 1990. "Both tissue and serum phospholipases release rat intestinal alkaline phosphatase." *Am. J. Physiol.* 259:G618-25.

Eliakim, R., DeSchryver-Kecskemeti, K., Nogee, L., Stenson, W. F., and Alpers, D. H. 1989. "Isolation and characterization of a small intestinal surfactant-like particle containing alkaline phosphatase and other digestive enzymes." *J. Biol. Chem.* 264:20614-9.

Eliakim, R., Mahmood, A., and Alpers, D. H. 1991. "Rat intestinal alkaline phosphatase secretion into lumen and serum is coordinately regulated." *Biochim. Biophys. Acta* 1091:1-8.

Englund, P. T. 1993. "The structure and biosynthesis of glycosyl phosphatidylinositol protein anchors." *Ann. Rev. Biochem.* 62:121-38.

Ermonval, M., Baychelier, F., and Fonta, C. 2015. "TNAP, an essential player in membrane lipid rafts of neuronal cells." *Subcell. Biochem.* 76:167-83.

Escola, J. M., Kleijmeer, M. J., Stoorvogel, W., Griffith, J. M., Yoshie, O., and Geuze, H. J. 1998. "Selective enrichment of tetraspanin proteins on the internal vesicles of multivesicular endosomes and on exosomes secreted by human B-lymphocytes." *J. Biol. Chem.* 273:20121-7.

Esposito, K., Ciotola, M., and Giugliano, D. 2006a. "Pioglitazone reduces endothelial microparticles in the metabolic syndrome." *Arterioscler. Thromb. Vasc. Biol.* 26:1926.

Esposito, K., Ciotola, M., Schisano, B., Gualdiero, R., Sardelli, L., Misso, L., Giannetti, G., Giugliano, D., Gualdiero, R., Sardelli, L., Misso, L., Giannetti, G., and Giugliano, D. 2006b. "Endothelial microparticles correlate with endothelial dysfunction in obese women." *J. Clin. Endocrinol. Metab.* 91:3676-9.

Esposito, K., Maiorino, M. I., Di Palo, C., Gicchino, M., Petrizzo, M., Bellastella, G., Saccomanno, F., and Giugliano, D. 2011. "Effects of pioglitazone versus metformin on circulating endothelial microparticles and progenitor cells in patients with newly diagnosed type 2 diabetes—a randomized controlled trial." *Diabetes Obes. Metab.* 13:439-45.

Evdokimovskaya, Y., Skarga, Y., Vrublevskaya, V., and Morenkov, O. 2010. "Secretion of the heat shock proteins HSP70 and HSC70 by baby hamster kidney (BHK-21) cells." *Cell Biol. Int.* 34:985-90.

Evdokimovskaya, Y., Skarga, Y., Vrublevskaya, V., and Morenkov, O. 2012. "Release of the glucose-regulated protein 94 by the baby hamster kidney cells." *Cell. Biochem. Funct.* 30:558-62.

Faerch, K., Hulmán, A., and Solomon, T. P. 2016. "Heterogeneity of pre-diabetes and type 2 diabetes: Implications for prediction, prevention and treatment responsiveness." *Curr. Diabetes Rev.* 12:30-41.

Fankhauser, N., and Mäser, P. 2005. "Identification of GPI anchor attachment signals by a Kohonen self-organizing map." *Bioinformatics* 21:1846-52.

Farese, R. V. 1990. "Lipid-derived mediators in insulin." *Proc. Soc. Exp. Biol. Med.* 195:312-24.

Faris, R., Moore, R. A., Ward, A., Race, B., Dorward, D. W., Hollister, J. R., Fischer, E. R., and Priola, S. A. 2017. "Cellular prion protein is present in mitochondria of healthy mice." *Sci. Rep.* 7:41556.

Faure, J., Lachenal, G., Court, M., Hirrlinger, J., Chatellard-Causse, C., Blot, B., Grange, J., Schoehn, G., Goldberg, Y., Boyer, V., Kirchhoff, F., Raposo, G., Garin, J., and Sadoul, R. 2006. "Exosomes are released by cultured cortical neurones." *Mol. Cell. Neurosci.* 31:642-8.

Feng, J., White, B., Tyurina, O. V., Guner, B., Larson, T., Lee, H. Y., Karlstrom, R. O., and Kohtz, J. D. 2004. "Synergistic and antagonistic roles of the Sonic hedgehog N- and C-terminal lipids." *Development* 131:4357-70.

Ferguson, M. A. J., and Williams, A. F. 1988. "Cell-surface anchoring of proteins *via* glycosylphosphatidylinositol structure." *Ann. Rev. Biochem.* 57:285-320.

Ferguson, M. A. J., Haldar, K., and Cross, G. A. M. 1985. "*Trypanosoma brucei* variant surface glycoprotein has a *sn*-1,2-dimyristyl glycerol membrane anchor at its COOH terminus." *J. Biol. Chem.* 260:4963-8.

Ferguson, M. A. J., Homans, S. W., Dwek, R. A., and Rademacher, T. W. 1988. "Glycosyl-phosphatidylinositol moiety that anchors *Trypanosoma brucei* variant surface glycoprotein to the membrane." *Science* 239:753-9.

Ferguson, M. A., Brimacombe, J. S., Brown, J. R., Crossman, A., Dix, A., Field, R. A., Güther, M. L., Milne, K. G., Sharma, D. K., and Smith, T. K. 1999. "The GPI biosynthetic pathway as a therapeutic target for African sleeping sickness." *Biochim. Biophys. Acta* 1455:327-40.

Fernandez-Messina, L., Ashiru, O., Boutet, P., Agüera-González, S., Skepper, J. N., Reyburn, H. T., and Valés-Gómez, M. 2010. "Differential mechanisms of shedding of the glycosylphosphatidylinositol (GPI)-anchored NKG2D ligands." *J. Biol. Chem.* 285:8543-51.

Fernandez-Messina, L., Reyburn, H. T., and Vales-Gomez, M. 2012. "Human NKG2D-ligans cell biology strategies to enssure immune recognition." *Front. Immunol.* 3:299.

Ferreira, A. C., Peter, A. A., Mendez, A. J., Jimenez, J. J., Mauro, L. M., Chirinos, J. A., Ghany, R., Virani, S, Garcia, S., Horstman, L. L., Purow, J., Jy, W., Ahn, Y. S., and de Marchena, E. 2004. "Postprandial hypertriglyceridemia increases circulating levels of endothelial cell microparticles." *Circulation* 110:3599-603.

Fevrier, B., Vilette, D., Archer, F., Loew, D., Faigle, W., Vidal, M., Laude, H., and Raposo, G. 2004. "Cells release prions in association with exosomes." *Proc. Natl. Acad. Sci. USA* 101:9683-8.

Fico, A., Maina, F., and Dono, R. 2011. "Fine-tuning of cell signaling by glypicans." *Cell. Mol. Life Sci.* 68:923-9.

Field, M. C. 1997. "Is there evidence for phospho-oligosaccharides as insulin mediators?" *Glycobiology* 7:161-8.

Fielding, P. E., and Fielding, C. J. 1996. "Intracellular transport of low density lipoprotein derived free cholesterol begins at clathrin-coated pits and terminates at cell surface caveolae." *Biochemistry* 35:14932-8.

Filmus, J., Capurro, M., and Rast, J. 2008. "Glypicans." *Genome Biol.* 9:224.

Findlay, J. B. C. 1987. "The isolation and labelling of membrane proteins and peptides." In *Biological Membranes: A Practical Approach,* edited by Findlay, J. B. C., and Evans, H. W., 179-217. Oxford: IRL Press.

Findlay, J. B. C., and Pappin, D. J. C. 1986. "The opsin family of proteins." *Biochem. J.* 238:625-42.

Findlay, J. B. C., Eliopoulos, E. E., and Finbow, M. 1990. "Molecular modelling of integral membrane proteins." *Biochem. Soc. Trans.* 18:838-40.

Finean, J. B., Bramley, T. A., and Coleman, R. 1971. "Lipid layer in cell membrane." *Nature* 229:114.

Fletcher, C. M., Harrison, R. A., Lachmann, P. J., and Neuhaus, D. 1993. "Sequence-specific 1H-NMR assignments and folding topology of human CD59." *Prot. Science* 2:2015-27.

Fletcher, C. M., Harrison, R. A., Lachmann, P. J., and Neuhaus, D. 1994. "Structure of a soluble, glycosylated form of the human complement regulatory protein CD59." *Structure* 2:185-99.

Flores-Borja, F., Kieszkievicz, J., Church, V., Francis-West, P. H., Schofield, J., Rademacher, T. W., and Lund, T. 2004. "Genetic regulation of mouse glycosylphosphatidylinositol-phospholipase D." *Biochimie* 86:275-82.

Fogel, R., Limson, J., and Seshia, A. A. 2016. "Acoustic biosensors." *Essays Biochem.* 60:101-10.

Fong, K. P., Barry, C., Tran, A. N., Traxler, E. A., Wannemacher, K. M., Tang, H. Y., Speicher, K. D., Blair, I. A., Speicher, D. W., Grosser, T., and Brass, L. F. 2011. "Deciphering the human platelet sheddome." *Blood* 117:e15-26.

Fonteles, M. C., Huang, L. C., and Larner, J. 1996. "Infusion of pH 2.0 D-chiro-inositol glycan insulin putative mediator normalizes plasma glucose in streptozotocin diabetic rats at a dose equivalent to insulin without inducing hypoglycaemia." *Diabetologia* 39:731-4.

Franssen, R., Young, S. G., Peelman, F., Hertecant, J., Sierts, J. A., Schimmel, A. W., Bensadoun, A., Kastelein, J. J., Fong, L. G., Dallinga-Thie, G. M., and Beigneux, A.

P. 2010. "Chylomicronemia with low postheparin lipoprotein lipase levels in the setting of GPIHBP1 defects." *Circ. Cardiovasc. Genet.* 3:169-78.

Franz, C., Boing, A. N., Hau, C. M., Montag, M., Strowitzki, T., Nieuwland, R., and Toth, B. 2013. "Procoagulant tissue factor-exposing vesicles in human seminal fluid." *J. Reprod. Immunol.* 98:45-51.

Freyssinet, J. M. 2003. "Cellular microparticles: What are they bad or good for?" *J. Thromb. Haemost.* 1:1655-62.

Freyssinet, J. M., and Toti, F. 2010. "Formation of procoagulant microparticles and properties." *Thromb. Res.* 125(Suppl 1):S46-8.

Frick, W., Bauer, A., Bauer, J., Wied, S., and Müller, G. 1998. "Structure-activity relationship of synthetic phosphoinositolglycans mimicking metabolic insulin action." *Biochemistry* 37:13421-36.

Fuchs, S., Frenzel, K., Hubert, C., Lyng, R., Muller, L., Michaud, A., Xiao, H. D., Adams, J. W., Capecchi, M. R., Corvol, P., Shur, B. D., and Bernstein, K. E. 2005. "Male fertility is dependent on dipeptidase activity of testis ACE." *Nat. Med.* 11:1140-2.

Fujihara, Y., and Ikawa, M. 2016. "GPI-AP release in cellular, developmental, and reproductive biology." *J. Lipid Res.* 57:538-45.

Fujii, T., Sakata, A., Nishimura, S., Eto, K., and Nagata, S. 2015. "TMEM16F is required for phosphatidylserine exposure and microparticle release in activated mouse platelets." *Proc. Natl. Acad. Sci. USA* 112:12800-5.

Fujimoto, T., Ohsaki, Y., Cheng, J., Suzuki, M., and Shinohara, Y. 2008. "Lipid droplets: a classic organelle with new outfits." *Histochem. Cell. Biol.* 130:263-79.

Fujita, M., and Kinoshita, T. 2010. "Structural remodeling of GPI anchors during biosynthesis and after attachment to proteins." *FEBS Lett.* 584:1670-7.

Fujita, M., and Kinoshita, T. 2012. "GPI-anchor remodeling: Potential functions of GPI-anchors in intracellular trafficking and membrane dynamics." *Biochim. Biophys. Acta* 1821:1050-8.

Fujita, M., Maeda, Y., Ra, M., Yamaguchi, Y., Taguchi, R., and Kinoshita, T. 2009. "GPI glycan remodeling by PGAP5 regulates transport of GPI-anchored proteins from the ER to the Golgi." *Cell* 139:352-65.

Fujita, M., Umemura, M., Yoko-o, T., and Jigami, Y. 2006. "PER1 is required for GPI-phospholipase A2 activity and involved in lipid remodeling of GPI-anchored proteins." *Mol. Biol. Cell* 17:5253-64.

Fujita, M., Watanabe, R., Jaensch, N., Romanova-Michaelides, M., Satoh, T., Kato, M., Riezman, H., Yamaguchi, Y., Maeda, Y., and Kinoshita, T. 2011. "Sorting of GPI-anchored proteins into ER exit sites by p24 proteins is dependent on remodeled GPI." *J. Cell Biol.* 194:61-75.

Fujiwara, S., and Amisaki, T. 2013. "Fatty acid binding to serum albumin: molecular simulation approaches." *Biochim. Biophys. Acta* 1830:5427-34.

Fukuoka, S., and Kobayashi, K. 2001. "Analysis of the C-terminal structure of urinary Tamm-Horsfall protein reveals that the release of the glycosyl phosphatidylinositol-anchored counterpart from the kidney occurs by phenylalanine-specific proteolysis." *Biochem. Biophys. Res. Commun.* 289:1044-8.

Fukushima, K., Ikehara, Y., Kanai, M., Kochibe, N., Kuroki, M., and Yamashita, K. 2003. "A beta-N-acetylglucosaminyl phosphate diester residue is attached to the glycosylphosphatidylinositol anchor of human placental alkaline phosphatase: a target of the channel-forming toxin aerolysin." *J. Biol. Chem.* 278:36296-303.

Futerman, A. H., Fiorini, R. M., Roth, E., Low, M. G., and Silman, I. 1985a. "Physicochemical behaviour and structural characteristics of membrane-bound acetylcholinesterase from *Torpedo* electric organ. Effect of phosphatidylinositol-specific phospholipase C." *Biochem. J.* 226:369-77.

Futerman, A. H., Low, M. G., Ackermann, K. E., Sherman, W. R., and Silman, I. 1985b. "Identification of covalently bound inositol in the hydrophobic membrane-anchoring domain of *Torpedo* acetylcholinesterase." *Biochem. Biophys. Res. Commun.* 129:312-7.

Gallet, A., Ruel, L., Staccini-Lavenant, L., and Thérond, P. P. 2006. "Cholesterol modification is necessary for controlled planar long-range activity of Hedgehog in *Drosophila* epithelia." *Development* 133:407-18.

Garaulet, M., Hernández-Morante, J. J., de Heredia, F. P., and Tébar, F. J. 2007. "Adiponectin, the controversial hormone." *Public Health Nutr.* 10:1145-50.

Garza, C., Butte, N. F., and Goldman, A. S. 1993. "Human milk and infant formula." In: *Textbook of Pediatric Nutrition*, edited by Suskind, R. M., and Lewinter-Suskind, L., 33. New York: Raven Press.

Gasser, O., and Schifferli, J. A. 2004. "Activated polymorphonuclear neutrophils disseminate anti-inflammatory microparticles by ectocytosis." *Blood* 104:2543-8.

Gasser, O., and Schifferli, J. A. 2005. "Microparticles released by human neutrophils adhere to erythrocytes in the presence of complement." *Exp. Cell Res.* 307:381-7.

Gaulton, G. N., and Pratt, J. C. 1994. "Glycosylated phosphatidylinositol molecules as second messengers." *Semin. Immunol.* 6:97-104.

Geisen, K. 1988. "Special pharmacology of the new sulfonylurea glimepiride." *Drug Res.* 38:1120-30.

Gerber, L. D., Kodukula, K., and Udenfriend, S. 1992. "Phosphatidylinositolglycan (PI-G) anchored membrane proteins. Amino acid requirements adjacent to the site of cleavage and PI-G attachment in the COOH-terminal signal peptide." *J. Biol. Chem.* 267:12168-73.

Ghosh, P., Sahoo, R., Vaidya, A., Cantel, S., Kavishwar, A., Goldfine, A., Herring, N., Bry, L., Chorev, M., and Halperin, J. A. 2013. "A specific and sensitive assay for blood levels of glycated CD59: A novel biomarker for diabetes." *Am. J. Hematol.* 88:670-6.

Ghugtyal, V., Vionnet, C., Roubaty, C., and Conzelmann, A. 2007. "CWH43 is required for the introduction of ceramides into GPI anchors in *Saccharomyces cerevisiae*." *Mol. Microbiol.* 67:1493-502.

Gibbings, D. J., Ciaudo, C., Erhardt, M., and Voinnet, O. 2009. "Multivesicular bodies associate with components of miRNA effector complexes and modulate miRNA activity." *Nat. Cell Biol.* 11:1143-9.

Gill, M. J. 2014. "The Possibilities of phenomenology for organizational research. *Organ.*" *Res. Meth.* 17:118-37.

Gillmor, C. S., Lukowitz, W., Brininstool, G., Sedbrook, J. C., Hamann, T., Poindexter, P., and Somerville, C. 2005. "Glycosylphosphatidylinositol-anchored proteins are required for cell wall synthesis and morphogenesis in *Arabidopsis*." *Plant Cell* 17:1128-40.

Gillon, A. D., Latham, C. F., and Miller, E. A. 2012. "Vesicle-mediated ER export of proteins and lipids." *Biochim. Biophys. Acta* 1821:1040-9.

Gin, P., Beigneux, A. P., Voss, C., Davies, B. S., Beckstead, J. A., Ryan, R. O., Bensadoun, A., Fong, L. G., and Young, S. G. 2011. "Binding preferences for GPIHBP1, a glycosylphosphatidylinositol-anchored protein of capillary endothelial cells." *Arterioscler. Thromb. Vasc. Biol.* 31:176-82.

Ginestra, A., Monea, S., Seghezzi, G., Dolo, V., Nagase, H., Mignatti, P., and Vittorelli, M. L. 1997. "Urokinase plasminogen activator and gelatinases are associated with membrane vesicles shed by human HT1080 fibrosarcoma cells." *J. Biol. Chem.* 272:17216-22.

Girouard, J., Frenette, G., and Sullivan, R. 2011. "Comparative proteome and lipid profiles of bovine epididymosomes collected in the intraluminal compartment of the caput and cauda epididymidis." *Int. J. Androl.* 34:e475-86.

Gmachl, M., Sagan, S., Ketter, S., and Kreil G. 1993. "The human sperm protein PH-20 has hyaluronidase activity." *FEBS Lett.* 336:545-8.

Goichot, B., Grunebaum, L., Desprez, D., Vinzio, S., Meyer, L., Schlienger, J. L., Lessard, M., and Simon, C. 2006. "Circulating procoagulant microparticles in obesity." *Diabetes Metab.* 32:82-5.

Gong, J., Sun, Z., and Li, P. 2009. "CIDE proteins and metabolic disorders." *Curr. Opin. Lipidol.* 20:121-6.

Gonzales, P. A., Pisitkun, T., Hoffert, J. D., Tchapyjnikov, D., Star, R. A., Kleta, R., Wang, N. S., and Knepper, M. A. 2009. "Large-scale proteomics and phosphoproteomics of urinary exosomes." *J. Am. Soc. Nephrol.* 20:363-79.

Gonzales, S., Lopez-Soto, A., Suarez-Alvarez, B., López-Vázquez, A., and López-Larrea, C. 2008. "NKG2D ligands key targets of the immune response." *Trends Immunol.* 29:397-403.

Gordon, V. M., Nelson, K. L., Buckley, J. T., Stevens, V. L., Tweten, R. K., Elwood, P. C., and Leppla, S. H. 1999. "*Chlostridium septicum* α-toxin uses glycosyl-phosphatidylinositol-anchored protein receptors." *J. Biol. Chem.* 274:27274-80.

Goswami, D., Gowrishankar, K., Bilgrami, S., Ghosh, S., Raghupathy, R., Chadda, R., Vishwakarma, R., Rao, M., and Mayor, S. 2008. "Nanoclusters of GPI-anchored proteins are formed by cortical actin-driven activity." *Cell* 135:1085-97.

Götze, H., Adelson, J. W., Hadorn, H. B., Portmann, R., and Troesch, V. 1972. "Hormone-elicited enzyme release by the small intestinal wall." *Gut* 13:471-6.

Grand, R. J. A. 1989. "Acylation of viral and eukaryotic proteins." *Biochem. J.* 258:625-38.

Gregorieff, A., Pinto, D., Begthel, H., Destrée, O., Kielman, M., and Clevers, H. 2005. "Expression pattern of Wnt signaling components in the adult intestine." *Gastroenterology* 129:626-38.

Griffiths, G. S., Galileo, D. S., Aravindan, R. G., and Martin-DeLeon, P. A. 2009. "Clusterin facilitates exchange of glycosyl phosphatidylinositol-linked SPAM1 between reproductive luminal fluids and mouse and human sperm membranes." *Biol. Reprod.* 81:562-70.

Griffiths, G. S., Galileo, D. S., Reese, K., and Martin-Deleon, P. A. 2008a. "Investigating the role of murine epididymosomes and uterosomes in GPI-linked protein transfer to sperm using SPAM1 as a model." *Mol. Reprod. Dev.* 75:1627-36.

Griffiths, G. S., Miller, K. A., Galileo, D. S., and Martin-DeLeon, P. A. 2008b. "SPAM1 is secreted by the estrous murine uterus and oviduct in a form which can bind to sperm during capacitation: acquisition enhances hyaluronic acid-binding ability and cumulus penetration efficiency." *Reproduction* 135:293-301.

Groh, V., Wu, J., Yee, C., and Spies, T. 2002. "Tumour-derived soluble MIC ligands impair expression of NKG2D and T-cell activation." *Nature* 419:734-8.

Gronewold, T. M. A. 2007. "Surface acoustic wave sensors in the bioanalytical field: recent trends and challenges." *Anal. Chim. Acta* 603:119-28.

Gronewold, T. M. A., Baumgartner, A., Hierer, J., Sierra, S., Blind, M., Schäfer, F., Blümer, J., Tillmann, T., Kiwitz, A., Kaiser, R., Zabe-Kühn, M., Quandt, E., and Famulok, M. 2009. "Kinetic binding analysis of aptamers targeting HIV-1 proteins by a combination of a microbalance array and mass spectrometry (MAMS)." *J. Prot. Res.* 8:3568-77.

Gronewold, T. M. A., Glass, S., Quandt, E., and Famulok M. 2005. "Monitoring complex formation in the blood-coagulation cascade using aptamer-coated SAW sensors." *Biosens. Bioelectron.* 20:2044-52.

Gruhl, F. J., Rapp, B. E., and Länge, K. 2013. "Biosensors for diagnostic applications." *Adv. Biochem. Eng. Biotechnol.* 133:115-48.

Guervilly, C., Lacroix, R., Forel, J. M., Roch, A., Camoin-Jau, L., Papazian, L., and Dignat-George, F. 2011. "High levels of circulating leukocyte microparticles are associated with better outcome in acute respiratory distress syndrome." *Crit. Care* 15:R31.

Gullotti, E., and Yeo, Y. 2009. "Extracellularly Activated Nanocarriers: A new paradigm of tumor targeted drug delivery." *Mol. Pharmaceut.* 6:1041-51.

Gumulec, J., Sochor, J., Hlavna, M., Sztalmachova, M., Krizkova, S., Babula, P., Hrabec, R., Rovny, A., Adam, V., Eckschlager, T., Kizek, R., and Masarik, M. 2012. "Caveolin-1 as a potential high-risk prostate cancer biomarker." *Oncology Rep.* 27:831-41.

Guther, M. L., Cardoso de Almeida, M. L., Rosenberry, T. L., and Ferguson, M. A. 1994. "The detection of phospholipase-resistant and -sensitive glycosyl-phosphatidylinositol membrane anchors by western blotting." *Anal. Biochem.* 219:249-55.

Guzman-Flores, J. M., Cortez-Espinosa. N., Cortés-Garcia, J. D., Vargas-Morales, J. M., Cataño-Cañizalez, Y. G., Rodríguez-Rivera, J. G., and Portales-Perez, D. P. 2015. "Expression of CD73 and A2A receptors in cells from subjects with obesity and type 2 diabetes mellitus." *Immunobiology* 220:976-84.

György, B., Szabo, T. G., Pasztoi, M., Pál, Z., Misják, P., Aradi, B., László, V., Pállinger, E., Pap, E., Kittel, A., Nagy, G., Falus, A., and Buzás, E. I. 2011. "Membrane vesicles, current state-of-the-art: emerging role of extracellular vesicles." *Cell. Mol. Life Sci.* 68:2667-88.

Hagaman, J. R., Moyer, J. S., Bachman, E. S., Sibony, M., Magyar, P. L., Welch, J. E., Smithies, O., Krege, J. H., and O'Brien, D. A. 1998. "Angiotensin-converting enzyme and male fertility." *Proc. Natl. Acad. Sci. USA* 95:2552-7.

Hakulinen, J., and Meri, S. 1994. "Expression and function of the complement membrane attack complex inhibitor CD59 protectin on human breast cancer cells." *Lab. Invest.* 71:820.

Hakulinen, J., and Meri, S. 1995. "Shedding and enrichment of the glycoplipid-anchored complement lysis inhibitor protectin (CD59) into milk fat globules." *Immunology* 85:495-501.

Haldar, K., Ferguson, M. A. J., and Cross, G. A. M. 1985. "Acylation of a *Plasmodium falciparum* merozoite surface antigen via *sn*-1,2-diacyl glycerol." *J. Biol. Chem.* 260:4969-74.

Hale, G., Xia, M. Q., Tighe, H. P., Dyer, M. J., and Waldmann, H. 1990. "The CAMPATH-1 antigen." *Tissue Antigens* 35:118-27.

Hambor, J. H., Kaplan, D. R., and Tycocinski, M. L. 1990. "CD8 functions as an inhibitory ligand in mediating the immunoregulatory activity of $CD8^+$ cells." *J. Immunol.* 145:1646-52.

Hamilton, K. K., Ji, Z., Rollins, S., and Sims, P. J. 1990. "Regulatory control of the terminal complement proteins at the surface of human endothelial cells: neutralization of a C5b-9 inhibitor by antibody to CD59." *Blood* 76:2572-7.

Hanada, K., Izawa, K., Nishijima, M., and Akamatsu, Y. 1993. "Sphingolipid deficiency induces hypersensitivity of CD14, a glycosyl phosphatidylinositol-anchored protein, to phosphatidylinositol-specific phospholipase C." *J. Biol. Chem.* 268:13820-3.

Hansen, G. H., Niels-Christiansen, L. L., Immerdal, L., Nystrøm, B. T., and Danielsen, E. M. 2007. "Intestinal alkaline phosphatase: Selective endocytosis from the enterocyte brush border during fat absorption." *Am. J. Physiol. Gastrointest. Liver Physiol.* 293:G1325-32.

Hansen, G. H., Rasmussen, K., Niels-Christiansen, L. L., and Danielsen, E. M. 2011. "Dietary free fatty acids form alkaline phosphatase-enriched microdomains in the intestinal brush border membrane." *Mol. Membr. Biol.* 28:136-44.

Hantke, K., and Braun, V. 1973. "Covalent binding of lipid to protein. Diglyceride and amide-linked fatty acid at the N-terminal end of the murein-lipoprotein of the Escherichia coli outer membrane." *Eur. J. Biochem.* 34:284-96.

Hanzal-Bayer, M. F., and Hancock, J. F. 2007. "Lipid rafts and membrane traffic." *FEBS Lett.* 581:2098-104.

Hara, T., Kuriyama, S., Kiyohara, H., Nagase, Y., Matsumoto, M., and Seya, T. 1992. "Soluble forms of membrane cofactor protein (CD46, MCP) are present in plasma, tears and seminal fluid in normal subject." *Clin. Exp. Immunol.* 89:490-4.

Hara, T., Matsumoto, M., Fukumori, Y., Miyagawa, S., Hatanaka, M., Kinoshita, T., Seya, T., and Akedo, H. 1993. "A monoclonal antibody against human decay accelerating factor (DAF, CD55), D17, which lacks reactivity with semen-DAF." *Immunol. Lett.* 37:145-52.

Haraba, R., Suica, V. I., Uyy, E., Ivan, L., and Antohe, F. 2011. "Hyperlipidemia stimulates the extracellular release of the nuclear high mobility group box 1 protein." *Cell. Tissue Res.* 346:361-8.

Harding, C., Heuser, J., and Stahl, P. 1983. "Receptor-mediated endocytosis of transferrin and recycling of the transferrin receptor in rat reticulocytes." *J. Cell Biol.* 97:329-39.

Hari, T., Kunze, H., Bohn, E., Brodbeck, U., and Bütikofer, P. 1996. "Subcellular distribution of glycosylphosphatidylinositol-specific phospholipase D in rat liver." *Biochem. J.* 320:315-9.

Häring, H. U. 2016. "Novel phenotypes of prediabetes?" *Diabetologia* 59:1806-18.

Harris, R., and Findlay, J. B. C. 1983. "Investigation of the organisation of the major proteins in bovine myelin membranes. Use of chemical probes and bifunctional crosslinking reagents." *Biochim. Biophys. Acta* 732:75-82.

Hartel, A. J., Glogger, M., Jones, N. G., Abuillan, W., Batram, C., Hermann, A., Fenz, S. F., Tanaka, M., and Engstler, M. 2016. "N-glycosylation enables high lateral mobility of GPI-anchored proteins at a molecular crowding threshold." *Nat. Commun.* 7:12870.

Hazenbos, W. L., Clausen, B. E., Takeda, J., and Kinoshita, T. 2004. "GPI-anchor deficiency in myeloid cells causes impaired FcγR effector functions." *Blood* 104:2825-31.

Hazenbos, W. L., Wu, P., Eastham-Anderson, J., Kinoshita, T., and Brown, E. J. 2011. "Impaired FcεR1 stability, signaling and effector functions in murine mast cells lacking glycosylphosphatidylinositol-anchored proteins." *Blood* 118:4377-83.

Heerwagen, M. J., Miller, M. R., Barbour, L. A., and Friedman, J. E. 2010. "Maternal obesity and fetal metabolic programming: a fertile epigenetic soil." *Am. J. Physiol. Regul. Integr. Comp. Physiol.* 299:R711-22.

Heidegger, M. 1967. *Sein und Zeit*, Tübingen: Niemeyer.

Heidegger, M. 1977. "Der Ursprung des Kunstwerkes." ["The origin of the artwork"]. In *Holzwege*, GA 5, Frankfurt am Main: Suhrkamp. S. 33.

Heidegger, M. 2009. *Leitgedanken zur Entstehung der Metaphysik, der neuzeitlichen Wissenschaft und der modernen Technik* [*Guidelines for the development of metaphysics, modern science and modern technology*], GA 76, Frankfurt am Main: Suhrkamp.

Heidegger, M. 2014. *Überlegungen VII-XI* [*Considerations VII-XI*], GA 95, Frankfurt am Main: Suhrkamp.

Heijnen, H. F. G., Schiel, A. E., Fijnheer, R., Geuze, H. J., and Sixma, J. J. 1999. "Activated platelets release two types of membrane vesicles: microvesicles by surface shedding and exosomes derived from exocytosis of multivesicular bodies and alpha-granules." *Blood* 94:3791-9.

Heilbronn, L. K., and Cambell, L. V. 2008. "Adipose tissue macrophages, low grade inflammation and insulin resistance in human obesity." *Curr. Pharm. Des.* 14:1225-30.

Heinemann, L., and Jacques, Y. 2009. "Oral Insulins and Buccal Insulins: a Critical Repraisal." *J. Diabetes Sci. Technol.* 3:568-84.

Helal, O., Defoort, C., Robert, S., Marin, C., Lesavre, N., Lopez-Miranda, J., Risérus, U., Basu, S., Lovegrove, J., McMonagle, J., Roche, H. M., Dignat-George, F., and Lairon, D. 2010. "Increased levels of microparticles originating from endothelial cells, platelets and erythrocytes in subjects with metabolic syndrome: relationship with oxidative stress." *Nutr. Metab. Cardiovasc. Dis.* 21:665-71.

Heller, M., Bieri, S., and Brodbeck, U. 1992. "A novel form of glycosyl-phosphatidylinositol-anchor converting activity with a specificity of a phospholipase D in mammalian liver membranes." *Biochim. Biophys. Acta* 1109:109-16.

Hemler, M. E. 2005. "Tetraspanin functions and associated microdomains." *Nat. Rev. Mol. Cell Biol.* 6:801-11.

Henderson, R., and Unwin, P. N. 1975. "Three-dimensional model of purple membrane obtained by electron microscopy." *Nature* 257:28-32.

Heneghan, H. M., Miller, N., and Kerin, M. J. 2010. "Role of microRNAs in obesity and the metabolic syndrome." *Obes. Rev.* 11:354-61.

Herder, C., Karakas, M., and Koenig, W. 2011. "Biomarkers for the prediction of type 2 diabetes and cardiovascular disease." *Clin. Pharmacol. Ther.* 90:52-66.

Herrera, M. B., Fonsato, V., and Gatti, S. 2010. "Human liver stem cell-derived microvesicles accelerate hepatic regeneration in hepatectomized rats." *J. Cell. Mol. Med.* 14:1605-18.

Ho, J. C., Cheung, S. T., Patil, M., Chen, X., and Fan, S. T. 2006. "Increased expression of glycosyl-phosphatidylinositol anchor attachment protein 1 (GPAA1) is associated with gene amplification in hepatocellular carcinoma." *Int. J. Cancer* 119:1330-7.

Hoener, M. C., and Brodbeck, U. 1992. "Phosphatidylinositol-glycan-specific phospholipase D is an amphiphilic glycoprotein that in serum is associated with high density lipoproteins." *Eur. J. Biochem.* 206:747-57.

Hoener, M. C., Bolli, R., and Brodbeck, U. 1993. "Glycosyl-phosphatidylinositol-specific phospholipase D. Interaction with and stimulation by apolipoprotein A-I." *FEBS Lett.* 327:203-6.

Hogan, M. C., Manganelli, L., Woolard, J. R., Masyuk, A. I., Masyuk, T. V., Tammachote, R., Huang, B. Q., Leontovich, A. A., Beito, T. G., Madden, B. J., Charlesworth, M. C., Torres, V. E., LaRusso, N. F., Harris, P. C., and Ward, C. J. 2009. "Characterization of PKD protein-positive exosome-like vesicles." *J. Am. Nephrol.* 20:278-88.

Holguin, M. H., Fredrick, L. R., Bernshaw, N. J., Wilcox, L. A., and Parker, C. J. 1989. "Isolation and characterization of a membrane protein from normal human erythrocytes that inhibits reactive lysis of the erythrocytes of paroxysmal nocturnal hemoglobinuria." *J. Clin. Invest.* 84:7-17.

Holvoet, P., De Keyzer, D., and Jacobs, D. R. 2008. "Oxidized LDL and the metabolic syndrome." *Future Lipidol.* 3:637-49.

Holvoet, P., Kritchevsky, S. B., Tracy, R. P., Mertens, A., Rubin, S. M., Butler, J., Goodpaster, B., and Harris, T. B. 2004. "The metabolic syndrome, circulating oxidized LDL, and risk of myocardial infarction in well-functioning elderly people in the health, aging, and body composition cohort." *Diabetes* 53:1068-73.

Homans, S. W., Ferguson, M. A., Dwek, R. A., Rademacher, T. W., Anand, R., and Williams, A. F. 1988. "Complete structure of the glycosyl phosphatidylinositol membrane anchor of rat brain Thy-1 glycoprotein." *Nature* 333:269-72.

Honigmann, A., Mueller, V., Ta, H., Schoenle, A., Sezgin, E., Hell, S. W., and Eggeling, C. 2014a. "Scanning STED-FCS reveals spatiotemporal heterogeneity of lipid interaction in the plasma membrane of living cells." *Nat. Commun.* 5:5412.

Honigmann, A., Sadeghi, S., Keller, J., Hell, S. W., Eggeling, C., and Vink, R. 2014b. "A lipid bound actin meshwork organizes liquid phase separation in model membranes." *Elife* 18,3:e01671.

Horstman, L. L., and Ahn, Y. S. 1999. "Platelet microparticles: A wide-angle perspective." *Crit. Rev. Oncol. Hematol.* 30:111-42.

Horstman, L. L., Jy, W., Jimenez, J. J., and Ahn, Y. S. 2003. "Endothelial microparticles as markers of endothelial dysfunction." *Front. Biosci.* 9:1118-35.

Horstman, L. L., Jy, W., Minagar, A., Bidot, C. J., Jimenez, J. J., Alexander, J. S., and Ahn, Y. S. 2007. "Cell-derived microparticles and exosomes in neuroinflammatory disorders." *Int. Rev. Neurobiol.* 79:227-68.

Howard, A., Berger, J., Familletti, P., and Udenfriend, S. 1987. "Characterization of the phosphatidylinositol-glycan membrane anchor of human placental alkaline phosphatase." *Proc. Natl. Acad. Sci. USA* 84:6055-9.

Hu, G., Drescher, K. M., and Chen, X. M. 2012. "Exosomal miRNAs: biological properties and therapeutic potential." *Front. Gene.* 3:56.

Huang, J. H., Getty, R. R., Chisari, F. V., Fowler, P., Greenspan, N. S., and Tykocinski, M. L. 1994. "Protein transfer of preformed MHC-peptide complexes sensitizes target cells to T cell cytolysis." *Immunity* 1:607-13.

Huber, J., Vales, A., Mitulovic, G., Blumer, M., Schmid, R., Witztum, J. L., Binder, B. R., and Leitinger, N. 2002. "Oxidized membrane vesicles and blebs from apoptotic cells contain biologically active oxidized phospholipids that induce monocyte-endothelial interactions." *Arterioscler. Thromb. Vasc. Biol.* 22:101-7.

Hugel, B., Martinez, M. C., Kunzelmann, C., and Freyssinet, J. M. 2005. "Membrane microparticles: two sides of the coin." *Physiology* 20:22-7.

Hunter, M. P., Ismail, N., Zhang, X., Aguda, B. D., Lee, E. J., Yu, L., Xiao, T., Schafer, J., Lee, M. L., Schmittgen, T. D., Nana-Sinkam, S. P., Jarjoura, D., and Marsh, C. B. 2008. "Detection of microRNA expression in human peripheral blood microvesicles." *PLoS One* 3:e3694.

Husserl, E. 1996. *Die Krisis der europäischen Wissenschaften und die transzendentale Phänomenologie: eine Einleitung in die phänomenologische Philosophie* [*The crisis of the European sciences and transcendental phenomenology: an introduction to phenomenological philosophy*]. Hamburg: Meiner.

Iero, M., Valenti, R., Huber, V., Filipazzi, P., Parmiani, G., Fais, S., and Rivoltini, L. 2008. "Tumour-released exosomes and their implications in cancer immunity." *Cell Death Differ.* 15:80-8.

Ikezawa, H. 2002. "Glycosylphosphatidylinositol (GPI)-anchored proteins." *Biol. Pharm. Bull.* 25:409-17.

Ikezawa, H., and Taguchi, R. 1981. "Phosphatidylinositol-specific phospholipase C from *Bacillus cereus* and *Bacillus thuringiensis*." *Methods Enzymol.* 71:731-41.

Ikezawa, H., Yamanegi, M., Taguchi, R., Miyashita, T., and Ohyabu, T. 1976. "Studies on phosphatidylinositol phosphodiesterase (phospholipase C type) of Bacillus cereus. I. purification, properties and phosphatase-releasing activity." *Biochim. Biophys. Acta* 450:154-64.

Ilangumaran, S., Arni, S., Poincelet, M., Theler, J. M., Brennan, P. J., Nasir-ud-Din, and Hoessli, D. C. 1995. "Integration of mycobacterial lipoarabinomannans into glycosylphosphatidylinositol-rich domains of lymphomonocytic cell plasma membranes." *I. Immunol.* 155:1334-42.

Ilangumaran, S., He, H. T., and Hoessli, D. C. 2000. "Microdomains in lymphocyte signalling: beyond GPI-anchored proteins." *Immunol. Today* 21:2-7.

Ilangumaran, S., Robinson, P. J., and Hoessli, D. C. 1996. "Transfer of exogenous glycosylphosphatidylinositol (GPI)-linked molecules to plasma membranes." *Trends Cell Biol.* 6:163-7.

Inal, J. M., Kosgodage, U., Azam, S., Stratton, D., Antwi-Baffour, S., and Lange, S. 2013. "Blood/plasma secretome and microvesicles." *Biochim. Biophys. Acta* 1834:2317-25.

Ingham, V., Williams, A., and Bate, C. 2014. "Glimepiride reduces CD14 expression and cytokine secretion from macrophages." *J. Neuroinflammation* 11:115-28.

Inglis, N. I., Krant, M. J., and Fishman, W. H. 1967. "Influence of a fat-enriched meal on human serum (L-phenylalanine-sensitive) 'intestinal' alkaline phosphatase." *Proc. Soc. Exp. Biol. Med.* 124:699-702.

Ioka, R. X., Kang, M. J., Kamiyama, S., Kim, D. H., Magoori, K., Kamataki, A., Ito, Y., Takei, Y. A., Sasaki, M., Suzuki, T., Sasano, H., Takahashi, S., Sakai, J., Fujino, T., and Yamamoto, T. T. 2003. "Expression cloning and characterization of a novel glycosylphosphatidylinositol-anchored high density lipoprotein-binding protein, GPIHBP1." *J. Biol. Chem.* 278:7344-9.

Iyer, H., Khedkar, A., and Verma, M. 2010. "Oral insulin – a review of current status." *Diabetes Obesity Metabol.* 12:179-85.

Jack, A., and Shallice, T. 2001. "Introspective physicalism as an approach to the science of consciousness." *Cognition* 79:161-96.

Jacquemin, C. 1991. "Glycosyl phosphatidylinositol in thyroid: cell signalling or protein anchor?" *Biochimie* 73:37-40.

Jäger, K., Meyer, P., Stieger, S., and Brodbeck, U. 1990. "Production and characterization of antibodies against the cross-reacting determinant of glycosyl-phosphatidylinositol-anchored acetylcholinesterase." *Biochim. Biophys. Acta* 1039:367-73.

Jain, A., and Holthuis, J. C. M. 2017. "Membrane contact sites, ancient and central hubs of cellular lipid logistics." *Biochim. Biophys. Acta* 1864:1450-8.

Janowska-Wieczorek, A., Wysoczynski, M., Kijowski, J., Marquez-Curtis, L., Machalinski, B., Ratajczak, J., and Ratajczak, M. Z. 2005. "Microvesicles derived from activated platelets induce metastasis and angiogenesis in lung cancer." *Int. J. Cancer* 113:752-60.

Jap, B. K. 1989. "Molecular design of PhoE porin and its functional consequences." *J. Mol. Biol.* 205:407-19.

Jaquenoud, M., Pagac, M., Signorell, A., Benghezal, M., Jelk, J., Bütikofer, P., and Conzelmann, A. 2008. "The Gup1 homologue of *Trypanosoma brucei* is a GPI glycosylphosphatidylinositol remodelase." *Mol. Microbiol.* 67:202-12.

Jarsch, I. K., Daste, F., and Gallop, J. L. 2016. "Membrane curvature in cell biology: An integration of molecular mechanisms." *J. Cell Biol.* 214:375-87.

Jensen, D., and Schekman, R. 2011. "COPII-mediated vesicle formation at a glance." *J. Cell Sci.* 124:1-4.

Jenuwein, T. 2006. "The epigenetic magic of histone lysine methylation." *FEBS J.* 273:3121-35.

Jiang, W. W., Zahurak, M., Zhou, Z. T., Park, H. L., Guo, Z. M., Wu, G. J., Sidransky, D., Trink, B., and Califano, J. A. 2007. "Alterations of GPI transamidase subunits in head and neck squamous carcinoma." *Mol. Cancer* 21:74.

Jimenez, J. J., Jy, W., Mauro, L. M., Horstman, L. L., and Ahn, Y. S. 2001. "Elevated endothelial microparticles in thrombotic thrombocytopenic purpura: findings from brain and renal microvascular cell culture and patients with active disease." *Br. J. Haematol.* 112:81-90.

Jimenez, J. J., Jy, W., Mauro, L. M., Soderland, C., Horstman, L. L., and Ahn, Y. S. 2003. "Endothelial cells release phenotypically and quantitatively distinct microparticles in activation and apoptosis." *Thromb. Res.* 109:175-80.

Johannes, L., and Mayor, S. 2010. "Induced domain formation in endocytic invagination, lipid sorting, and scission." *Cell* 142:507-10.

Johansen Taber, K. A., and Dickinson, B. D. 2015. "Genomic-based tools for the risk assessment, management, and prevention of type 2 diabetes." *Appl. Clin. Genet.* 8:1-8.

Johnstone, R. M. 2003. "Revisiting the road to the discovery of exosomes." *Blood Cells Mol. Dis.* 34:214-9.

Johnstone, R. M., Adam, M., Hammond, J. R., Orr, L., and Turbide, C. 1987. "Vesicle formation during reticulocyte maturation. Association of plasma membrane activities with released vesicles (exosomes)." *J. Biol. Chem.* 262:9412-20.

Johnstone, R. M., Mathew, A., Mason, A. B., and Teng, K. 1991. "Exosome formation during maturation of mammalian and avian reticulocytes: evidence that exosome release is a major route for externalisation of obsolete membrane proteins." *J. Cell. Pathol.* 147:27-36.

Jones, D. R., and Varela-Nieto, I. 1999. "Diabetes and the role of inositol-containing lipids in insulin signaling." *Mol. Med.* 5:505-14.

Joop, K., Berckmans, R. J., Nieuwland, R., Berkhout, J., Romijn, F. P., Hack, C. E., and Sturk, A. 2001. "Microparticles from patients with multiple organ dysfunction syndrome and sepsis support coagulation through multiple mechanisms." *Thromb. Haemost.* 85:810-20.

Jung, Y., McCall, S. J., Li, Y. X., and Diehl, A. M. 2007. "Bile ductules and stromal cells express Hedgehog ligands and/or Hedgehog target genes in primary biliary cirrhosis." *Hepatology* 45:1091-6.

Kalsi, K., Lawson, C., Dominquez, M., Taylor, P., Yacoub, M. H., and Smolenski, R. T. 2002. "Regulation of ecto-5'-nucleotidase by TNF-alpha in human endothelial cells." *Mol. Cell. Biochem*. 232:113-9.

Kamps, M. P., Buss, J. E., and Sefton, B. M. 1985. "Mutation of NH2-terminal glycine of p60src prevents both myristoylation and morphological transformation." *Proc. Natl. Acad. Sci. USA* 82:4625-8.

Kapitza, C., Zijlstra, E., Heinemann, L., Castelli, M. C., Riley, G., and Heise, T. 2010. "Oral insulin: A comparison with subcutaneous regular human insulin in patients with type 2 diabetes." *Diabetes Care* 33:1288-90.

Karlsson, M., Lundin, S., Dahlgren, U., Kahu, H., Pettersson, I., and Telemo, E. 2001. "'Tolerosomes' are produced by intestinal epithelial cells." *Eur. J. Immunol.* 31:2892-900.

Karnchanaphanurach, P., Mirchev, R., Ghiran, I., Asara, J. M., Papahadjopoulos-Sternberg, B., Nicholson-Weller, A., and Golan, D. E. 2009. "C3b deposition on human erythrocytes induces the formation of a membrane skeleton-linked protein complex." *J. Clin. Invest.* 119:788-801.

Karnik, S. S., Sakmar, T. P., Chen, M. B., and Khorana, H. G. 1988. "Cysteine residues 110 and 187 are essential for the formation of correct structure in bovine rhodopsin." *Proc. Natl. Acad. Sci. USA* 82:4625-8.

Kataoka, K., Matsumoto, T., and Yokoyama, M. 2000. "Doxorubicin-loaded poly(ethylene glycol)-poly(beta-benzyl-L-aspartate) copolymer micelles: Their pharmaceutical characteristics and biological significance." *J. Control Rel*. 64:143-53.

Kawagoe, K., Kitamura, D., Okabe, M., Taniuchi, I., Ikawa, M., Watanabe, T., Kinoshita, T., and Takeda, J. 1996. "Glycosylphosphatidylinositol-anchor-deficient mice: implications for clonal dominance of mutant cells in paroxysmal nocturnal hemoglobinuria." *Blood* 87:3600-6.

Keiding, N. R. 1966. "Intestinal alkaline phosphatase in human lymph and serum." *Scand. J. Clin. Lab. Invest*. 18:134-40.

Keller, S., Ridinger, J., Rupp, A. K., Janssen, J. W., and Altevogt, P. 2011. "Body fluid derived exosomes as a novel template for clinical diagnostics." *J. Transl. Med*. 9:86.

Kim, H. K., Song, K. S., Park, Y. S., Kang, Y. H., Lee, Y. J., Lee, K. R., Kim, H. K., Ryu, K. W., Bae, J. M., and Kim, S. 2003. "Elevated levels of circulating platelet microparticles, VEGF, IL-6 and RANTES in patients with gastric cancer. Possible role of a metastasis predictor." *Eur. J. Cancer* 39:184-91.

Kinoshita, T., and Fujita, M. 2016. "Biosynthesis of GPI-anchored proteins: special emphasis on GPI lipid remodeling." *J. Lipid Res*. 57:6-24.

Kinoshita, T., Fujita, M., and Maeda, Y. 2008. "Biosynthesis, remodelling and functions of mammalian GPI-anchored proteins: Recent progress." *J. Biochem.* 144:287-94.

Kinoshita, T., Maeda, Y., and Fujita, M. 2013. "Transport of glycosyl-phosphatidylinositol-anchored proteins from the endoplasmic reticulum." *Biochim. Biophys. Acta* 1833:2473-8.

Kinoshita, T., Medof, M. E., Hong, K., and Nussenzweig, V. 1986. "Membrane-bound C4b interacts endogenously with complement receptor CR1 of human red cells." *J. Exp. Med.* 164:1377-88.

Kirchhoff, C., and Hale, G. 1996. "Cell-to-cell transfer of glycosylphosphatidylinositol-anchored membrane proteins during sperm maturation." *Mol. Hum. Reprod.* 2:177-84.

Kirchhoff, C., Pera, I., Derr, P., Yeung, C. H., and Cooper, T. 1997. "The molecular biology of the sperm surface. Post-testicular membrane remodelling." *Adv. Exp. Med. Biol.* 424:221-32.

Kirpotin, D. B., Drummond, D. C., Shao, Y., Shalaby, M. R., Hong, K., Nielsen, U. B., Marks, J. D., Benz, C. C., and Park, J. W. 2006. "Antibody targeting of long-circulating lipidic nanoparticles does not increase tumor localization but does increase internalization in animal models." *Cancer Res.* 66:6732-40.

Kleerekoper, M., Horne, M., Cornish, C. J., and Posen, S. 1970. "Serum alkaline phosphatase after fat ingestion: an immunological study." *Clin. Sci.* 38:339-45.

Knaus, A., Pantel, J. T., Pendziwiat, M., Hajjir, N., Zhao, M., Hsieh, T. C., Schubach, M., Gurovich, Y., Fleischer, N., Jäger, M., Köhler, S., Muhle, H., Korff, C., Møller, R. S., Bayat, A., Calvas, P., Chassaing, N., Warren, H., Skinner, S., Louie, R., Evers, C., Bohn, M., Christen, H. J., van den Born, M., Obersztyn, E., Charzewska, A., Endziniene, M., Kortüm, F., Brown, N., Robinson, P. N., Schelhaas, H. J., Weber, Y., Helbig, I., Mundlos, S., Horn, D., and Krawitz, P. M. 2018. "Characterization of glycosylphosphatidylinositol biosynthesis defects by clinical features, flow cytometry, and automated image analysis." *Genome Med.* 10:3. doi: 10.1186/s13073-017-0510-5.

Kobayashi, T., Okamoto, H., Yamada, J., Setaka, M., and Kwan, T. 1984. "Vesiculation of platelet plasma membranes. Dilauroylglycerophosphocholine-induced shedding of a platelet plasma membrane fraction enriched in acetylcholinesterase activity." *Biochim. Biophys. Acta* 778:210-8.

Koga, H., Sugiyama, S., Kugiyama, K., Fukushima, H., Watanabe, K., Sakamoto, T., Yoshimura, M., Jinnouchi, H., and Ogawa, H. 2006. "Elevated levels of remnant lipoproteins are associated with plasma platelet microparticles in patients with type-2 diabetes mellitus without obstructive coronary artery disease." *Eur. Heart J.* 27:817-23.

Kogure, K., Nakamura, C., Okuda, O., Hayashi, K., and Ueno, M. 1997. "Effect of dicetylphosphate or stearic acid on spontaneous transfer of protein from influenza virus-infected cells to dimyristoylphosphatidylcholine liposomes." *Biochim. Biophys. Acta* 1329:174-82.

Kogure, T., Lin, W. L., Yan, I. K., Braconi, C., and Patel, T. 2011. "Intercellular nanovesicle-mediated microRNA transfer: A mechanism of environmental modulation of hepatocellular cancer cell growth." *Hepatology* 54:1237-48.

Kolberg, J. A., Jorgensen, T., Gerwien, R. W., Hamren, S., McKenna, M. P., Moler, E., Rowe, M. W., Urdea, M. S., Xu, X. M., Hansen, T., Pedersen, O., and Borch-Johnsen, K. 2009. "Development of a type 2 diabetes risk model from a panel of serum biomarkers from the Inter99 cohort." *Diab. Care* 32:1207-12.

Kolowos, W., Gaipl, U. S., Sheriff, A., Voll, R. E., Heyder, P., Kern, P., Kalden, J. R., and Herrmann, M. 2005. "Microparticles shed from different antigen-presenting cells display an individual pattern of surface molecules and a distinct potential of allogeneic T-cell activation." *Scand. J. Immunol.* 61:226-33.

Kondoh, G., Gao, X. H., Nakano, Y., Koike, H., Yamada, S., Okabe, M., and Takeda, J. 1999. "Tissue-inherent fate of GPI revealed by GPI-anchored GFP transgenesis." *FEBS Lett.* 458:299-303.

Kondoh, G., Tojo, H., Nakatani, Y., Komazawa, N., Murata, C., Yamagata, K., Maeda, Y., Kinoshita, T., Okabe, M., Taguchi, R., and Takeda, J. 2005. "Angiotensin-converting enzyme is a GPI-anchored protein releasing factor crucial for fertilization." *Nat. Med.* 11:160-6.

Kooyman, D. L., Byrne, G. W., and Logan, J. S. 1998. "Glycosyl phosphatidylinositol anchor." *Exp. Nephrol.* 6:148-51.

Kooyman, D. L., Byrne, G. W., McClellan, S., Nielsen, D., Tone, M., Waldmann, H., Coffman, T. M., McCurry, K. R., Platt, J. L., and Logan, J. S. 1995. "*In vivo* transfer of GPI-linked complement restriction factors from erythrocytes to the endothelium." *Science* 269:89-92.

Korn, E. D. 1955. "Clearing factor, a heparin-activated lipoprotein lipase. I. Isolation and characterization of the enzyme from normal rat heart." *J. Biol. Chem.* 215:1-14.

Kouzayha, A., and Besson, F. 2005. "GPI-alkaline phosphatase insertion into phosphatidylcholine monolayers: phase behavior and morphology changes." *Biochem. Biophys. Res. Commun.* 333:1315-21.

Kovac, V., Hafner-Bratkovic, I., and Curin Serbec, V. 2016. "Anchorless forms of prion protein - Impact of truncation on structure destabilization and prion protein conversion." *Biochem. Biophys. Res. Commun.* 481:1-6.

Krawitz, P. M., Schweiger, M. R., Rodelsperger, C., Marcelis, C., Kölsch, U., Meisel, C., Stephani, F., Kinoshita, T., Murakami, Y., Bauer, S., Isau, M., Fischer, A., Dahl, A., Kerick, M., Hecht, J., Köhler, S., Jäger, M., Grünhagen, J., de Condor, B. J., Doelken, S., Brunner, H. G., Meinecke, P., Passarge, E., Thompson, M. D., Cole, D. E., Horn, D., Roscioli, T., Mundlos, S., and Robinson, P. N. 2010. "Identity-bydescent filtering of exome sequence data identifies PIGV mutations in hyperphosphatasia mental retardation syndrome." *Nat. Genet.* 42:827-9.

Kreuger, J., Perez, L., Giraldez, A. J., and Cohen, S. M. 2004. "Opposing activities of Dally-like glypican at high and low levels of Wingless morphogen activity." *Dev. Cell* 7:503-11.

Kuchel, P. W., Campbell, D. G., Barclay, A. N., and Williams, A. F. 1978. "Molecular weights of the Thy-1 glycoproteins from rat thymus and brain in the presence and absence of deoxycholate." *Biochem. J.* 169:411-7.

Kurtz, T. A., Fineberg, N. S., Considine, R. V., and Deeg, M. A. 2004. "Insulin resistance is associated with increased serum levels of glycosylphosphatidylinositol-specific phospholipase D." *Metabolism* 53:138-9.

Kusumi, A., Shirai, Y. M., Koyama-Honda, I., Suzuki, K. G., and Fujiwara, T. K. 2010. "Hierarchical organization of the plasma membrane: investigations by single-molecule tracking *vs.* fluorescence correlation spectroscopy." *FEBS Lett.* 584:1814-23.

Kusumi, A., Suzuki, K. G., Kasai, R. S., Ritchie, K., and Fujiwara, T. K. 2011. "Hierarchical mesoscale domain organization of the plasma membrane." *Trends Biochem. Sci.* 36:604-15.

Kuypers, F. A., Larkin, S. K., Emeis, J. J., and Allison, A. C. 2007. "Interaction of an annexin V homodimer (Diannexin) with phosphatidylserine on cell surfaces and consequent antithrombotic activity." *Thromb. Haemost.* 97:478-86.

Labani-Motlagh, A., Israelsson, P., Ottander, U., Lundin, E., Nagaev, I., Nagaeva, O., Dehlin, E., Baranov, V., and Mincheva-Nilsson, L. 2015. "Differential expression of ligands for NKG2D and DNAM-1 receptors by epithelial ovarian cancer-derived exosomes and its influence on NK cell cytotoxicity." *Tumour Biol.* 37:5455-66.

Lallanne, E., Honys, D., Johnson, A., Borner, G. H., Lilley, K. S., Dupree, P., Grossniklaus, U., and Twell, D. 2004. "SETH1 and SETH2, two components of the glycosylphosphatidylinositol anchor biosynthetic pathway, are required for pollen germination and tube growth in Arabidopsis." *Plant Cell* 16:229-40.

Lamaze, C., Tardif, N., Dewulf, M., Vassilopoulos, S., and Blouin, C. M. 2017. "The caveolae dress code: structure and signaling." *Curr. Opin. Cell Biol.* 47:117-25.

Lampen, J. O., and Nielson, J. B. 1984. "N-terminal glyceride-cysteine modification of membrane penicillinases in gram-positive bacteria." *Meth. Enzymol.* 106:365-8.

Landi, A. P. G., Wilson, A. B., Davies, A., Lachmann, P. J., Ferriani, V. P., Seilly, D. J., and Assis-Pandochi, A. I. 2003. "Determination of CD59 protein in normal human serum by enzyme immunoassay, using octyl-glucoside detergent to release glycosyl-phosphatidyl-CD59 from lipid complex." *Immunol. Lett.* 90:209-13.

Larkin, M., Watts, S., and Clifton, E. 2006. "Giving voice and making sense in Interpretative Phenomenological Analysis." *Qualit. Res. Psychol.* 3:102-20.

Larner, J., Brautigan, D. L., and Thorner M. O. 2010. "D-*chiro*-inositol glycans in insulin signaling and insulin resistance." *Mol. Med.* 16:543-52.

Lasch, F., and Schönbrunner, E. 1936. "Weitere Untersuchungen über die enterale Insulinresorption, über den Schutz des Insulins gegen Trypsin." ["Further studies on enteral insulin absorption on the protection of insulin against trypsin"]. *Naunyn Schmiedeberg Arch. Pharmacol.* 182:452-8.

Lathia, C. D. 2002. "Biomarkers and surrogate endpoints: How and when they might impact drug development." *Disease Markers* 18:83-90.

Latour, B. 2010. *Eine neue Soziologie für eine neue Gesellschaft*. [*A new sociology for a new society*]. Frankfurt am Main: Suhrkamp. S. 237.

Lauc, G., and Heffer-Lauc, M. 2006. "Shedding and uptake of gangliosides and glycosylphosphatidylinositol-anchored proteins." *Biochim. Biophys. Acta* 1760:584-602.

Laulagnier, K., Motta, C., Hamdi, S., Roy, S., Fauvelle, F., Pageaux, J. F., Kobayashi, T., Salles, J. P., Perret, B., Bonnerot, C., and Record, M. 2004. "Mast cell- and dendritic cell-derived exosomes display a specific lipid composition and an unusual membrane organization." *Biochem. J.* 380:161-71.

Lauressergues, E., Bert, E., Duriez, P., Hum, D., Majd, Z., Staels, B., and Cussac, D. 2012. "Does endoplasmic reticulum stress participate in APD-induced hepatic metabolic dysregulation?" *Neuropharmacology* 62:784-96.

Lawrence, J. C. Jr., James, C., and Hiken, J. F. 1986. "Control of glycogen synthase by insulin and isoproterenol in rat adipocytes. Changes in the distribution of phosphate in the synthase subunit in response to insulin and ß-adrenergic receptor activation." *J. Biol. Chem.* 261:669-77.

Lawson, C., Vicencio, J. M., Yellon, D. M., and Davidson, S. M. 2016. "Microvesicles and exosomes: new players in metabolic and cardiovascular disease." *J. Endocrinol.* 228:R57-71.

Lazar, D. F., Knez, J. J., Medof, M. E., Cuatrecasas, P., and Saltiel, A. R. 1994. "Stimulation of glycogen synthesis by insulin in human erythroleukemia cells requires the synthesis of glycosyl-phosphatidylinositol." *Proc. Natl. Acad. Sci. USA* 91:9665-9.

Le Lay, S., and Dugail, I. 2009. "Connecting lipid droplet biology and the metabolic syndrome." *Progr. Lipid Res.* 48:191-5.

Le Lay, S., Blouin, C. M., Hajduch, E., and Dugail, I. 2008. "Filling up adipocytes with lipids. Lessons from caveolin-1 deficiency." *Biochim. Biophys. Acta* 1791:514-8.

Le Lay, S., Hajduch, E., Lindsay, M. R., Le Lièpvre, X., Thiele, C., Ferré, P., Parton, R. G., Kurzchalia, T., Simons, K., and Dugail, I. 2006. "Cholesterol-induced caveolin targeting to lipid droplets in adipocytes: A role for caveolar endocytosis." *Traffic* 7:549-61.

Le Lay, S., Krief, S., Farnier, C., Lefrère, I., Le Liepvre, X., Bazin, R., Ferré, P., and Dugail, I. 2001. "Cholesterol, a cell size-dependent signal that regulates glucose metabolism and gene expression in adipocytes." *J. Biol. Chem.* 276:16904-10.

Lear, J. D., Wasserman, Z. R., and DeGrado, W. F. 1988. "Synthetic amphiphilic peptide models for protein ion channels." *Science* 240:1177-81.

Lee, E. S., Na, K., and Bae, Y. H. 2005. "Super pH-sensitive multifunctional polymeric micelle." *Nano Lett.* 5:325-9.

Lee, H. C., Shoda, R., Krall, J. A., Foster, J. D., Selhub, J., and Rosenberry, T. L. 1992. "Folate binding protein from kidney brush border membranes contains components characteristic of a glycoinositol phospholipid anchor." *Biochemistry* 31:3236-43.

Lee, J. M., Han, J. J., Altwerger, G., Kohn, E. C. 2011. "Proteomics and biomarkers in clinical trials for drug development." *J. Proteomics* 74:2632-41.

Lee, S., Kim, Y. I., and Kim, K. B. 2013. "Comparative study of binding constants from Love wave surface acoustic wave and surface plasmon resonance biosensors using kinetic analysis." *J. Nanosci. Nanotechnol.* 13:7319-24.

Lee, T. H., D'Asti, E., Magnus, N., Al-Nedawi, K., Meehan, B., and Rak, J. 2001. "Microvesicles as mediators of intracellular communication in cancer – the emerging science of cellular 'debris'." *Semin. Immunopathol.* 33:455-67.

Legler, D. F., Doucey, M. A., Schneider, P., Chapatte, L., Bender, F. C., and Bron, C. 2005. "Differential insertion of GPI-anchored GFPs into lipid rafts of live cells." *FASEB J.* 19:73-5.

Lehto, M. T., and Sharom, F. J. 1998. "Release of the glycosylphosphatidylinositol-anchored enzyme ecto-5'-nucleotidase by phospholipase C: catalytic activation and modulation by the lipid bilayer." *Biochem. J.* 332:101-9.

Lehto, M. T., and Sharom, F. J. 2002. "PI-specific phospholipase C cleavage of a reconstituted GPI-anchored protein: modulation by the lipid bilayer." *Biochemistry* 41:1398-408.

Lehto, M. T., Honkanen, E., Teppo, A. M., and Meri, S. 1995. "Urinary excretion of protectin (CD59), complement SC5b-9 and cytokines in membranous glomerulonephritis." *Kidney Int.* 47:918-26.

Lehto, T., and Meri, S. 1993. "Interactions of soluble CD59 with the terminal complement complexes: CD59 and C9 compete for a nascent epitope on C8." *J. Immunol.* 151:4941-9.

Leidlich, S. D., Drapp, D. A., and Orlean, P. 1994. "A conditionally lethal yeast mutant blocked at the first step in glycosyl phosphatidylinositol anchor synthesis." *J. Biol. Chem.* 269:10193-6.

Leifer, D., and Henderson, R. 1983. "Three-dimensional structure of orthorhombic purple membrane at 6.5 Å resolution." *J. Mol. Biol.* 163:451-66.

Leisle, L., Parkin, E. T., Turner, A. J., and Hooper, N. M. 2005. "Angiotensin-converting enzyme as a GPIase: a critical reevaluation." *Nat. Med.* 11:1139-40.

Leroyer, A. S., Tedgui, A., and Boulanger, C. M. 2008. "Microparticles and type 2 diabetes." *Diabetes Metab*. 34:S27-31.

Li, C., Hancock, M. A., Sehgal, P., Zhou, S., Reinhardt, D. P., and Philip, A. 2016. "Soluble CD109 binds TGF-ß and antagonizes TGF-ß signalling and responses." *Biochem. J.* 473:537-47.

Li, D., Zhang, Y., Xu, L., Zhou, L., Wang, Y., Xue, B., Wen, Z., Li, P., and Sang, J. 2010. "Regulation of gene expression by FSP27 in white and brown adipose tissue." *BMC Genomics* 11:446.

Lim, J. M., Wollaston-Hayden, E. E., Teo, C. F., Hausman, D., and Wells, L. 2014. "Quantitative secretome and glycome of primary human adipocytes during insulin resistance." *L. Clin. Proteomics* 11:20.

Lin, Q., Gao, Z., and Alarcon, R. M. 2009. "A role of miR-27 in the regulation of adipogenesis." *FEBS Lett*. 276:2348-58.

Lin, Y. H., Mi, F. L., Chen, C. T., Chang, W. C., Peng, S. F., Liang, H. F., and Sung, H. W. 2007. "Preparation and characterization of nanoparticles shelled with chitosan for oral insulin delivery." *Biomacromolecules* 8:146-52.

Ling, C., and Groop, L. 2009. "Epigenetics: a molecular link between environmental factors and type 2 diabetes." *Diabetes* 58:2718-25.

Lingwood, D., and Simons, K. 2010. "Lipid rafts as a membrane-organizing principle." *Science* 327:46-50.

Lipardi, C., Nitsch, L., and Zurzolo, C. 2000. "Detergent-insoluble GPI-anchored proteins are apically sorted in Fischer rat thyroid cells, but interference with cholesterol or sphingolipids differentially affects detergent insolubility and apical sorting." *Mol. Biol. Cell* 11:531-42.

Lisanti, M. P., Caras, L. W., Davitz, M. A., and Rodriguez-Boulan, E. 1989. "A glycophospholipid membrane anchor acts as an apical targeting signal in polarized epithelial cells." *J. Cell Biol*. 109:2145-56.

Liu, B., Chen, J. S., Cao, M., Gu, S. L., Liao, C., Li, D. Z., and Zhong, H. Z. 2004. "Platelet characteristic antigens of CD34+ cells in cryopreserved cord blood. A study of platelet-derived microparticles in transplant processing." *Vox Sang*. 87:96-104.

Liu, B., Liao, C., Chen, J., Gu, S., Wu, S., and Xu, Z. 2003. "Significance of increasing adhesion of cord blood hematopoietic cells and a new method: platelet microparticles." *Am. J. Hematol*. 74:216-7.

Liu, D., Liu, F., and Song, Y. K. 1995. "Recognition and clearance of liposomes containing phosphatidylserine are mediated by serum opsonin." *Biochim. Biophys. Acta* 1235:140-6.

Liu, S. Q., Tong, Y. W., and Yang, Y. Y. 2005. "Incorporation and *in vitro* release of doxorubicin in thermally sensitive micelles made from poly(*N*-isopropylacrylamide-*co-N,N*-dimethylacrylamide)-*b*-poly(D,L-lactide-co-glycolide) with varying compositions." *Biomaterials* 26:5064-74.

Liu, T., Li, R., Pan, T., Liu, D., Petersen, R. B., Wong, B. S., Gambetti, P., and Sy, M. S. 2002. "Intercellular transfer of the cellular prion protein." *J. Biol. Chem.* 277:47671-8.

Liu, Y., Xiang, X., Zhuang, X., Zhang, S., Liu, C., Cheng, Z., Michalek, S., Grizzle, W., and Zhang, H. G. 2010. "Contribution of MyD88 to the tumor exosome-mediated induction of myeloid derived suppressor cells." *Am. J. Pathol.* 176:2490-9.

Long, K. E., Yomtovian, R., Kida, M., Knez, J. J., and Medof, M. E. 1993. "Time-dependent loss of surface complement regulatory activity during storage of donor blood." *Transfusion* 33:294-300.

Lookene A., Savonen, R., and Olivecrona, G. 1997. "Interaction of lipoproteins with heparan sulfate proteoglycans and with lipoprotein lipase. Studies by surface plasmon resonance technique." *Biochemistry* 36:5267-75.

Lopez-Cobo, S., Campos-SIlva, C., and Vales-Gomez, M. 2016. "Glycosyl-phosphatidyl-inositol (GPI)-anchors and metalloproteases: Their roles in the regulation of exosome composition and NKG2D-mediated immune recognition." *Front. Cell Dev. Biol.* 4:97.

Lotta, L. A., Abbasi, A., Sharp, S. J., Sahlqvist, A. S., Waterworth, D., Brosnan, J. M., Scott, R. A., Langenberg, C., and Wareham, N. J. 2015. "Definitions of metabolic health and risk of future type 2 diabetes in BMI categories: A systematic review and network meta-analysis." *Diabetes Care* 38:2177-87.

Low, M. G. 1989. "Glycosyl-phosphatidylinositol: a versatile anchor for cell surface proteins." *FASEB J.* 3:1600-8.

Low, M. G. 1990. "The glycosyl-phosphatidylinositol anchor of membrane proteins." *Biochim. Biophys. Acta* 988:427-54.

Low, M. G., and Finean, J. B. 1977a. "Non-lytic release of acetylcholinesterase from erythrocytes by a phosphatidylinositol-specific phospholipase C." *FEBS Lett.* 82: 143-6.

Low, M. G., and Finean, J. B. 1977b. "Release of alkaline phosphatase from membranes by a phosphatidylinositol-specific phospholipase C." *Biochem. J.* 167:281-4.

Low, M. G., and Finean, J. B. 1977c. "Modification of erythrocyte membranes by a purified phosphatidylinositol-specific phospholipase C (*Staphylococcus aureus*)." *Biochem. J.* 162:235-40.

Low, M. G., and Finean, J. B. 1978. "Specific release of plasma membrane enzymes by a phosphatidylinositol-specific phospholipase C." *Biochim. Biophys. Acta* 508:565-70.

Low, M. G., and Huang, K. S. 1991. "Factors affecting the ability of glycosylphosphatidylinositol-specific phospholipase D to degrade the membrane anchors of cell surface proteins." *Biochem. J.* 279:483-93.

Low, M. G., and Prasad, A. R. 1988. "A phospholipase D specific for the phosphatidylinositol anchor of cell-surface proteins is abundant in plasma." *Proc. Natl. Acad. Sci. USA* 85:980-4.

Low, M. G., and Saltiel, A. R. 1988. "Structural and functional roles of glycosylphosphatidylinositol in membranes." *Science* 239:268-75.

Low, M. G., and Zilversmit, D. B. 1980. "Role of phosphatidylinositol in attachment of alkaline phosphatase to membranes." *Biochemistry* 19:3913-8.

Lunkes, G. I., Lunkes, D. S., Morsch, V. M., Mazzanti, C. M., Morsch, A. L., Miron, V. R., and Schetinger, M. R. 2004. "NTPDase and 5'-nucleotidase activities in rats with alloxan-induced diabetes." *Diabetes Res. Clin. Pract.* 65:1-6.

Lunkes, G. I., Lunkes, D., Stefanello, F., Morsch, A., Morsch, V. M., Mazzanti, C. M., and Schetinger, M. R. 2003. "Enzymes that hydrolyze adenine nucleotides in diabetes and associated pathologies." *Thromb. Res.* 109:189-94.

Luo, M., Oza-Frank, R., Narayan, K. M., Gokulakrishnan, K., and Mohan, V. 2010. "Serum total adiponectin is associated with impaired glucose tolerance in Asian Indian females but not in males." *J. Diabetes Sci. Technol.* 4:645-51.

Lynch, S. F., and Ludlam, C. A. 2007. "Plasma microparticles and vascular disorders." *Br. J. Haematol.* 137:36-48.

Lyons, T. J., and Basu, A. 2012. "Biomarkers in diabetes: hemoglobin A1c, vascular and tissue markers." *Transl. Res.* 159:303-12.

MacDonald, C., Stamnes, M. A., Katzmann, D. J., and Piper, R. C. 2015. "Tetraspan cargo adaptors usher GPI-anchored proteins into multivesicular bodies." *Cell Cycle* 14:3673-8.

Mack, M., Kleinschmidt, A., Brühl, H., Klier, C., Nelson, P. J., Cihak, J., Plachý, J., Stangassinger, M., Erfle, V., and Schlöndorff, D. 2000. "Transfer of the chemokine receptor CCR5 between cells by membrane-derived microparticles: a mechanism for cellular human immunodeficiency virus 1 infection." *Nat. Med.* 6:769-75.

Macrae, J. L., Acosta-Serrano, A., Morrice, N. A., Mehlert, A., and Ferguson, M. A. 2005. "Structural characterization of NETNES, a novel glycoconjugate in *Trypanosoma cruzi* epimastigotes." *J. Biol. Chem.* 280:12201-11.

Maeda, Y., Tashima, Y., Houjou, T., Fujita, M., Yoko-o, T., Jigami, Y., Taguchi, R., and Kinoshita, T. 2007. "Fatty acid remodeling of GPI-anchored proteins is required for their raft association." *Mol. Biol. Cell* 18:1497-506.

Magee, A. I., and Courtneidge, S. A. 1985. "Two classes of fatty acid acylated proteins exist in eukaryotic cells." *EMBO J.* 4:1137-44.

Mahmood, A., Engle, M. J., and Alpers, D. H. 2002. "Secreted intestinal surfactant-like particles interact with cell membranes and extracellular matrix proteins in rats." *J. Physiol.* 542:237-44.

Mahmood, A., Yamagishi, F., Eliakim, R., DeSchryver-Kecskemeti, K., Gramlich, T. L., and Alpers, D. H. 1994. "A possible role for rat intestinal surfactant-like particles in transepithelial triacylglycerol transport." *J. Clin. Invest.* 93:70-80.

Malek, T. R., Fleming, T. J., and Codias, E. K. 1994. "Regulation of T lymphocyte function by glycosylphosphatidylinositol (GPI)-anchored proteins." *Semin. Immunol.* 6:105-13.

Mallat, Z., Benamer, H., Hugel, B., Benessiano, J., Steg, P. G., Freyssinet, J. M., and Tedgui, A. 2000. "Elevated levels of shed membrane microparticles with procoagulant potential in the peripheral circulating blood of patients with acute coronary syndromes." *Circulation* 29:841-3.

Mamot, C., Drummond, D. C., Noble, C. O., Kallab, V., Guo, Z., Hong, K., Kirpotin, D. B., and Park, J. W. 2005. "Epidermal growth factor receptor-targeted immunoliposomes significantly enhance the efficacy of multiple anticancer drugs *in vivo*." *Cancer Res.* 65:11631-8.

Mančekk-Keber, M., Frank-Bertoncelj, M., Hafner-Bratkovič, I., Smole, A., Zorko, M., Pirher, N., Hayer, S., Kralj-Iglič, V., Rozman, B., Ilc, N., Horvat, S., and Jerala, R. 2015. "Toll-like receptor 4 senses oxidative stress mediated by the oxidation of phospholipids in extracellular vesicles." *Sci. Signal.* 8(381):ra60.

Mann, K. J., Hepworth, M. R., Raikwar, N. S., Deeg, M. A., and Sevlever, D. 2004. "Effect of glycosylphosphatidylinositol (GPI)-phospholipase D overexpression on GPI metabolism." *Biochem. J.* 378:641-8.

Mann, R. S., and Culi, J. 2005. "Developmental biology: morphogens hitch a greasy ride." *Nature* 435:30-3.

Mapstone, M., Cheema, A. K., Fiandaca, M. S., Zhong, X., Mhyre, T. R., MacArthur, L. H., Hall, W. J., Fisher, S. G., Peterson, D. R., Haley, J. M., Nazar, M. D., Rich, S. A., Berlau, D. J., Peltz, C. B., Tan, M. T., Kawas, C. H., and Federoff, H. J. 2014. "Plasma phospholipids identify antecedent memory impairment in older adults." *Nat. Med.* 20:415-8.

Marques, M. B., Weller, P. F., Parsonnet, J., Ransil, B. J., and Nicolson-Weller, A. 1989. "Phosphatidylinositol-specific phospholipase C, a possible virulence factor of *Staphylococcus aureus*." *J. Clin. Microbiol.* 27:2451-4.

Marshall, K. E., Hughson, A., Vascellari, S., Priola, S. A., Sakudo, A., Onodera, T., and Baron, G. S. 2017. "PrP knockout cells expressing transmembrane PrP resist prion infection." *J. Virol.* 91(2):e01686-16.

Martin, C., and Zhang, Y. 2005. "The diverse functioins of histone lysine methylation." *Nat. Rev. Mol. Cell Biol.* 6:838-49.

Martin, S., and Parton, R. G. 2006. "Lipid droplets: a unified view of a dynamic organelle." *Nat. Rev. Mol. Cell Biol.* 7:373-8.

Martin, S., and Pol, A. 2005. "Caveolin, cholesterol, and lipid bodies." *Sem. Cell. Dev. Biol.* 16:163-74.

Martin-DeLeon, P. A. 2011. "Germ-cell hyaluronidases: their roles in sperm function." *Int. J. Androl.* 34:e306-18.

Martin-DeLeon, P. A. 2015. "Epididymosomes: transfer of fertility-modulating proteins to the sperm surface." *Asian J. Androl.* 17:720-5.

Martinez, M. C., and Andriantsitohaina, R. 2011. "Microparticles in angiogenesis: therapeutic potential." *Circ. Res.* 109:110-9.

Martinez, M. C., Kunzelmann, C., and Freyssinet, J. M. 2004. "Plasma membrane remodelling and cell stimulation." *Med. Sci.* 20:189-95.

Marzesco, A. M., Janich. P., Wilsch-Brauninger. M., Dubreuil, V., Langenfeld, K., Corbeil, D., and Huttner, W. B. 2005. "Release of extracellular membrane particles carrying the stem cell marker prominin-1 (CD133) from neural progenitors and other epithelial cells." *J. Cell Sci.* 118:2849-58.

Masella, R., Cantafora, A., Guidoni, L., Luciani, A. M., Mariutti, G., Rosi, A., and Viti, V. 1989. "Characterization of vesicles, containing an acylated oligopeptide, released by human colon adenocarcinoma cells. NMR and biochemical studies." *FEBS Lett.* 246:25-9.

Maslowska, M., Wang, H. W., and Cianflone, K. 2005. "Novel roles for acylation stimulating protein/C3adesArg: a review of recent *in vitro* and *in vivo* evidence." *Vitam. Horm.* 70:309-32.

Masuishi, Y., Nomura, A., Okayama, A., Kimura, Y., Arakawa, N., and Hirano, H. 2013. "Mass spectrometric identification of glycosylphosphatidylinositol-anchored peptides." *J. Prot. Res.* 12:4617-26.

Matas-Rico, E., van Veen, M., and Moolenaar, W. H. 2016b. "Neuronal differentiation through GPI-anchor cleavage." *Cell Cycle* 16:388-9.

Matas-Rico, E., van Veen, M., Leyton-Puig, D., van den Berg, J., Koster, J., Kedziora, K. M., Molenaar, B., Weerts, M. J. A., de Rink, I., Medema, R. H., Giepmans, B. N. G., Perrakis, A., Jalink, K., Versteeg, R., and Moolenaar, W. H. 2016a. "Glycerophosphodiesterase GDE2 promotes neuroblastome differentiation through glypican release and is a marker of clinical outcome." *Cancer Cell* 30:548-62.

Mathiowitz, E. 2008. "Drug delivery system." *Toxicol. Pathol.* 36:6-22.

Mathiowitz, E., Jacob, J. S., Jong, Y. S., Carino, G. P., Chickering, D. E., Chaturvedi, P., Santos, C. A., Vijayaraghavan, K., Montgomery, S., Bassett, M., and Morrell, C. 1997. "Biologically erodable microspheres as potential oral drug delivery systems." *Nature* 386:410-4.

Mathivanan, S., Fahner, C. J., and Simpson, R. J. 2012. "ExoCarta 2012: database of exosomal proteins, RNA and lipids." *Nucleic Acids Res.* 40(Database issue):D1241–4.

Mathivanan, S., Ji, H., and Simpson, R. J. 2010. "Exosomes: extracellular organelles important in intercellular communication." *J. Proteom.* 73:1907-20.

Mato, J. M. 1989. "Insulin mediators revisited." *Cell. Signal.* 1:143-6.

Matsumara, Y. 2008. "Polymeric micellar delivery systems in oncology." *Jpn. J. Clin. Oncology* 38:793-802.

Maydan, G., Noyman, I., Har-Zahav, A., Neriah, Z. B., Pasmanik-Chor, M., Yeheskel, A., Albin-Kaplanski, A., Maya, I., Magal, N., Birk, E., Simon, A. J., Halevy, A., Rechavi, G., Shohat, M., Straussberg, R., and Basel-Vanagaite, L. 2011. "Multiple congenital anomalies-hypotonia-seizures syndrome is caused by a mutation in PIGN." *J. Med. Genet.* 48:383-9.

Mayor, S., and Riezman, H. 2004. "Sorting GPI-anchored proteins." *Nat. Rev. Mol. Cell Biol.* 5:110-20.

McConville, M. J., and Ferguson, M. A. 1993. "The structure, biosynthesis and function of glycosylated phosphatidylinositols in the parasitic protozoa and higher eukaryotes." *Biochem. J.* 294:305-24.

McCurry, K. R., Kooyman, D. L., Alvarado, C. G., Cotterell, A. H., Martin, M. J., Logan, J. S., and Platt, J. L. 1995. "Human complement regulatory proteins protect swine-to-primate cardiac xenografts from humoral injury." *Nat. Med.* 1:423-7.

McDowell, M. A., Ransom, D. M., and Bangs, J. D. 1998. "Glycosylphosphatidylinositol-dependent secretory transport in *Trypanosoma brucei*." *Biochem. J.* 335:681-9.

McGwire, G. B., Becker, R. P., and Skidgel, R. A. 1999. "Carboxypeptidase M, a glycosylphosphatidylinositol-anchored protein, is localized on both the apical and basolateral domains of polarized Madin-Darby canine kidney cells." *J. Biol. Chem.* 274:31632-40.

McHugh, R. S., Ahmed, A. N., Wang, Y. C., Sell, K. W., and Selvaraj, P. 1995. "Construction, purification and functional reconstitution on tumor cells of a glycolipid anchored human B7-1 (CD80)." *Proc. Natl. Acad. Sci. USA* 92:8059-63.

McHugh, R. S., Nagarajan, S., Wang, Y. C., Sell, K. W., and Selvaraj, P. 1999. "Protein transfer of glycosyl-phosphatidylinositol-B7-1 into tumor cell membranes: a novel approach to tumor immunotherapy." *Cancer Res.* 59:2433-7.

McLendon, P. M., Fichter, K. M., and Reineke, T. M. 2010. "Poly(glycoamidoamine) vehicles promote pDNA uptake through multiple routes and efficient gene expression via caveolae-mediated endocytosis." *Mol. Pharmaceut.* 7:738-50.

McVey, M., Tabuchi, A., and Kuebler, W. M. 2012. "Microparticles and acute lung injury." *Am. J. Physiol. Lung Cell. Mol. Physiol.* 303:L364-81.

Medof, M. E., Kinoshita, T., and Nussenzweig, V. 1984. "Inhibition of complement activation on the surface of cells after incorporation of decay-accelerating factor (DAF) into their membranes." *J. Exp. Med.* 160:1558-78.

Medof, M. E., Kinoshita, T., Silber, R., and Nussenzweig, V. 1985. "Amelioration of lytic abnormalities of paroxysmal nocturnal hemoglobinuria with decay-accelerating factor." *Proc. Natl. Acad. Sci. USA* 82:2980-4.

Medof, M. E., Nagarajan, S., and Tykocinski, M. L. 1996. "Cell-surface engineering with GPI-anchored proteins." *FASEB J.* 10:574-86.

Medof, M. E., Walter, E. I., Roberts, W. L., Haas, R., and Rosenberry, T. L. 1986. "Decay accelerating factor of complement is anchored to cells by a C-terminal glycolipid." *Biochemistry* 25:6740-7.

Meisinger, C., Peters, A., and Linseisen, J. 2016. "From the MONICA-project via KORA to the NAKO-study: Practical Utility of Epidemiological Studies in Augsburg Region." *Gesundheitswesen* 78:84-90.

Mensa-Wilmot, K., Morris, J. C., al Qahtani, A., and Englund, P. T. 1995. "Purification and use of recombinant glycosylphosphatidylinositol-phospholipase C." *Methods Enzymol.* 250:641-55.

Meri, S., Morgan, B. P., Davies, A., Daniels, R. H., Olavesen, M. G., Waldmann, H., and Lachmann, P. J. 1990. "Human protectin (CD59), an 18000-20000 MW complement lysis restricting factor, inhibits C5b-8 catalyzed insertion of C9 into lipid bilayer." *Immunology* 71:1-9.

Meri, S., Väkevä, A., and Laari, T. 1991. "Soluble forms of CD59-antigen: distribution in body fluids and functional activity." *Compl. Inflamm.* 8:193.

Merleau-Ponty, M. 1966. *Phänomenologie der Wahrnehmung.* [*Phenomenology of perception*]. Berlin: Walter De Gruyter.

Mesiha, M., Plakogiannis, F., and Vejosoth, S. 1994. "Enhanced oral absorption of insulin from desolvated fatty acid-sodium glycocholate emulsions." *Int. J. Pharm.* 111:213-6.

Mesri, M., and Altieri, D. C. 1999. "Leukocyte microparticles stimulate endothelial cell cytokine release and tissue factor induction in a JNK1 signaling pathway." *J. Biol. Chem.* 274:23111-8.

Michael, A., Bajracharya, S. D., Yuen, P. S., Zhou, H., Star, R. A., Illei, G. G., and Alevizos, I. 2010. "Exosomes from human saliva as a source of microRNA biomarkers." *Oral Dis.* 16:34-8.

Milhiet, P. E., Giocondi, M. C., Baghdadi, O., Ronzon, F., Roux, B., and Le Grimellec, C. 2002. "Spontaneous insertion and partitioning of alkaline phosphatase into model lipid rafts." *EMBO Rep.* 3:485-90.

Miller, J. L. 1998. "Release and extracellular transit of glycosylphosphatidylinositol proteins." *J. Lab. Clin. Med.* 131:115-23.

Millington, P. F., and Finean, J. B. 1962. Electron microscope studies of the structure of the microvilli on principal epithelial cells of rat jejunum after treatment in hypo- and hypertonic saline. *J. Cell Biol.* 14:125-39.

Minagar, A., Jy, W., Jimenez, J. J., and Ahn, Y. S. 2001. "Elevated plasma endothelial microparticles in multiple sclerosis." *Neurology* 56:1319-24.

Miranda, K. C., Bond, D. T., McKee, M., Skog, J., Păunescu, T. G., Da Silva, N., Brown, D., and Russo, L. M. 2010. "Nucleic acids within urinary exosomes/microvesicles are potential biomarkers for renal disease." *Kidney Int.* 78:191-9.

Misek, D. E., and Saltiel, A. R. 1992. "An inositol phosphate glycan derived from a *Trypanosoma brucei* glycosyl-phosphatidylinositol mimics some of the metabolic actions of insulin." *J. Biol. Chem.* 267:16266-76.

Misek, D. E., and Saltiel, A. R. 1994. "An inositol phosphate glycan derived from a *Trypanosoma brucei* glycosyl phosphatidylinositol promotes protein dephosphorylation in rat epididymal adipocytes." *Endocrinology* 135:1869-75.

Mishra, R., Grzybek, M., Niki, T., Hirashima, M., and Simons, K. 2010. "Galectin-9 trafficking regulates apical-basal polarity in Madin-Darby canine kidney epithelial cells." *Proc. Natl. Acad. Sci. USA* 107:17633-8.

Mitchell, P. J., Welton, J., Staffurth, J., Court, J., Mason, M. D., Tabi, Z., and Clayton, A. 2009. "Can urinary exosomes act as treatment response markers in prostate cancer?" *J. Transl. Med.* 7:4.

Mittelbrunn, M., Gutierrez-Vazquez, C., Villarroya-Beltri, C., González, S., Sánchez-Cabo, F., González, M. Á., Bernad, A., and Sánchez-Madrid, F. 2011. "Unidirectional transfer of microRNA-loaded exosomes from T cells to antigen-presenting cells." *Nat. Commun.* 2:282.

Moldovan, N. I., Radu, A. N., and Simionescu, N. 1987. "Endothelial cell plasma membrane obtained by chemically induced vesiculation." *Exp. Cell Res.* 170:499-510.

Molino, D., Nola, S., Lam, S. M., Verraes, A., Proux-Gillardeaux, V., Boncompain, G., Perez, F., Wenk, M., Shui, G., Danglot, L., and Galli, T. 2015. "Role of tetanus neurotoxin insensitive vesicle-associated membrane protein in membrane domains transport and homeostasis." *Cell. Logist.* 5(1):e1025182.

Montecalvo, A., Larregina, A. T., Shufesky, W. J., Stolz, D. B., Sullivan, M. L., Karlsson, J. M., Baty, C. J., Gibson, G. A., Erdos, G., Wang, Z., Milosevic, J., Tkacheva, O. A., Divito, S. J., Jordan, R., Lyons-Weiler, J., Watkins, S. C., and Morelli, A. E. 2012. "Mechanism of transfer of functional microRNAs between mouse dendritic cells *via* exosomes." *Blood* 119:756-66.

Moon, Y. G., Lee, H. J., Kim, M. R., Myung, P. K., Park, S. Y., and Sok, D. E. 1999. "Conversion of glycosylphosphatidylinositol (GPI)-anchored alkaline phosphatase by GPI-PLD." *Arch. Pharm. Res.* 22:249-54.

Moran, P., and Caras, I. W. 1991. "Fusion of sequence elements from non-anchored proteins to generate a fully functional signal for glycophosphatidylinositol membrane anchor attachment." *J. Cell Biol.* 115:1595-600.

Morel, O., Hugel, B., Jesel, L., Lanza, F., Douchet, M. P., Zupan, M., Chauvin, M., Cazenave, J. P., Freyssinet, J. M., and Toti, F. 2004. "Sustained elevated amounts of circulating procoagulant membrane microparticles and soluble GPV after acute myocardial infarction in diabetes mellitus." *Thromb. Haemost.* 91:345-53.

Morel, O., Toti, F., Hugel, B., and Freyssinet, J. M. 2004. "Cellular microparticles: A disseminated storage pool of bioactive vascular effectors." *Curr. Opin. Hematol.* 11:156-64.

Morgan, B. P., van den Berg, C. W., Davies, E. V., Hallett, M. B., and Horejsi, V. 1993. "Cross-linking of CD59 and of other glycosyl phosphatidylinositol-anchored molecules on neutrophils triggers cell activation via tyrosine kinase." *Eur. J. Immunol.* 23:2841-50.

Moskovtsev, S. I., Jarvi, K., Legare, C., Sullivan, R., and Mullen, J. B. 2007. "Epididymal P34H protein deficiency in men evaluated for infertility." *Fertil. Steril.* 88:1455-7.

Movahedi, S., and Hooper, N. M. 1997. "Insulin stimulates the release of the glycosylphosphatidylinositol-anchored membrane dipeptidase from 3T3-L1 adipocytes through the action of a phospholipase C." *Biochem. J.* 326:531-7.

Mukasa, R., Umeda, M., Endo, T., Kobata, A., and Inoue, K. 1995. "Characterization of glycosylphosphatidylinositol (GPI)-anchored NCAM on mouse skeletal muscle cell line C2C12: the structure of the GPI glycan and release during myogenesis." *Arch. Biochem. Biophys.* 318:182-90.

Müller, G. 2000. "The molecular mechanism of the insulin-mimetic/sensitizing activity of the antidiabetic sulfonylurea drug Amaryl." *Mol Med.* 6:907-33.

Müller, G. 2002. "Dynamics pf plasma membrane microdomains and cross-talk to the insulin signalling cascade." *FEBS Lett.* 531:81-7.

Müller, G. 2010a. "Oral protein therapy for the future? Transport of glycolipid-modified therapeutic proteins." *Pharmacology* 86:92-116.

Müller, G. 2010b. "Oral delivery of protein drugs – Driver of personalized medicine?" *Curr. Issues Mol. Biol.* 13:13-24.

Müller, G. 2010c. "Let's shift lipid burden – from large to small adipocytes." *Eur. J. Pharmacol.* 656:1-4.

Müller, G. 2010d. "Control of lipid storage and cell size between adipocytes by vesicle-associated glycosylphosphatidylinositol-anchored proteins." *Arch. Physiol. Biochem.* 117:23-43.

Müller, G. 2011a. "Novel applications for glycosylphosphatidylinositol-anchored proteins in pharmaceutical and industrial biotechnology." *Mol. Membr. Biol.* 28:187-205.

Müller, G. 2011b. "Let's shift lipid burden--from large to small adipocytes." *Eur. J. Pharmacol.* 656:1-4.

Müller, G. 2012a. "Personalised strategies for the diagnosis and therapy of type II diabetes and obesity." *Immunol. Endocr. Metabol. Agents Med. Chem.* 12:80-109.

Müller, G. 2012b. "Microvesicles/exosomes as potential novel biomarkers of metabolic diseases." *Diabetes Metab. Syndr. Obes.* 5:247-82.

Müller, G. 2013. "Adipose tissue-derived microvesicles and exosomes and the diagnosis of metabolic diseases." In *Adipocytes – Biology, Regulation and Health Impact*, edited by Weber, M. J., and Hoffmann, J. M., 65-102. New York: Nova Biomedical.

Müller, G. A. 2016. *Zivilisationskrankheiten – Für eine gesamtwissenschaftliche Prädiktion und Prävention* [*Civilization Diseases - For Overall Scientific Prediction and Prevention*]. Frankfurt am Main: PL Academic Research. p. 205-219.

Müller, G. A. 2017. *Gesundheit – Vom (bio)technologischen, (eigen)verantwortlichen, fairen und realistischen Umgang* [*From (bio) technological, (own) responsible, fair and realistic handling*]. Frankfurt am Main: PL Academic Research. p. 287-318.

Müller, G. A., Herling, A. W., Stemmer, K., Lechner, A., and Tschöp, M. 2018. "*Chip-based sensing of glycosylphosphatidylinositol-anchored proteins in complex with phospholipids and correlation to metabolic states*." In press.

Müller, G., and Bandlow, W. 1993. "Glucose induces lipolytic cleavage of a glycolipidic plasma membrane anchor in yeast." *J. Cell Biol.* 122:325-36.

Müller, G., and Bandlow, W. 1994. "Lipolytic membrane release of two phosphatidy-linositol-anchored cAMP receptor proteins in yeast alters their ligand-binding parameters." *Arch. Biochem. Biophys.* 308:504-14.

Müller, G., and Frick, W. 1999. "Signalling via caveolin: involvement in the cross-talk between phosphoinositolglycans and insulin." *Cell. Mol. Life Sci.* 56:945-70.

Müller, G., and Geisen, K. 1996. "Characterization of the molecular mode of action of the sulfonylurea, glimepiride, at adipocytes." *Horm. Metab. Res.* 28:469-87.

Müller, G., and Tschöp, M. 2015. "Biomarkers for stress-induced diseases." *FEBS J.* 282(Suppl. 1):274-275.

Müller, G., and Tschöp, M. 2016. "Biosensing of extracellular complexes of glycosylphosphatidylinositol-anchored proteins and phospholipids as biomarkers for type II diabetes." *Biomed. Tech.* 61:S38.

Müller, G., and Welte, S. 2002. "Lipid raft domains are the targets for the insulin-independent blood glucose-decreasing activity of the sulfonylurea glimepiride." *Rec. Res. Develop. Endocrinol.* 3:401-23.

Müller, G., and Wied, S. 1993. "The sulfonylurea drug, glimepiride, stimulates glucose transport, glucose transporter translocation, and dephosphorylation in insulin-resistant rat adipocytes *in vitro*." *Diabetes* 42:1852-67.

Müller, G., and Zimmermann, R. 1987. "Import of honeybee prepromelittin into the endoplasmic reticulum: structural basis for independence of SRP and docking protein." *EMBO J.* 6:2099-107.

Müller, G., Dearey, E. A., and Pünter, J. 1993. "The sulphonylurea drug, glimepiride, stimulates release of glycosylphosphatidylinositol-anchored plasma-membrane proteins from 3T3 adipocytes." *Biochem. J.* 289:509-21.

Müller, G., Dearey, E. A., Korndörfer, A., and Bandlow, W. 1994a. "Stimulation of a glycosylphosphatidylinositol-specific phospholipase by insulin and the sulfonylurea, glimepiride, in rat adipocytes depends on increased glucose transport." *J. Cell Biol.* 126:1267-76.

Müller, G., Gross, E., Wied, S., and Bandlow, W. 1996. "Glucose-induced sequential processing of a glycosyl-phosphatidylinositol-anchored ectoprotein in *Saccharomyces cerevisiae*." *Mol. Cell. Biol.* 16:442-56.

Müller, G., Hanekop, N., Wied, S., and Frick, W. 2002. "Cholesterol depletion blocks redistribution of lipid raft components and insulin-mimetic signaling by glimepiride and phosphoinositolglycans in rat adipocytes." *Mol. Med.* 8:120-36.

Müller, G., Herling, A., and Tschöp, M. 2017a. "A chip-based biosensor for the detection of glycosylphosphatidylinositol-anchored proteins in serum as stress biomarkers." *Biomed. Eng. Biomed. Tech.* 62(S1):S358.

Müller, G., Herling, A., and Tschöp, M. 2017b. "Glycosylphosphatidylinositol-anchored proteins in serum are correlated with distinct metabolic states of obese rats." *Adipositas* 11:A19.

Müller, G., Jung, C., Straub, J., Wied, S., and Kramer, W. 2009a. "Induced release of membrane vesicles and exosomes from rat adipocytes containing lipid droplet, lipid raft and glycosylphosphatidylinositol-anchored proteins." *Cell. Signal.* 21:324-38.

Müller, G., Jung, C., Wied, S., and Biemer-Daub, G. 2009b. "Induced translocation of glycosylphosphatidylinositol-anchored proteins from lipid droplets to adiposomes in rat adipocytes." *Br. J. Pharmacol.* 158:749-70.

Müller, G., Jung, C., Wied, S., Biemer-Daub, G., and Frick, W. 2010a. "Transfer of the glycosylphosphatidylinositol-anchored 5'-nucleotidase CD73 from adiposomes into rat adipocytes stimulates lipid synthesis." *Br. J. Pharmacol.* 160:878-91.

Müller, G., Jung, C., Wied, S., Welte, S., and Frick, W. 2001a. "Insulin-mimetic signaling by the sulfonylurea glimepiride and phosphoinositolglycans involves distinct mechanisms for redistribution of lipid raft components." *Biochemistry* 40:14603-20.

Müller, G., Jung, C., Wied, S., Welte, S., Jordan, H., and Frick, W. 2001b. "Redistribution of glycolipid raft domain components induces insulin-mimetic signaling in rat adipocytes." *Mol. Cell. Biol.* 21:4553-67.

Müller, A., Klöppel, C., Smith-Valentine, M., Van Houten, J., and Simon, M. 2012a. "Selective and programmed cleavage of GPI-anchored proteins from the surface membrane by phospholipase C." *Biochim. Biophys. Acta* 1818:117-24.

Müller, G., Over, S., Wied, S., and Frick, W. 2008a. "Association of (c)AMP-degrading glycosylphosphatidylinositol-anchored proteins with lipid droplets is induced by palmitate, $H_2O_2$ and the sulfonylurea drug, glimepiride, in rat adipocytes." *Biochemistry* 47:12774-87.

Müller, G., Schneider, M., Biemer-Daub, G., and Wied, S. 2011a. "Microvesicles released from rat adipocytes and harboring glycosylphosphatidylinositol-anchored proteins transfer RNA stimulating lipid synthesis." *Cell. Signal.* 23:1207-23.

Müller, G., Schneider, M., Biemer-Daub, G., and Wied, S. 2011b. "Upregulation of lipid synthesis in small rat adipocytes by microvesicle-associated CD73 from large adipocytes." *Obesity* (Silver Spring) 19:1531-44.

Müller, G., Schneider, M., Gassenhuber, J., and Wied, S. 2012b. "Release of exosomes and microvesicles harbouring specific RNAs and glycosylphosphatidylinositol-anchored proteins from rat and human adipocytes is controlled by histone methylation." *Am. J. Mol. Biol.* 2:187-209.

Müller, G., Schubert, K., Fiedler, F., and Bandlow, W. 1992. "The cAMP-binding ectoprotein from *Saccharomyces cerevisiae* is membrane-anchored by glycosyl-phosphatidylinositol." *J. Biol. Chem.* 267:25337-46.

Müller, G., Schulz, A., Dearey, E. A., Wetekam, E. M., Wied, S., and Frick, W. 2010b. "Synthetic phosphoinositolglycans regulate lipid metabolism between rat adipocytes *via* release of GPI-protein-harbouring adiposomes." *Arch. Physiol. Biochem.* 116:97-115.

Müller G., Schulz, A., Wied, S., and Frick, W. 2005. "Regulation of lipid raft proteins by glimepiride- and insulin-induced glycosylphosphatidylinositol-specific phospholipase C in rat adipocytes." *Biochem. Pharmacol.* 69:761-80.

Müller, G., Wetekam, E. M., Jung, C., and Bandlow W. 1994b. "Membrane association of lipoprotein lipase and a cAMP-binding ectoprotein in rat adipocytes." *Biochemistry* 33:12149-59.

Müller, G., Wied, S., Crecelius, A., Kessler, A., and Eckel, J. 1997. "Phosphoinositol-glycan-peptides from yeast potently induce metabolic insulin actions in isolated rat adipocytes, cardiomyocytes and diaphragms." *Endocrinology* 138:3459-75.

Müller, G., Wied, S., Dearey, E. A., and Biemer-Daub, G. 2011c. "Glycosyl-phosphatidylinositol-anchored proteins coordinate lipolysis inhibition between large and small adipocytes." *Metabolism* 60:1021-37.

Müller, G., Wied, S., Dearey, E. A., Wetekam, E. M., and Biemer-Daub, G. 2010c. "Lipid storage in large and small rat adipocytes by vesicle-associated glycosylphosphatidylinositol-anchored proteins." In *Results and Problems in Cell Differentiation*: *Sensory and Metabolic Control of Energy Balance* Vol. 52, edited by W. Richter, U. Beisiegel, H. Joost, and P. Meyerhof, 27-34. Berlin, Heidelberg, New York: Springer Press.

Müller, G., Wied, S., Jung, C., and Over, S. 2008b. "Hydrogen peroxide-induced translocation of glycolipid-anchored (c)AMP-hydrolases to lipid droplets mediates inhibition of lipolysis in rat adipocytes." *Br. J. Pharmacol.* 154:901-13.

Müller, G., Wied, S., Jung, C., and Over, S. 2008c. "Translocation of glycosyl-phosphatidylinositol-anchored proteins to lipid droplets and inhibition of lipolysis in rat adipocytes is mediated by reactive oxygen species." *Br. J. Pharmacol.* 154:901-13.

Müller, G., Wied, S., Jung, C., Frick, W., and Biemer-Daub, G. 2010d. "Inhibition of lipolysis by adiposomes containing glycosylphosphatidylinositol-anchored Gce1 protein in rat adipocytes." *Arch. Physiol. Biochem.* 116:28-41.

Müller, G., Wied, S., Over, S., and Frick, W. 2008d. "Inhibition of lipolysis by palmitate, $H_2O_2$ and the sulfonylurea drug, glimepiride, in rat adipocytes depends on cAMP degradation by lipid droplets." *Biochemistry* 47:1259-73.

Müller, G., Wied, S., Straub, J., and Jung, C. 2008e. "Coordinated regulation of esterification and lipolysis by palmitate, $H_2O_2$ and the anti-diabetic sulfonylurea drug, glimepiride, in rat adipocytes." *Eur. J. Pharmacol.* 597:6-18.

Müller, G., Wied, S., Walz, N., and Jung, C. 2008f. "Translocation of glycosylphosphatidylinositol-anchored proteins from plasma membrane microdomains to lipid droplets in rat adipocytes is induced by palmitate, $H_2O_2$ and the sulfonylurea drug, glimepiride." *Mol. Pharmacol.* 73:1513-29.

Müller, G., Wied, S., Wetekam, E. M., Crecelius, A., Unkelbach, A., and Pünter, J. 1994c. "Stimulation of glucose utilization in 3T3 adipocytes and rat diaphragm in vitro by the sulphonylureas, glimepiride and glibenclamide, is correlated with modulations of the cAMP regulatory cascade." *Biochem. Pharmacol.* 48:985-96.

Müller, B., Wilcke, A., Boulesteix, A. L., Brauer, J., Passarge, E., Boltze, J., and Kirsten, H. 2016. "Improved prediction of complex diseases by common genetic markers: state of the art and further perspectives." *Hum. Genet.* 135:259-72.

Muniz, M., and Riezman, H. 2016. "Trafficking of glycosylphosphatidylinositol anchored proteins from the endoplasmic reticulum to the cell surface." *J. Lipid Res.* 57:352-60.

Muniz, M., Nuoffer, C., Hauri, H. P., and Riezman, H. 2000. "The Emp24 complex recruits a specific cargo molecule into endoplasmic reticulum-derived vesicles." *J. Cell Biol.* 148:925-30.

Munteanu, M., Ratziu, V., and Poynard, T. 2011. "FibroStic: a large confirmatory study for non-invasive biomarkers accuracy, if correctly interpreted." *J. Hepatol.* 55:233.

Muralidharan-Chari, V., Clancy, J. W., Sedgwick, A., and D'Souza-Schorey, C. 2010. "Microvesicles: mediators of extracellular communication during cancer progression." *J. Cell Sci.* 123:1603-11.

Muralidharan-Chari, V., Clancy, J., Plou, C., Romao, M., Chavrier, P., Raposo, G., and D'Souza-Schorey, C. 2009. "ARF6-regulated shedding of tumor cell-derived plasma membrane microvesicles." *Curr. Biol.* 19:1875-85.

Musri, M. M., Carmona, M. C., Hanzu, F. A., Kaliman, P., Gomis, R., and Párrizas, M. 2010a. "Histone demethylase LSD1 regulates adipogenesis." *J. Biol. Chem.* 285:30034-41.

Musri, M. M., Gomis, R., and Parrizas, M. 2010b. "A chromatin perspective of adipogenesis." *Organogenesis* 6:15-23.

Nagamune, K., Nozaki, T., Maeda, Y., Ohishi, K., Fukuma, T., Hara, T., Schwarz, R. T., Sutterlin, C., Brun, R., Riezman, H., and Kinoshita, T. 2000. "Critical roles of glycosylphosphatidylinositol for *Trypanosoma brucei*." *Proc. Natl. Acad. Sci. USA* 97:10336-41.

Nagarajan, S., Anderson, M., Ahmed, S. N., Sell, K. W., and Selvaraj, P. 1995. "Purification and optimization of functional reconstitution on the surface of leukemic cell lines of GPI-anchored Fc gamma receptor III." *J. Immunol. Methods* 184:241-51.

Nagase, H., Visse, R., and Murphy, G. 2006. "Structure and function of matrix metalloproteinases and TIMPs." *Cardiovasc. Res*. 69:562-73.

Naghibalhossaini, F., and Ebadi, P. 2006. "Evidence for CEA release from human colon cancer cells by an endogenous GPI-PLD enzyme." *Cancer Lett*. 28:158-67.

Nagpal, J. K., Dasgupta, S., Jadallah, S., Chae, Y. K., Ratovitski, E. A., Toubaji, A., Netto, G. J., Eagle, T., Nissan, A., Sidransky, D., and Trink, B. 2008. "Profiling the expression pattern of GPI transamidase complex subunits in human cancer." *Mod. Pathol.* 21:979-91.

Nakamura, M., Tsujii, K., Katsuragi, Y., Kurihara, K., and Sunamoto, J. 1994. "Taste receptor proteins directly extracted by liposome from intact epithelium of bullfrog tongue." *Biochem. Biophys. Res. Commun*. 201:415-22.

Nakayasu, E. S., Yashunsky, D. V., Nohara, L. L., Torrecilhas, A. C., Nikolaev, A. V., and Almeida, I. C. 2009. "GPIomics: global analysis of glycosylphosphatidylinositol-anchored molecules of *Trypanosoma cruzi*." *Mol. Syst. Biol*. 5:261.

Naylor, S. 2003. "Biomarkers: Current perspectives and future prospects." *Expert Rev. Mol. Diagn*. 3:525-9.

Naylor, S. 2004. "Genomic and proteomic sample preparation: higher throughput solutions. 26-27th April, 2004, World Trade Center, Boston, MA, USA." *Expert. Rev. Proteomics* 1:11-6.

Nelson, K. L., Raja, S. M., and Buckley, J. T. 1997. "The glycosylphosphatidylinositol-anchored surface glycoprotein Thy-1 is a receptor for the channel-forming toxin aerolysin." *J. Biol. Chem*. 272:12170-4.

Neumann, S., Harterink, M., and Sprong, H. 2007. "Hitch-hiking between cells on lipoprotein particles." *Traffic* 8:331-8.

Nicholson, T. B., and Stanners, C. P. 2006. "Specific inhibition of GPI-anchored protein function by homing and self-association of specific GPI anchors." *J. Cell Biol.* 175:647-59.

Nielsen, M. H., Irvine, H., Vedel, S., Raungaard, B., Beck-Nielsen, H., and Handberg, A. 2016. "The impact of lipoprotein-associated oxidative stress on cell-specific microvesicle release in patients with familial hypercholesterolemia." *Oxid. Med. Cell. Longev*. 2016:2492858.

Nieuwland, R., Berckmans, R. J., McGregor, S., Böing, A. N., Romijn, F. P., Westendorp, R. G., Hack, C. E., and Sturk, A. 2000. "Cellular origin and procoagulant properties of microparticles in meningococcal sepsis." *Blood* 95:930-5.

Nishino, N., Tamori, Y., Tateya, S., Böing, A. N., Romijn, F. P., Westendorp, R. G., Hack, C. E., and Sturk, A. 2008. "FSP27 contributes to efficient energy storage in murine white adipocytes by promoting the formation of unilocular lipid droplets." *J. Clin. Invest.* 118:2808-21.

Noda, M. 1984. "Primary structure of Electrophorus electricus sodium channel deduced from cDNA sequence." *Nature* 312:121-7.

Nomura, S. 2001. "Function and clinical significance of platelet-derived microparticles." *Int. J. Hematol.* 74:397-404.

Nomura, S., Shouzu, A., Omoto, S., Hayakawa, T., Kagawa, H., Nishikawa, M., Inada, M., Fujimura, Y., Ikeda, Y., and Fukuhara, S. 1998. "Effect of cilostazol on soluble adhesion molecules and platelet-derived microparticles in patients with diabetes." *Thromb. Haemost.* 80:388-92.

Nomura, S., Suzuki, M., Katsura, K., Xie, G. L., Miyazaki, Y., Miyake, T., Kido, H., Kagawa, H., and Fukuhara, S. 1995. "Platelet-derived microparticles may influence the development of atherosclerosis in diabetes mellitus." *Atherosclerosis* 116:235-40.

Nomura, S., Tandon, N. N., Nakamura, T., Cone, J., Fukuhara, S., and Kambayashi, J. 2001. "High-shear-stress-induced activation of platelets and microparticles enhances expression of cell adhesion molecules in THP-1 and endothelial cells." *Atherosclerosis* 158:277-87.

Nose, M., Katoh, M., Okada, N., Kyogoku, M., and Okada, H. 1990. "Tissue distribution of HRF20, a novel factor preventing the membrane attack of homologous complement, and its predominant expression on endothelial cells *in vivo*." *Immunology* 70:145-9.

Nosjean, O., and Roux, B. 2003. "Anchoring of glycosylphosphatidylinositol-proteins to liposomes." *Methods Enzymol.* 372:216-32.

Nosjean, O., Briolay, A., and Roux, B. 1997. "Mammalian GPI proteins: sorting, membrane residence and functions." *Biochim. Biophys. Acta* 1331:153-86.

Notohamiprodjo, S., Djafarzadeh, R., Rieth, N., Hofstetter, M., Jaeckel, C., and Nelson, P. J. 2012. "Cell surface engineering of renal cell carcinoma with glycosylphosphatidylinositol-anchored TIMP-1 blocks TGF-ß1 activation and reduces regulatory ID gene expression." *Biol. Chem.* 393:1463-70.

Nozaki, M., Ohishi, K., Yamada, N., Kinoshita, T., Nagy, A., and Takeda, J. 1999. "Developmental abnormalities of glycosylphosphatidylinositol-anchor-deficient embryos revealed by Cre/loxP system." *Lab. Invest.* 79:293-9.

Nuoffer, C., Jenö, P., Conzelmann, A., and Riezman, H. 1991. "Determinants for glycophospholipid anchoring of the Saccharomyces cerevisiae GAS1 protein to the plasma membrane." *Mol. Cell. Biol.* 11:27-37.

O'Brien, P. J., and Zatz, M. 1984. "Acylation of bovine rhodopsin by ($^3$H)palmitic acid." *J. Biol. Chem.* 259:5054-7.

O'Brien, P. J., St. Jules, R. S., Ready, T. S., Bazan, N. G., and Zatz, M. 1987. "Acylation of disc membrane rhodopsin may be nonenzymatic." *J. Biol. Chem.* 262:5210-5.

Ogawa, R., Tanaka, C., Sato, M., Nagasaki, H., Sugimura, K., Okumura, K., Nakagawa, Y., and Aoki, N. 2010. "Adipocyte-derived microvesicles contain RNA that is transported into macrophages and might be secreted into blood circulation." *Biochem. Biophys. Res. Commun.* 398:723-9.

Ohmichi, M., Decker, S. J., and Saltiel, A. R. 1992. "Nerve growth factor stimulates the tyrosine phosphorylation of a 38-kDa protein that specifically associates with the src homology domain of phospholipase C-gamma 1." *J. Biol. Chem.* 267:21601-6.

Ohsaki, Y., Cheng, J., Suzuki, M., Shinohara, Y., Fujita, A., and Fujimoto, T. 2009. "Biogenesis of cytoplasmic lipid droplets: from the lipid ester globule in the membrane to the visible structure." *Biochim. Biophys. Acta* 1791:399-407.

Ohyabu, T., Taguchi, R., and Ikezawa, H. 1978. "Studies on phosphatidylinositol phosphodiesterase (phospholipase C type) of *Bacillus cereus* II. *In vivo* and immunochemical studies of phosphatase-releasing activity." *Arch. Biochem. Biophys.* 190:1-7.

Oiki, S., Danho, W., and Montal, M. 1988. "Channel protein engineering: synthetic 22-mer peptide from the primary structure of the voltage-sensitive sodium channel forms ionic channels in lipid bilayers." *Proc. Natl. Acad. Sci. USA* 85:2392-7.

Okada, N., Harada, R., Fujita, T., and Okada, H. 1989. "A novel membrane glycoprotein capable of inhibiting membrane attack by homologous complement." *Int. Immunol.* 1:205-8.

Okumura, Y., Ishitobi, M., Sobel, M., Akiyoshi, K., and Sunamoto, J. 1994. "Transfer of membrane proteins from human platelet to liposomal fraction by interaction with liposomes containing an artificial boundary lipid." *Biochim. Biophys. Acta* 1194:335-40.

Olivecrona, G., Ehrenborg, E., Semb, H., Makoveichuk, E., Lindberg, A., Hayden, M. R., Gin, P., Davies B. S., Weinstein, M. M., Fong, L. G., Beigneux, A. P., Young, S. G., Olivecrona, T., and Hernell, O. 2010. "Mutation of conserved cysteines in the Ly6 domain of GPIHBP1 in familial chylomicronemia." *J. Lipid Res.* 51:1535-45.

Olivecrona, T., Hultin, M., Bergö, M., and Olivecrona, G. 1997. "Lipoprotein lipase: regulation and role in lipoprotein metabolism." *Proc. Nutr. Soc.* 56:723-9.

Olofsson, S. O., Asp, L., and Boren, J. 1999. "The assembly and secretion of apolipoprotein B-containing lipoproteins." *Curr. Opin. Lipidol.* 10:341-6.

Omaetxebarria, M. J., Hagglund, P., Elortza, F., Hooper, N. M., Arizmendi, J. M., and Jensen, O. N. 2006. "Isolation and characterization of glycosylphosphatidylinositol-anchored peptides by hydrophilic interaction chromatography and MALDI tandem mass spectrometry." *Anal. Chem.* 78:3335-41.

Omary, M. B., and Trowbridge, I. S. 1981. "Covalent binding of fatty acid to the transferrin receptor in cultured human cells." *J. Biol. Chem.* 356:4715-8.

Omenetti, A., and Diehl, A. M. 2008. "The adventures of sonic hedgehog in development and repair II." *Am. J. Physiol. Gastrointest. Liver Physiol*. 294:G595-8.

Omoto, S., Nomura, S., Shouzu, A., Nishikawa, M., Fukuhara, S., and Iwasaka, T. 2002. "Detection of monocyte-derived microparticles in patients with type II diabetes mellitus." *Diabetologia* 45:550-5.

O'Neill, S., Bohl, M., Gregersen, S., Hermansen, K., and O'Driscoll, L. 2016. "Blood-based biomarkers for metabolic syndrome." *Trends Endocrinol. Metab*. 27:363-74.

Orlean, P., and Menon, A. K. 2007. "Thematic review series: lipid posttranslational modifications. GPI anchoring of protein in yeast and mammalian cells, or: how we learned to stop worrying and love glycophospholipids." *J. Lipid Res*. 48:993-1011.

Ott, P., Hope, M. J., Verkleij, A. J., Roelofsen, B., Brodbeck, U., and van Deenen, L. L. 1981. "Effect of dimyristoylphosphatidylcholine on intact erythrocytes. Release of spectrin-free vesicles without ATP depletion." *Biochim. Biophys. Acta* 641:79-87.

Ovchinnikov, Yu. A., Abdulaev, N. G., and Bogachuk, A. S. 1988. "Two adjacent cysteine resdiues in the C-terminal cytoplasmic fragment of bovine rhodopsin are palmitylated." *FEBS Lett.* 230:1-5.

Oxley, D., and Bacic, A. 1999. "Structure of the glycosylphosphatidylinositol anchor of an arabinogalactan protein from *Pyrus communis* suspension-cultured cells." *Proc. Natl. Acad. Sci. USA* 96:14246-51.

Oyarzún, C., Salinas, C., Gómez, D., Jaramillo, K., Pérez, G., Alarcón, S., Podestá, L., Flores, C., Quezada, C., and San Martín, R. 2015. "Increased levels of adenosine and ecto 5'-nucleotidase (CD73) activity precede renal alterations in experimental diabetic rats." *Biochem. Biophys. Res. Commun*. 468:354-9.

Paladino, S., Lebreton, S., and Zurzolo, C. 2015. "Trafficking and membrane organization of GPI-anchored proteins in health and diseases." *Curr. Top. Membr.* 75:269-303.

Paladino, S., Pocard, T., Catino, M. A., and Zurzolo, C. 2006. "GPI-anchored proteins are directly targeted to the apical surface in fully polarized MDCK cells." *J. Cell Biol.* 172:1023-34.

Paladino, S., Sarnataro, D., Pillich, R., Tivodar, S., Nitsch, L., and Zurzolo, C. 2004. "Protein oligomerisation modulates raft partitioning and apical sorting of GPI-anchored proteins." *J. Cell Biol*. 167:699-709.

Palin, K. J., Philipps, A. J., and Ning, A. 1986. "The oral absorption of cefoxitin from oil and emulsion vesicles in rats." *Int. J. Pharm*. 33:99-104.

Pan, B. T., and Johnstone, R. M. 1983. "Fate of the transferrin receptor during maturation of sheep reticulocytes *in vitro*: selective externalization of the receptor." *Cell* 33:967-78.

Pan, B. T., Teng, K., Wu, C., Adam, M., and Johnstone, R. M. 1985. "Electron microscopic evidence for externalization of the transferrin receptor in vesicular form in sheep reticulocytes." *J. Cell Biol.* 101:942-8.

Panakova, D., Sprong, H., Marois, E., Thiele, C., and Eaton, S. 2005. "Lipoprotein particles are required for Hedgehog and Wingless signalling." *Nature* 435:58-65.

Pap, E., Pallinger, E., Pasztoi, M., and Falus, A. 2009. "Highlights of a new type of intercellular communication: microvesicle-based information transfer." *Inflamm. Res.* 58:1-8.

Pappin, D. J., and Findlay, J. B. 1984. "Sequence variability in the retinal-attachment domain of mammalian rhodopsins." *Biochem. J.* 217:260-4.

Park, S. W., Choi, K., Kim, I. C., Lee, H. H., Hooper, N. M., and Park, H. S. 2001. "Endogenous glycosylphosphatidylinositol-specific phospholipase C releases renal dipeptidase from kidney proximal tubules *in vitro*." *Biochem. J.* 353:339-44.

Park, S. W., Choi, K., Lee, H. B., Park, S. K., Turner, A. J., Hooper, N. M., and Park, H. S. 2002. "Glycosyl-phosphatidylinositol (GPI)-anchored renal dipeptidase is released by a phospholipase C *in vivo*." *Kidney Blood Press. Res.* 25:7-12.

Parkin, E. T., Watt, N. T., Turner, A. J., and Hooper, N. M. 2004. "Dual mechanisms for shedding of the cellular prion protein." *J. Biol. Chem.* 279:11170-8.

Parra, P., Serra, F., and Palou, A. 2010. "Expression of adipose microRNAs is sensitive to dietary conjugated linoleic acid treatment in mice." *PLoS ONE* 5:e13005.

Pasca di Magliano, M., and Hebruk, M. 2003. "Hedgehog signalling in cancer formation and maintenance." *Nat. Rev. Cancer* 3:903-11.

Patel, R., Al-Dossary, A. A., Stabley, D. L., Barone, C., Galileo, D. S., Strehler, E. E., and Martin-DeLeon, P. A. 2013. "Plasma membrane $Ca^{2+}$ATPase 4 in murine epididymis: secretion of splice variants in the luminal fluid and a role in sperm maturation." *Biol. Reprod.* 89:1-11.

Patton, S., and Huston, G. E. 1986. "A method for isolation of milk fat globules." *Lipids* 21:170-4.

Paulik, M. G., and Bertozzi, C. R. 2008. "The glycosylphosphatidylinositol anchor: a complex membrane-anchoring structure for proteins." *Biochemistry* 47:6991-7000.

Pavlou, A. K., and Reichert, J. M. 2004. "Recombinant protein therapeutics: Success rates, market trends and values to 2010." *Nat. Biotechnol.* 22:1513-9.

Pearce, E. J., Hall, B. F., and Sher, A. 1990. "Host-spepcific evasion of the alternative complement pathway by schistosomes correlates with the presence of a phospholipase C-sensitive surface molecule resembling human decay accelerating factor." *J. Immunol.* 144:2751-6.

Pellman, D., Garber, E. A., Cross, F. R., and Hanafusa, H. 1985. "An N-terminal peptide from p60src can direct myristylation and plasma membrane localization when fused to heterelogous proteins." *Nature* 314:374-7.

Pereira-Lancha, L. O., and Campos-Ferraz, P. L. 2012. "Obesity: considerations about etiology, metabolism, and the use of experimental models." *Diabetes Metab. Syndr. Obes.* 5:75-87.

Perez-Hernandez, J., and Cortes, R. 2015. "Extracellular vesicles as biomarkers of systemic lupus erythematosus." *Dis. Markers* 613536.

Peters, P. J., Borst, J., Oorschot, V., Fukuda, M., Krähenbühl, O., Tschopp, J., Slot, J. W., and Geuze, H. J. 1991. "Cytotoxic T lymphocyte granules are secretory lysosomes containing both perforin and granzymes." *J. Exp. Med.* 173:1099-109.

Pettengill, M., Robson, S., Tresenriter, M., Millán, J. L., Usheva, A., Bingham, T., Belderbos, M., Bergelson, I., Burl, S., Kampmann, B., Gelinas, L., Kollmann, T., Bont, L., and Levy, O. 2013. "Soluble ecto-5'-nucleotidase (5'-NT), alkaline phosphatase, and adenosine deaminase (ADA1) activities in neonatal blood favor elevated extracellular adenosine." *J. Biol. Chem.* 288:27315-26.

Piccin, A., Murphy, W. G., and Smith, O. P. 2007. "Circulating microparticles: pathophysiology and clinical implications." *Blood Rev.* 21:157-71.

Pinaud, F., Michalet, X., Iyer, G., Margeat, E., Moore, H. P., and Weiss, S. 2009. "Dynamic partitioning of a glycosyl-phosphatidylinositol-anchored protein in glycosphingolipid-rich microdomains imaged by single-quantum dot tracking." *Traffic* 10:691-712.

Pisitkun, T., Shen, R. F., and Knepper, M. A. 2004. "Identification and proteomic profiling of exosomes in human urine." *Proc. Natl. Acad. Sci. USA* 101:13368-73.

Poisson, G., Chauve, C., Chen, X., and Bergeron, A. 2007. "FragAnchor: a large-scale predictor of glycosylphosphatidylinositol anchors in eukaryote protein sequences by qualitative scoring." *Genomics Proteomics Bioinformatics* 5:121-30.

Pol, A., Martin, S., Fernandez, M. A., Ingelmo-Torres, M., Ferguson, C., Enrich, C., and Parton, R. G. 2005. "Cholesterol and fatty acids regulate dynamic caveolin trafficking through the Golgi complex and between the cell surface and lipid bodies." *Mol. Biol. Cell* 16:2091-105.

Porter, J. A., Young, K. E., and Beachy, P. A. 1996. "Cholesterol modification of hedgehog signaling proteins in animal development." *Science* 274:255-9.

Poynard, T., Munteanu, M., Deckmyn, O., Ngo, Y., Drane, F., Castille, J. M., Housset, C., Ratziu, V., and Imbert-Bismut, F. 2012. "Validation of liver fibrosis biomarker (FibroTest) for assessing liver fibrosis progression: proof of concept and first application in a large population." *J. Hepatol.* 57:541-8.

Pradhan, A. D., Manson, J. E., Rifai, N., Buring, J. E., and Ridker, P. M. 2001. "C-reactive protein, interleukin 6, and risk of developing type 2 diabetes mellitus." *JAMA* 286:327-34.

Premkumar, D. R. D., Fukuoka, Y., Sevlever, D., Brunschwig, E., Rosenberry, T. L., Tykocinski, M. L., and Medof, M. E. 2001. "Properties of exogenously added GPI-anchored proteins following their incorporation into cells." *J. Cell. Biochem.* 82:234-45.

Preston, R. A., Jy, W., Jimenez, J. J., Mauro, L. M., Horstman, L. L., Valle, M., Aime, G., and Ahn, Y. S. 2003. "Effects of severe hypertension on endothelial and platelet microparticles." *Hypertension* 41:211-7.

Puri, V., Konda, S., Ranjit, S., Aouadi, M., Chawla, A., Chouinard, M., Chakladar, A., and Czech, M. P. 2007. "Fat-specific protein 27, a novel lipid droplet protein that enhances triglyceride storage." *J. Biol. Chem.* 282:34213-8.

Puri, V., Ranjit, S., Konda, S., Nicoloro, S. M., Straubhaar, J., Chawla, A., Chouinard, M., Lin, C., Burkart, A., Corvera, S., Perugini, R. A., and Czech, M. P. 2008. "Cidea is associated with lipid droplets and insulin sensitivity in humans." *Proc. Natl. Acad. Sci. USA* 105:7833-8.

Qazi, K. R., Torregrosa Paredes, P., Dahlberg, B., Grunewald, J., Eklund, A., and Gabrielsson, S. 2010. "Proinflammatory exosomes in bronchoalveolar lavage fluid of patients with sarcoidosis." *Thorax* 65:1016-24.

Qin, X., Goldfine, A., Krumrei, N., Grubissich, L., Acosta, J., Chorev, M., Hays, A. P., and Halperin, J. A. 2004. "Glycation inactivation of the complement regulatory protein CD59: A possible role in the pathogenesis of the vascular complications of human diabetes." *Diabetes* 53:2653-61.

Queiroz, K. C. S., Tio, R. A., Zeebregts, C. J., Bijlsma, M. F., Zijlstra, F., Badlou, B., de Vries, M., Ferreira, C. V., Spek, C. A., Peppelenbosch, M. P., and Rezaee, F. 2010. "Human plasma very low density lipoprotein carries Indian Hedgehog." *J. Proteome Res.* 9:6052-9.

Rabesandrata, H., Toutant, J. P., Reggio, H., and Vidal, M. 1998. "Decay-Accelerating factor (CD55) and membrane inhibitor of reactive lysis (CD59) are released within exosomes during in vitro maturation of reticulocytes." *Blood* 91:2573-80.

Rabinowits, G., Gercel-Taylor, C., Day, J. M., Taylor, D. D., and Kloecker, G. H. 2009. "Exosomal microRNA: a diagnostic marker for lung cancer." *Clin. Lung Cancer* 10:42-6.

Rademacher, T. W., Edge, C. J., and Dwek, R. A. 1991. "Dropping anchor with the lipophosphoglycans." *Curr. Biol.* 1:41-2.

Radons, J., and Multhoff, G. 2005. "Immunostimulatory functions of membrane-bound and exported heat shock protein 70." *Exerc. Immunol. Rev.* 11:17-33.

Raggi, M. C., Djafarzadeh, R., Muenchmeier, N., Hofstetter, M., Jahn, B., Rieth, N., and Nelson, P. J. 2009. "Peritumoral administration of GPI-anchored TIMP-1 inhibits colon carcinoma growth in Rag-2 gamma chain-deficient mice." *Biol. Chem.* 390:893-7.

Raghupathy, R., Anilkumar, A. A., Polley, A., Singh, P. P., Yadav, M., Johnson, C., Suryawanshi, S., Saikam, V., Sawant, S. D., Panda, A., Guo, Z., Vishwakarma, R. A.,

Rao, M., and Mayor, S. 2015. "Transbilayer lipid interactions mediate nanoclustering of lipid-anchored proteins." *Cell* 161:581-94.

Raikwar, N. S., Bowen-Deeg, R. F., Du, X. S., Low, M. G., and Deeg, M. A. 2010. "Glycosylphosphatidylinositol-specific phospholipase D improves glucose tolerance." *Metabolism* 59:1413-20.

Rajendran, L., and Simons, K. 2005. "Lipid rafts and membrane dynamics." *J. Cell Sci.* 118:1099-102.

Raposo, G., Nijman, H. W., Stoorvogel, W., Liejendekker, R., Harding, C. V., Melief, C. J., and Geuze, H. J. 1996. "B lymphocytes secrete antigen-presenting vesicles." *J. Exp. Med.* 183:1161-72.

Ratajczak, J., Miekus, K., and Kucia, M. 2006. "Embryonic stem cell-derived microvesicles reprogram hematopoietic progenitors: evidence for horizontal transfer of mRNA and protein delivery." *Leukemia* 20:847-56.

Ratnoff, O. D., Gleich, G. J., Shurin, S. B., Kazura, J., Everson, B., and Embury, P. 1993. "Inhibition of the activation of hageman factor (factor XII) by eosinophils and eosinophilic constituents." *Am. J. Hematol.* 42:138-45.

Ratziu, V., Voiculescu, M., and Poynard, T. 2012. "Touching some firm ground in the epidemiology of NASH." *J. Hepatol.* 56:23-5.

Ray S., Britschgi, M., Herbert, C., Takeda-Uchimura, Y., Boxer, A., Blennow, K., Friedman, L. F., Galasko, D. R., Jutel, M., Karydas, A., Kaye, J. A., Leszek, J., Miller, B. L., Minthon, L., Quinn, J. F., Rabinovici, G. D., Robinson, W. H., Sabbagh, M. N., So, Y. T., Sparks, D. L., Tabaton, M., Tinklenberg, J., Yesavage, J. A., Tibshirani, R., and Wyss-Coray, T. 2007. "Classification and prediction of clinical Alzheimer's diagnosis based on plasma signaling proteins." *Nat. Med.* 13:1359-62.

Raymond, F., Datta, H., and Moss, D. 1991. "Alkaline phosphatase isoforms in bile and serum and their generation from cells *in vitro*." *Biochim. Biophys. Acta* 1074:217-22.

Razani, B., and Lisanti, M. P. 2001. "Caveolins and caveolae: molecular and functional relationships." *Exp. Cell Res.* 271:36-44.

Record, M., Subra, C., and Silvente-Poirot, M. 2011. "Exosomes as intercellular signalosomes and pharmacological effectors." *Biochem. Pharmacol.* 81:1171-82.

Redman, C. A., Thomas-Oates, J. E., Ogata, S., Ikehara, Y., and Ferguson, M. A. 1994. "Structure of the glycosylphosphatidylinositol membrane anchor of human placental alkaline phosphatase." *Biochem. J.* 302:861-5.

Redman, C. W., and Sargent, I. L. 2007. "Microparticles and immunomodulation in pregnancy and pre-eclampsia." *J. Reprod. Immunol.* 76:61-7.

Reed, G. L. 2004. "Platelet secretory mechanism." *Semin. Thromb. Hemost.* 30:441–50.

Reid, K., Flowers, P., and Larkin, M. 2005. "Exploring lived experience: An introduction to Interpretative Phenomenological Analysis." *The Psychologist* 18:20-3.

Reilly, M. P., Lehrke, M., Wolfe, M. L., Rohatgi, A., Lazar, M. A., and Rader, D. J. 2005. "Resistin is an inflammatory marker of atherosclerotic in humans." *Circulation* 111:932-9.

Reya, T., and Clevers, H. 2005. "Wnt signalling in stem cells and cancer." *Nature* 434:843-50.

Rhee, S. G., Suh, P. G., Ryu, S. H., and Lee, S. Y. 1989. "Studies of inositol phospholipid-specific phospholipase C." *Science* 244:546-50.

Riakwar, N. S., Bowen-Deeg, R. F., Du, X. S., Low, M. G., and Deeg, M. A. 2010. Glycosylphosphatidylinositol-specific phospholipase D improves glucose tolerance. *Metabolism* 59:1413-20.

Ridker, P. M., Buring, J. E., Cook, N. R., and Rifai, N. 2003. "C-reactive protein, the metabolic syndrome, and risk of incident cardiovascular events: an 8-year follow-up of 14 719 initially healthy American women." *Circulation* 107:391-7.

Rifkin, M. R., and Landsberger, F. R. 1990. "*Trypanosome* variant surface glycoprotein transfer to target membranes: a model for the pathogenesis of trypanosomiasis." *Proc. Natl. Acad. Sci. USA* 87:801-5.

Rios, J. J., Shastry, S., Jasso, J., Hauser, N., Garg, A., Bensadoun, A., Cohen, J. C., and Hobbs, H. H. 2011. "Deletion of GPIHBP1 causing severe chylomicronemia." *J. Inherit. Metab. Dis.* 35:531-40.

Roberts, W. L., Myyher, J. J., Kuksis, A., Low, M. G., and Rosenberry, T. L. 1988. "Lipid analysis of the glycoinositol phospholipid membrane anchor of human erythrocyte acetylcholinesterase. Palmitoylation of inositol results in resistance to phosphatidylinositol-specific phospholipase C." *J. Biol. Chem.* 263:18766-75.

Robertson, J. D. 1964. "Unit membranes: A review with recent new studies of experimental alterations and a new subunit structure in synaptic membranes." In *Cellular Membranes in Development*, edited by M. Locke, 1-81. New York, and London: Academic Press.

Robinson, P. J. 1991a. "Signal transduction by GPI-anchored membrane proteins." *Cell Biol. Intern. Rep.* 15:761-7.

Robinson, P. J. 1991b. "Phosphatidylinositol membrane anchors and T-cell activation." *Immunol. Today* 12:35-41.

Robinson, P. J. 1997. "Signal transduction *via* GPI-anchored membrane proteins." *Adv. Exp. Med. Biol.* 419:365-70.

Robinson, P. J., Millrain, M., Antoniou, J., Simpson, E, and Mellor, A. L. 1989. "A glycophospholipid anchor is required for Qa-2-mediated T cell activation." *Nature* 342:85-7.

Roddy, J., Clark, I., Hazleman, B. L., Compston, D. A., and Scolding, N. J. 1994. "Cerebrospinal fluid concentrations of the complement MAC inhibitor CD59 in multiple sclerosis and patients with other neurological disorders." *J. Neurol.* 241:557-60.

Rollins, S. A., and Sims, P. J. 1990. "The complement-inibitory activity of CD59 resides in its capacity to block incorporation of C9 into membrane C5b-9." *J. Immunol.* 144:3478-83.

Rollins, S. A., Zhao, J., Ninomiya, H., and Sims, P. J. 1991. "Inhibition of homologous complement by CD59 is mediated by a species-selective recognition conferred through binding to C8 within C5b-8 or C9 within C5b-9." *J. Immunol.* 146:2345-51.

Romero, G., and Larner, J. 1993. "Insulin mediators and the mechanism of insulin action." *Adv. Pharmacol.* 24:21-50.

Romero, G., Luttrell, L., Rogol, A., Zeller, K., Hewlett, E., and Larner, J. 1988. "Phosphatidylinositol-glycan anchors of membrane proteins: potential precursors of insulin mediators." *Science* 240:509-11.

Rondinone, C. M. 2006. "Adipocyte-derived hormones, cytokines, and mediators." *Endocrine* 29:81-90.

Ronquist, G., and Brody, I. 1985. "The prostasome: its secretion and function in man." *Biochim. Biophys. Acta* 822:203-18.

Ronzon, F., Desbat, B., Chauvet, J. P., and Roux, B. 2002. "Behavior of a GPI-anchored protein in phospholipid monolayers at the air-water interface." *Biochim. Biophys. Acta* 1560:1-13.

Ronzon, F., Morandat, S., Roux, B., and Bortolato, M. 2004. "Insertion of a glycosylphosphatidylinositol-anchored enzyme into liposomes." *J. Membr. Biol.* 197:169-77.

Rooney, I. A., Atkinson, J. P., Krul, E. S., Schonfeld, G., Polakoski, K., Saffitz, J. E., and Morgan, B. P. 1993. "Physiologic relevance of the membrane attack complex inhibitory protein CD59 in human seminal plasma: CD59 is present on extracellular organelles (prostasomes), binds cell membranes, and inhibits complement-mediated lysis." *J. Exp. Med.* 177:1409-20.

Rooney, I. A., Davies, A., and Morgan, B. P. 1992. "Membrane attack complex (MAC)-mediated damage to spermatozoa: protection of the cells by the presence on their membranes of MAC inhibitory proteins." *Immunology* 75:499-506.

Rooney, I. A., Heuser, J. E., and Atkinson, J. P. 1996. "GPI-anchored complement regulatory proteins in seminal plasma. An analysis of their physical condition and the mechanisms of their binding to exogenous cells." *J. Clin. Invest.* 97:1675-86.

Root, K. T., Plucinsky, S. M., and Glover, K. J. 2015. "Recent progress in the topology, structure, and oligomerization of caveolin: a building block of caveolae." *Curr. Top. Membr.* 75:305-36.

Rosse, W. F. 2001. "New insights into paroxysmal nocturnal hemoglobinuria." *Curr. Opin. Hematol.* 8:61-7.

Rowsen, M. L., Dileep, K. J., and Sharma, C. P. 2000. "Beta cyclodextrin-insulin-encapsulated chitosan/alginate matrix: Oral delivery system." *J. Appl. Polym. Sci.* 75:1089-96.

Rudd, P. M., Morgan, B. O., Wormald, M. R., Harvey, D. J., van den Berg, C. W., Davis, S. J., Ferguson, M. A., and Dwek, R. A. 1997. "The glycosylation of the complement regulatory protein, human erythrocyte CD59." *J. Biol. Chem.* 272:7229-44.

Rupert, D. L., Claudio, V., Lässer, C., and Bally, M. 2017. "Methods for the physical characterization and quantification of extracellular vesicles in biological samples." *Biochim. Biophys. Acta* 1861:3164-79.

Sabatier, F., Darmon, P., Hugel, B., Combes, V, Sanmarco, M., Velut, J. G., Arnoux, D., Charpiot, P., Freyssinet, J. M., Oliver, C., Sampol, J., and Dignat-George, F. 2002. "Type 1 and type 2 diabetic patients display different patterns of cellular microparticles." *Diabetes* 51:2840-5.

Sack, T. L., Gum, J. R., Lowe, M. G., and Kim, Y. S. 1988. "Release of carcinoembryonic antigen from human colon cancer cells by phosphatidylinositol-specific phospholipase C." *J. Clin. Invest.* 82:586-93.

Saha, S., Anilkumar, A. A., and Mayor, S. 2016. "GPI-anchored protein organization and dynamics at the cell surface." *J. Lipid Res.* 57:159-75.

Saha, S., Lee, I. H., Polley, A., Groves, J. T., Rao, M., and Mayor, S. 2015. "Diffusion of GPI-anchored proteins is influenced by the activity of dynamic cortical actin." *Mol. Biol. Cell* 26:4033-45.

Saitakis, M., and Gizeli, E. 2012. "Acoustic sensors as a biophysical tool for probing cell attachment and cell/surface interactions." *Cell. Mol. Life Sci.* 69:357-71.

Saltiel, A. R. 1990. "Second messengers of insulin action." *Diabetes Care* 13:244-56.

Saltiel, A. R., and Cuatrecasas, P. 1988. "In search of a second messenger for insulin." *Am. J. Physiol.* 255:C1-11.

Saltiel, A. R., Fox, J. A., Sherline, P., and Cuatrecasas, P. 1986. "Insulin-stimulated hydrolysis of a novel glycolipid generates modulators of cAMP phosphodiesterase." *Science* 233:967-72.

Saltiel, A. R., Osterman, D. G., Darnell, J. C., Sorbara-Cazan, L. R., Chan, B. L., Low, M. G., and Cuatrecasas, P. 1988. "The function of glycosyl phosphoinositides in hormone action." *Philos. Trans. R. Soc. Lond. B. Biol. Sci.* 320:345-58.

Samuel, V. T., and Shulman, G. I. 2012. "Mechanisms for insulin resistance: common threads and missing links." *Cell* 148:852-71.

Saqr, H. E., Pearl, D. K., and Yates, A. J. 1993. "A review and predictive models of ganglioside uptake by biological membranes." *J. Neurochem.* 61:395-411.

Sargiacomo, M., Sudol, M., Tang, Z. L., and Lisanti, M. P. 1993. "Signal transducing molecules and GPI-linked proteins form a caveolin-enriched insoluble complex in MDCK cells." *J. Cell Biol.* 122:789-807.

Sarnataro, D., Paladino, S., Campana, V., Grassi J, Nitsch, L., and Zurzolo, C. 2002. "PrPC is sorted to the basolateral membrane of epithelial cells independently of its association with rafts." *Traffic* 3:810-21.

Sato, K., Noda, Y., and Yoda, K. 2007. "Pga1 is an essential component of glycosylphosphatidylinositol-mannosyltransferase II of *Saccharomyces cerevisiae*." *Mol. Biol. Cell* 18:3472-85.

Satpute-Krishnan, P., Ajinkya, M., Bhat, S., Itakura, E., Hegde, R. S., and Lippincott-Schwartz, J. 2014. "ER stress-induced clearance of misfolded GPI-anchored proteins via the secretory pathway." *Cell* 158:522-33.

Satta, N., Freyssinet, J. M., and Toti, F. 1997. "The significance of human monocyte thrombomodulin during membrane vesiculation and after stimulation by lipopolysaccharide." *Br. J. Haematol.* 96:534-42.

Sawada, R., Ohashi, K., Anaguchi, H., Okazaki, H., Hattori, M., Minato, N., and Naruto, M. 1990. "Isolation and expression of the full-length cDNA encoding CD59 antigen of human lymphocytes." *DNA Cell Biol.* 9:213-20.

Schneider, P., Ferguson, M. A. J., McConville, M. J., Mehlert, A., Homans, S. W., and Bordier, C. 1990. "Structure of the glycosyl-phosphatidylinositol membrane anchor of the *Leishmania major* Promastigote surface protease." *J. Biol. Chem.* 1990:16955-64.

Schofield, L., and Hackett, F. 1993. "Signal transduction in host cells by a glycosylphosphatidylinositol toxin of malaria parasites." *J. Exp. Med.* 177:145-53.

Schorey, J. S., and Bhatnagar, S. 2008. "Exosome function: from tumor immunology to pathogen biology." *Traffic* 9:871-81.

Schulze, M. B., Weikert, C., Pischon, T., Bergmann, M. M., Al-Hasani, H., Schleicher, E., Fritsche, A., Häring, H. U., Boeing, H., and Joost, H. G. 2009. "Use of multiple metabolic and genetic markers to improve the prediction of type 2 diabetes: the EPIC-Potsdam study." *Diab. Care* 32:2116-9.

Schwendeman, S. P., Cardamone, M., and Klibanov, A. 1996. "Stability of proteins and their delivery from biodegradable polymer microspheres." In *Microparticulate Systems for the Delivery of Proteins and Vaccines*, edited by Cohen, S., and Bernstein, H., 1-49. New York: Marcel Dekker.

Sefton, B. M., Trowbridge, I. S., Cooper, J. A., and Scolnick, E. M. 1982. "The transforming proteins of Rous sarcoma virus, Harvey sarcoma virus and Abelson virus contain tightly bound lipid." *Cell* 31:465-74.

Segawa, K., Suzuki, J., and Nagata, S. 2014. "Flippases and scramblases in the plasma membrane." *Cell Cycle* 13:2990-1.

Sengupta, P., Jovanovic-Talisman, T., Skoko, D., Renz, M., Veatch, S. L., and Lippincott-Schwartz, J. 2011. "Probing protein heterogeneity in the plasma membrane using PALM and pair correlation analysis." *Nat. Methods* 8:969-75.

Senior, A. E. 1988. "ATP synthesis by oxidative phosphorylation." *Physiological Reviews* 68:177-231.

Seong, J., Wang, Y., Kinoshita, T., and Maeda, Y. 2013. "Implications of lipid moiety in oligomerization and immunoreactivities of GPI-anchored proteins." *J. Lipid Res.* 54:1077-91.

Setzer, F., Oberle, V., Bläss, M., Möller, E., Russwurm, S., Deigner, H. P., Claus, R. A., Bauer, M., Reinhart, K., and Lösche, W. 2006. "Platelet-derived microvesicles induce differential gene expression in monocytic cells: a DNA microarray study." *Platelets* 17: 571-6.

Sevcsik, E., Brameshuber, M., Fölser, M., Weghuber, J., Honigmann, A., and Schütz, G. J. 2015. "GPI-anchored proteins do not reside in ordered domains in the live cell plasma membrane." *Nat. Commun.* 6:6969.

Sezgin, E., Davis, S. J., and Eggeling, C. 2015. "Membrane Nanoclusters - Tails of the unexpected." *Cell* 161:433-4.

Sezgin, E., Kaiser, H. J., Baumgart,T., Schwille, P., Simons, K, and Levental, I. 2012. "Elucidating membrane structure and protein behavior using giant plasma membrane vesicles." *Nat. Protoc.* 7:1042-51.

Shao, H., Chung, J., Lee, K., Balaj, L., Min, C., Carter, B. S., Hochberg, F. H., Breakefield, X. O., Lee, H., and Weissleder, R. 2015. "Chip-based analysis of exosomal mRNA mediating drug resistance in glioblastoma." *Nat. Commun.* 6:6999.

Sharma, P., Varma, R., Sarasij, R. C., Gousset, K., Krishnamoorthy, G., Rao, M., and Mayor, S. 2004. "Nanoscale organization of multiple GPI-anchored proteins in living cell membranes." *Cell* 116:577-89.

Sharom, F. J., and Radeva, G. 2004. "GPI-anchored protein cleavage in the regulation of transmembrane signals." *Subcell. Biochem.* 37:285-315.

Sharom, F. J., Lorimer, I., and Lamb, M. P. 1985. "Reconstitution of lymphocyte 5'-nucleotidase in lipid bilayers: behaviour and interaction with concanavalin A." *Can. J. Biochem. Cell Biol.* 63:1049-57.

Shaskin, P. N., Shashkina, E. F., Fernqvist-Forbes E., Zhou, Y. P., Grill, V., and Katz, A. 1997. "Insulin mediators in man: effects of glucose ingestion and insulin resistance." *Diabetologia* 40:557-63.

Shen, M. M. 2007. "Nodal signaling: developmental roles and regulation." *Development* 134:1023-34.

Shin, A. S., and Camilleri, M. 2013. "Diagnostic assessment of diabetic gastroparesis." *Diabetes* 62:2667-73.

Shin, D. J., Chuoy, H. E., and Hong, Y. 2005. "Use of *Clostridium septicum* alpha toxins for isolation of various glycosylphosphatidylinositol-deficient cells." *J. Microbiol.* 43:266-71.

Shin, D. J., Lee, J. J., Choy, H. E., and Hong, Y. 2004. "Generation and characterization of *Clostridium septicum* alpha toxin mutants and their use in diagnosing paroxysmal nocturnal hemoglobinuria." *Biochem. Biophys. Res. Commun.* 324:753-60.

Shiri-Sverdlov, R., Wouters, K., van Gorp, P. J., Gijbels, M. J., Noel, B., Buffat, L., Staels, B., Maeda, N., van Bilsen, M., and Hofker, M. H. 2006. "Early diet-induced non-alcoholic steatohepatitis in APOE2 knock-in mice and its prevention by fibrates." *J. Hepatol.* 44:732-41.

Shukla, S. D., Coleman, R., Finean, J. B., and Mitchell, R. H. 1980. "Selective release of plasma-membrane enzymes from rat hepatocytes by a phosphatidylinositol-specific phospholipase C." *Biochem. J.* 187:277-80.

Sikorska, N., Lemus, L., Aguilera-Romero, A., Manzano-Lopez, J., Riezman, H., Muñiz, M., and Goder, V. 2016. "Limited ER quality control for GPI-anchored proteins." *J. Cell Biol.* 213:693-704.

Siljander, P. R. 2011. "Platelet-derived microparticles – an updated perspective." *Thromb. Res.* 127:S30-3.

Simak, J., and Gelderman, M. P. 2006. "Cell membrane microparticles in blood and blood products: Potentially pathogenic agents and diagnostic markers." *Transfusion Med. Rev.* 20:1-26.

Simon, M. C., and Kusch, J. 2013. "Communicative functions of GPI-anchored surface proteins in unicellular eukaryotes." *Crit. Rev. Microbiol.* 39:70-8.

Simons, K., and Gerl, M. J. 2010. "Revitalizing membrane rafts: new tools and insights." *Nat. Rev. Mol. Cell Biol.* 11:688-99.

Simons, K., and Ikonen, E. 1997. "Functional rafts in cell membranes." *Nature* 387:569-72.

Simons, K., and Toomre, D. 2000. "Lipid rafts and signal transduction." *Nat. Rev. Mol. Cell Biol.* 1:31-9.

Simons, M., and Raposo, G. 2009. "Exosomes - vesicular carriers for intercellular communication." *Curr. Opin. Cell Biol.* 21:575-81.

Simpson, R. J., Lim, J. W., Moritz, R. L., and Mathivanan, S. 2009. "Exosomes: proteomic insights and diagnostic potential." *Expert Rev. Proteomics* 6:267-83.

Sims, P. J., Rollins, S. A., and Wiedmer, T. 1989. "Regulatory control of complement on blood platelets. Modulation of platelet procoagulant responses by a membrane inhibitor of the C5b-9 complex." *J. Biol. Chem.* 264:19228-35.

Singer, S. J., and Nicolson, G. L. 1972. "The fluid mosaic model of the structure of cell membranes." *Science* 175:720-31.

Singh, R., and Lillard, Jr. J. W. 2009. "Nanoparticle-based targeted drug delivery." *Exp. Mol. Pathol.* 86:215-23.

Skog, J., Würdinger, T., van Rijn, S., Meijer, D. H, Gainche, L., Sena-Esteves, M., Curry, W. T. Jr., Carter B. S., Krichevsky, A. M., and Breakefield, X. O. 2008. "Glioblastoma microvesicles transport RNA and proteins that promote tumour growth and provide diagnostic biomarkers." *Nat. Cell Biol.* 10:1470-6.

Slein, M. W., and Logan, G. F. 1960. "Mechanism of action of the toxin of *Bacillus anthracis*. I. Effect *in vivo* on some blood serum components." *J. Bacteriol.* 80:77-85.

Slein, M. W., and Logan, G. F. 1962. "Mechanism of action of the toxin of *Bacillus anthracis* II.: Alkaline phosphatasemia produced by culture filtrates of various bacilli." *J. Bacteriol.* 83:359-69.

Slein, M. W., and Logan, G. F. 1963. "Partial purification and properties of two phospholipases of *Bacillus cereus*." *J. Bacteriol.* 85:369-81.

Slein, M. W., and Logan, G. F. 1965. "Characterization of the phospholipases of *Bacillus cereus* and their effects on erythrocytes, bone and kidney cells." *J. Bacteriol.* 90:69-81.

Sloand, E. M., Maciejewski, J. P., Dunn, D., Moss, J., Brewer, B., Kirby, M., and Young, N. S. 1998. "Correction of the PNH defect by GPI-anchored protein transfer." *Blood* 92:4439-45.

Smith, J. A. 1996. "Beyond the divide between cognition and discourse: Using interpretative phenomenological analysis in health psychology." *Psychol. Health* 11:261-71.

Smith, J. A. 2007. "Hermeneutics, human sciences and health: Linking theory and practice." *Internat. J. Qualit. Stud. Health Well-Being* 2:3-11.

Smith, J. A. 2011. "Evaluating the contribution of interpretative phenomenological analysis." *Health Psychol. Rev.* 5:9-27.

Soussan, E., Cassel, S., Blanzat, M., and Rico-Lattes, I. 2009. "Drug delivery by soft matter: Matrix and vesicular carriers." *Angew. Chem. Int. Ed.* 48:274-88.

Spannhoff, A., Hauser, A. T., Heinke, R., Sippl, W., and Jung, M. 2009. "The emerging therapeutic potential of histone methyltransferase and demethylase inhibitors." *Chem Med Chem.* 4:1568-82.

Spear, P., Wu, M. R., Sentman, M. L., and Sentman, C. L. 2013. "NKG2D ligands as therapeutic targets." *Cancer Immun.* 13:8.

Speare, J. O., Offerdahl, D. K., Hasenkrug, A., Carmody, A. B., and Baron, G. S. 2010. "GPI anchoring facilitates propagation and spread of misfolded Sup35 aggregates in mammalian cells." *EMBO J.* 29:782-94.

Spillane, K. M., Ortega-Arroyo, J., de Wit, G., Eggeling, C., Ewers, H., Wallace, M. I., and Kukura, P. 2014. "High-speed single-particle tracking of GM1 in model membranes reveals anomalous diffusion due to interleaflet coupling and molecular pinning." *Nano. Lett.* 14:5390-7.

Stadelmann, B., Bütikofer, P., König, A., and Brodbeck, U. 1997. "The C-terminus of glycosylphosphatidylinositol-specific phospholipase D is essential for biological activity." *Biochim. Biophys. Acta* 1355:107-13.

Staiger, H., Machicao, F., Fritsche, A., and Häring, H. U. 2009. "Pathomechanisms of type 2 diabetes genes." *Endocr. Rev.* 30:557-85.

Staudt, E., Ramasamy, P., Plattner, H., and Simon, M. 2016. ”Differential subcellular distribution of four phospholipase C isoforms and secretion of GPI-PLC activity.” *Biochim. Biophys. Acta* 1858:3157-68.

Staufenbiel, M., and Lazarides, E. 1986. ”Ankyrin is fatty acid acylated in erythrocytes.” *Proc. Natl. Acad. Sci. USA* 83:318-22.

Stefan, N., Fritsche, A., Schick, F., and Häring, H. U. 2016. “Phenotypes of prediabetes and stratification of cardiometabolic risk.” *Lancet Diabetes Endocrinol.* 4:789-98.

Stefanova, I., Hilgert, I., Kristofova, H., Brown, R., Low, M. G., and Horejsí, V. 1989. “Characterization of a broadly expressed human leucocyte surface antigen MEM-43 anchored in membrane through phosphatidylinositol.” *Mol. Immunol.* 26:153-61.

Stefanova, I., Horejsi, V., Ansotegui, I. J., Knapp, W., and Stockinger, H. 1991. “GPI-anchored cell-surface molecules complexed to protein tyrosine kinases.” *Science* 254:1016-9.

Stein, W. D., and Danielli, J. F. 1956. “Structure and function in red cell permeability.” *Disc. Faraday Soc.* 21:238-51.

Stephenson, F. A. 1988. “Understanding the GABAA receptor: a chemically gated ion channel.” *Biochem. J.* 249:21-32.

Stieger, A., Cardoso de Almeida, M. L., Blatter, M. C., Brodbeck, U., and Bordier, C. 1986. “The membrane-anchoring systems of vertebrate acetylcholinesterase and variant surface glycoproteins of African trypanosomes share a common antigenic determinant.” *FEBS Lett.* 199:182-6.

Stieger, S., Diem, S., Jakob, A., and Brodbeck, U. 1991. “Enzymatic properties of phosphatidylinositol-glycan-specific phospholipase C from rat liver and phosphatidylinositol-glycan-specific phospholipase D from rat serum.” *Eur. J. Biochem.* 197:67-73.

Stinson, R. A., and Hamilton, B. A. 1994. “Human liver plasma membranes contain an enzyme activity that removes membrane anchor from alkaline phosphatase and converts it to a plasma-like form.” *Clin. Biochem.* 27:49-55.

Stöger, R. 2008. “Epigenetics and obesity.” *Pharmacogenomics* 9:1851-60.

Stoorvogel, W., Kleijmeer, M. J., Geuze, H. J., and Raposo, G. 2002. “The biogenesis and functions of exosomes.” *Traffic* 3:321-30.

Strating, J. R., and Martens, G. J. 2009. “The p24 family and selective transport processes at the ER-Golgi interface.” *Biol. Cell* 101:495-509.

Subra, C., Grand, D., Laulagnier, K., Stella, A., Lambeau, G., Paillasse, M., De Medina, P., Monsarrat, B., Perret, B., Silvente-Poirot, S., Poirot, M., and Record, M. 2010. “Exosomes account for vesicle-mediated transcellular transport of activatable phospholipases and prostaglandins.” *J. Lipid Res.* 51:2105-20.

Sugita, Y., Ito, K., Shiozuka, K., Suzuki, H., Gushima, H., Tomita, M., and Masuho, Y. 1994. “Recombinant soluble CD59 inhibits reactive haemolysis with complement.” *Immunology* 82:34-41.

Sugita, Y., Nakano, Y., and Tomita, M. 1988. "Isolation from human erythrocytes of a new membrane protein which inhibits the formation of complement transmembrane channels." *J. Biochem*. 104:633-7.

Sullivan, R., Legare, C., Villeneuve, M., Foliguet, B., and Bissonnette, F. 2006. "Levels of P34H, a sperm protein of epididymal origin, as a predictor of conventional *in vitro* fertilization outcome." *Fertil. Steril*. 85:1557-9.

Sunamoto, J., Mori, Y., and Sato, T. 1992. "Direct transfer of membrane proteins from B16 melanoma cell to artificial cell liposome." *Proc. Jpn. Acad*. 68B:69-74.

Sussman, N. L., Shubin, R. A., Kyuwa, S., Perlmutter, D. H., DeSchryver-Kecskemeti, K., and Alpers, D. H. 1989. "Intestinal alkaline phosphatase is secreted bidirectionally from villous enterocytes." *Am. J. Physiol*. 257:G14-23.

Suzuki, K. G. 2015. "New insights into the organization of plasma membrane and its role in signal transduction." *Int. Rev. Cell. Mol. Biol*. 317:67-96.

Suzuki, K. G. N., Fujiwara, T. K., Sanematsu, F., Iino, R., Edidin, M., and Kusumi, A. 2007. "GPI-anchored receptor clusters transiently recruit Lyn and G alpha for temporary cluster immobilization and Lyn activation: single-molecule tracking study 1." *J. Cell Biol*. 177:717-30.

Suzuki, K. G. N., Kasai, R. S., Hirosawa, K. M., Nemoto, Y. L., Ishibashi, M., Miwa, Y., Fujiwara, T. K., and Kusumi, A. 2012. "Transient GPI-anchored protein homodimers are units for raft organisation and function." *Nat. Chem. Biol*. 8:774-83.

Suzuki, K., and Okumura, Y. 2000a. "GPI-linked proteins do not transfer spontaneously from erythocytes to liposomes. New aspects of reorganization of the cell membrane." *Biochemistry* 39:9477-85.

Suzuki, K., and Okumura, Y. 2000b. "Mechanism of selective release of membrane proteins from human erythrocytes in the presence of liposomes." *Arch. Biochem. Biophys*. 379:344-52.

Suzuki, K., Okumura, Y., and Sato, T. 1995. "Membrane protein transfer from human erythrocyte ghosts to liposomes containing an artificial boundary lipid." *Proc. Jpn. Acad*. 71B:93-7.

Suzuki, K., Okumura, Y., and Sunamoto, J. 1999. "Induction of acetylcholinesterase release from erythrocytes in the presence of liposomes." *J. Biochem*. 125:876-82.

Suzuki, S., Suzuki, C., Hinokio, Y., Ishigaki, Y., Katagiri, H., Kanzaki, M., Azev, V. N., Chakraborty, N., and d'Alarcao, M. 2014. "Insulin-mimicking bioactivities of acylated inositol glycans in several mouse models of diabetes with or without obesity." *PLoS ONE* 9:e100466.

Tachado, S. D., and Schofield, L. 1994. "Glycosylphosphatidylinositol toxin of Trypanosoma brucei regulates IL-1 alpha and TNF-alpha expression in macrophages by protein tyrosine kinase mediated signal transduction." *Biochem. Biophys. Res. Commun*. 205:984-91.

Taguchi, R., and Ikezawa, H. 1978. "Phosphatidylinositol-specific phospholipase C from *Clostridium novyi* type A." *Arch. Biochem. Biophys.* 186:196-201.

Taguchi, R., Asahi, Y., and Ikezawa, H. 1980. "Purification and properties of phosphatidylinositol-specific phospholipase C of *Bacillus thuringiensis*." *Biochim. Biophys. Acta* 619:48-57.

Taguchi, R., Hamakawa, N., Harada Nishida, M., Fukui, T., Nojima, K., and Ikezawa, H. 1994. "Microheterogeneity in glycosylphosphatidylinositol anchor structures of bovine liver 5'-nucleotidase." *Biochemistry* 33:1017-22.

Takami, N., Ogata, S., Oda, K., Misumi, Y., and Ikehara, Y. 1988. "Biosynthesis of placental alkaline phosphatase and its post-translational modification by glycophospholipid for membrane-anchoring." *J. Biol. Chem.* 263:3016-21.

Takeda, J., and Kinoshita, T. 1995. "GPI-anchor biosynthesis." *Trends Biochem. Sci.* 20:367-71.

Takeda, J., Miyata, T., Kawagoe, K., Iida, Y., Endo, Y., Fujita, T., Takahashi, M., Kitani, T., and Kinoshita, T. 1993. "Deficiency of the GPI anchor caused by a somatic mutation of the *PIG-A* gene in paroxysmal nocturnal hemoglobinuria." *Cell* 73:703-11.

Takida, S., Maeda, Y., and Kinoshita, T. 2008. "Mammalian GPI-anchored proteins require p24 proteins for their efficient transport from the ER to the plasma membrane." *Biochem. J.* 409:555-62.

Takka, S., and Acarturk, F. 1999. "Calcium alginate microparticles for oral administration. I: Effect of sodium alginate type on drug release and drug entrapment efficiency." *J. Microencapsul. Micro Nano Carriers* 16:275-90.

Tan, K. T., Tayebjee, M. H., Lim, H. S., and Lip, G. Y. 2005. "Clinically apparent atherosclerotic disease in diabetes is associated with an increase in platelet microparticle levels." *Diabet. Med.* 22:1657-62.

Tanaka, S., Maeda, Y., Tashima, Y., and Kinoshita, T. 2004. "Inositol deacylation of glycosylphosphatidylinositol-anchored proteins is mediated by mammalian PGAP1 and yeast Bst1p." *J. Biol. Chem.* 279:14256-63.

Tanaka, Y., Okada, Y., and Hirokawa, N. 2005. "FGF-induced vesicular release of Sonic hedgehog and retinoic acid in leftward nodal flow is critical for left-right determination." *Nature* 435:172-7.

Taron, B. W., Colussi, P. A., Wiedman, J. M., Orlean, P., and Taron, C. H. 2004. "Human Smp3p adds a fourth mannose to yeast and human glycosyl-phosphatidylinositol precursors *in vivo*." *J. Biol. Chem.* 279:36083-92.

Tarutani, M., Itami, S., Okabe, M., Orlean, P., and Taron, C. H. 1997. "Tissue specific knock-out of the mouse *Pig-a* gene reveals important roles for GPI-anchored proteins in skin development." *Proc. Natl. Acad. Sci. USA* 94:7400-5.

Tashima, Y., Taguchi, R., Murata, C., Ashida, H., Kinoshita, T., and Maeda Y. 2006. "PGAP2 is essential for correct processing and stable expression of GPI-anchored proteins." *Mol. Biol. Cell* 17:1410-20.

Taylor, D. D., and Gercel-Taylor, C. 2008. "MicroRNA signatures of tumor-derived exosomes as diagnostic biomarkers of ovarian cancer." *Gynecol. Oncol.* 110:13-21.

Tellier, E., Canault, M., Rebsomen, L., Bonardo, B., Juhan-Vague, I., Nalbone, G., and Peiretti, F. 2006. "The shedding activity of ADAM17 is sequestered in lipid rafts." *Exp. Cell Res.* 312:3969-80.

Thery, C. 2011. "Exosomes: secreted vesicles and intercellular communications." *F1000 Biol. Rep.* 3:15.

Thery, C., Boussac, M., Veron, P., Ricciardi-Castagnoli, P., Raposo, G., Garin, J., and Amigorena, S. 2001. "Proteomic analysis of dendritic cell-derived exosomes: a secreted subcellular compartment distinct from apoptotic vesicles." *J. Immunol.* 166:7309-18.

Thery, C., Regnault, A., Garin, J., Wolfers, J., Zitvogel, L., Ricciardi-Castagnoli, P., Raposo, G., and Amigorena S. 1999. "Molecular characterization of dendritic cell-derived exosomes: selective accumulation of the heat shock protein hsc73." *J. Cell Biol.* 147:599-610.

Thery, C., Zitvogel, L., and Amigorena, S. 2002. "Exosomes: composition, biogenesis and function." *Nat. Rev. Immunol.* 2:569-79.

Thomas, J. R., Dwek, R. A., and Rademacher, T. W. 1990. "Structure, biosynthesis, and function of glycosylphosphatidylinositols." *Biochemistry* 29:5413-22.

Tkach, M., and Thery, C. 2016. "Communication by extracellular vesicles: Where we are and where we need to go." *Cell* 164:1226-32.

Tozeren, A., Sung, K. L., Sung, L. A., Dustin, M. L., Chan, P. Y., Springer, T. A., and Chien, S. 1992. "Micromanipulation of adhesion of a Jurkat cell to a planar bilayer membrane containing lymphocyte function-associated antigen 3 molecules." *J. Cell Biol.* 116:997-1006.

Traister, A., Shi, W., and Filmus, J. 2007. "Mammalian Notum induces the release of glypicans and other GPI-anchored proteins from the cell surface." *Biochem. J.* 410:503-11.

Tramontano, A. F., Lyubarova, R., Tsiakos, J., Palaia, T., Deleon, J. R., and Ragolia, L. 2010. "Circulating endothelial microparticles in diabetes mellitus." *Mediators Inflamm.* 2010:250476.

Trams, E. G., Lauter, C. J., Salem, N., and Heine U. 1981. "Exfoliation of membrane ecto-enzymes in the form of microvesicles." *Biochim. Biophys. Acta* 645:63-70.

Treitz, G., Gronewold, T. M. A., Quandt, E., and Zabe-Kühn, M. 2008. "Combination of a SAW-biosensor with MALDI mass spectrometric analysis." *Biosens. Bioelectron.* 23:1496-502.

Trowell, S. C., Hines, E. R., Herit, A. J., and Rickards, R. W. 1994. "Characterization of a juvenile hormone binding lipophorin from the blowfly *Lucilla cuprina*." *Comp. Biochem. Physiol. B. Biochem. Mol. Biol.* 109:339-57.

Tschöp, M. H., Speakman, J. R., Arch, J. R., Auwerx, J., Brüning, J. C., Chan, L., Eckel, R. H., Farese, R. V. Jr., Galgani, J. E., Hambly, C., Herman, M. A., Horvath, T. L., Kahn, B. B., Kozma, S. C., Maratos-Flier, E., Müller, T. D., Münzberg, H., Pfluger, P. T., Plum, L., Reitman, M. L., Rahmouni, K., Shulman, G. I., Thomas, G., Kahn, C. R., and Ravussin, E. 2011. A guide to analysis of mouse energy metabolism. *Nat. Methods* 9:57-63.

Turner, A. J., and Hooper, N. M. 2002. "The angiotensin-converting enzyme gene family: genomics and pharmacology." *Trends Pharmacol. Sci.* 23:177-83.

Tykocinski, M. L., Kaplan, D. R., and Medof, M. E. 1996. "Antigen-presenting cell engineering. The molecular toolbox." *Amer. J. Pathol.* 148:1-16.

Ueda, Y., Yamaguchi, R., Ikawa, M., Okabe, M., Morii, E., Maeda, Y., and Kinoshita, T. 2007. "PGAP1 knock-out mice show otocephaly and male infertility." *J. Biol. Chem.* 282:30373-80.

Uhrich, K. E., Cannizzaro, S. M., Langer, R. S., and Shakesheff, K. M. 1999. "Polymeric systems for controlled drug release." *Chem. Rev.* 99:3181-98.

Umemura, M., Fujita, M., Yoko, O. T., Fukamizu, A., and Jigami, Y. 2007. "*Saccharomyces cerevisiae* CWH43 is involved in the remodeling of the lipid moiety of GPI anchors to ceramides." *Mol. Biol. Cell* 18:4304-16.

Unwin, P. N. T., and Ennis, P. D. 1984. "Two configurations of a channel-forming membrane protein." *Nature* 307:609-13.

Unwin, P. N. T., and Zampighi, G. 1980. "Structure of the junction between communicating cells." *Nature* 283:545-9.

Ussar, S., Bezy, O., Blüher, M., and Kahn, C. R. 2012. "Glypican-4 enhances insulin signaling *via* interaction with the insulin receptor and serves as a novel adipokine." *Diabetes* 61:2289-98.

Uyy, E., Ivan, L., Boteanu, R. M., Suica, V. I., and Antohe, F. 2013. "High-fat diet alters protein composition of detergent-resistant membrane microdomains." *Cell Tissue Res.* 354:771-81.

Väkevä, A., Jauhiainen, M., Ehnholm, C., Lehto, T., and Meri, S. 1994. "High-density lipoproteins can act as carriers of glycophosphoinositol lipid-anchored CD59 in human plasma." *Immunology* 82:28-33.

Väkevä, A., Lehto, T., Takala, A., and Meri, S. 2000. "Detection of a soluble form of the complement membrane attack complex inhibitor CD59 in plasma after acute myocardial infarction." *Scand. J. Immunol.* 52:411-4.

Valadi, H., Ekström, K., Bossios, A., Sjöstrand, M., Lee, J. J., and Lötvall, J. O. 2007. "Exosome-mediated transfer of mRNAs and microRNAs is a novel mechanism of genetic exchange between cells." *Nat. Cell Biol.* 9:654-9.

Van den Berg, C. W., Cinek, T., Hallett, M. B., Horejsi, V., and Morgan, B. P. 1995. "Exogenous glycosyl phosphatidylinositol-anchored CD59 associates with kinases in membrane clusters on U937 cells and becomes Ca(2+)-signaling competent." *J. Cell Biol.* 131:669-77.

van der Vusse, G. J. 2009. "Albumin as fatty acid transporter." *Drug Metab. Pharmacokinet.* 24:300-7.

van Niel, G., Raposo, G., Candalh, C., Boussac, M., Hershberg, R., Cerf-Bensussan, N., and Heyman, M. 2001. "Intestinal epithelial cells secrete exosome-like vesicles." *Gastroenterology* 121:337-49.

van Veen, M., Matas-Rico, E., van de Wetering, K., Leyton-Puig, D., Kedziora, K. M., De Lorenzi, V., Stijf-Bultsma, Y., van den Broek, B., Jalink, K., Sidenius, N., Perrakis, A., and Moolenaar, W. H. 2017. "Negative regulation of urokinase receptor activity by a GPI-specific phospholipase C in breast cancer cells." *Elife* 6pii:e23649.

VanWijk, M. J., Nieuwland, R., Boer, K., van der Post, J. A., VanBavel, E., and Sturk, A. 2002. "Microparticle subpopulations are increased in preeclampsia: possible involvement in vascular dysfunction?" *Am. J. Obstet. Gynecol.* 187:450-6.

VanWjik, M. J., VanBavel, E., Sturk, A., and Nieuwland, R. 2003. "Microparticles in cardiovascular diseases." *Cardiovasc. Res.* 59:277-87.

Varela-Nieto, I., León, Y., and Caro, H. N. 1996. "Cell signalling by inositol phosphoglycans from different species." *Comp. Biochem. Physiol. B Biochem. Mol. Biol.* 115:223-41.

Varma, R., and Mayor, S. 1998. "GPI-anchored proteins are organized in submicron domains at the cell surface." *Nature* 394:798-801.

Vega, V. L., Rodriguez-Silva, M., Frey, T., Gehrmann, M., Diaz, J. C., Steinem, C., Multhoff, G., Arispe, N., and De Maio, A. 2008. "Hsp70 translocates into the plasma membrane after stress and is released into the extracellular environment in a membrane-associated form that activates macrophages." *J. Immunol.* 180:4299-307.

Vidal, M., Mangeat, P., and Hoekstra, D. 1997. "Aggregation reroutes molecules from a recycling to a vesicle-mediated secretion pathway during reticulocyte maturation." *J. Cell Sci.* 110:1867-77.

Vila, M. C., Cozza, E. N., Lima, C., Ramirez M. I., and De Lederkremer, R. M. 1995. "An inositol phosphoglycan from *Trypanosoma cruzi* inhibits ACTH action in calf adrenocortical cells." *Cell. Signal.* 7:331-9.

Villar, A. V., Alonso, A., Pañeda, C., Varela-Nieto, I., Brodbeck, U., and Goñi, F. M. 1999. "Towards the *in vitro* reconstitution of caveolae. Asymmetric incorporation of glycosylphosphatidylinositol (GPI) and gangliosides into liposomal membranes." *FEBS Lett.* 457:71-4.

Vlassov, A. V., Magdaleno, S., Setterquist, R., and Conrad, R. 2012. "Exosomes: current knowledge of their composition, biological functions, and diagnostic and therapeutic potentials." *Biochim. Biophys. Acta* 1820:940-8.

Voiculescu, I., and Nordin, A. N. 2012. "Acoustic wave based MEMS devices for biosensing applications." *Biosens. Bioelectron.* 33:1-9.

von Toerne, C., Huth, C., de Las Heras Gala, T., Kronenberg, F., Herder, C., Koenig, W., Meisinger, C., Rathmann, W., Waldenberger, M., Roden, M., Peters, A., Thorand, B., and Hauck, S. M. 2016. "MASP1, THBS1, GPLD1 and ApoA-IV are novel biomarkers associated with prediabetes: the KORA F4 study." *Diabetologia* 59:1882-92.

Voss, C. V., Davies, B. S., Tat, S., Gin, P., Fong, L. G., Pelletier, C., Mottler, C. D., Bensadoun, A., Beigneux, A. P., and Young, S. G. 2011. "Mutations in lipoprotein lipase that block binding to the endothelial cell transporter GPIHBP1." *Proc. Natl. Acad. Sci. USA* 108:7980-4.

Waldhauer, I., Goehlsdorf, D., Gieseke, F., Weinschenk, T., Wittenbrink, M., Ludwig, A., Stevanovic, S., Rammensee, H. G., and Steinle, A. 2008. "Tumor-associated MICA is shed by ADAM proteases." *Cancer Res.* 68:6368-76.

Walter, E. I., Ratnoff, W. D., Long, K. E., Kazura, J. W., and Medof, M. E. 1992. "Effect of glycoinositolphospholipid anchor lipid groups on functional properties of decay-accelerating factor protein in cells." *J. Biol. Chem.* 267:1245-52.

Wang, G. H., Zhou, X. M., Bai, Y., Yin, X. M., Yang, L. F., and Zhao, D. 2011. "Hsp70 binds to PrPC in the process of PrPC release via exosomes from THP-1 monocytes." *Cell Biol. Int.* 35:553-8.

Wang, H., and Eckel, R. H. 2009. "Lipoprotein lipase: from gene to obesity." *Am. J. Physiol. Endocrinol. Metab.* 297:E271-88.

Wang, X., Jansen, G., Fan, J., Kohler, W. J., Ross, J. F., Schornagel, J., and Ratnam, M. 1996. "Variant GPI structure in relation to membrane-associated functions of a murine folate receptor." *Biochemistry* 35:16305-12.

Wang, X., Strizich, G., Hu, Y., Wang, T., Kaplan, R. C., and Qi, Q. 2016. "Genetic markers of type 2 diabetes: Progress in genome-wide association studies and clinical application for risk prediction." *J. Diabetes* 8:24-35.

Warkentin, T. E., Hayward, C. P., Boshkov, L. K., Santos, A. V., Sheppard, J. A., Bode, A. P., and Kelton, J. G. 1994. "Sera from patients with heparin-induced thrombocytopenia generate platelet-derived microparticles with procoagulant activity: an explanation for the thrombotic complications of heparin-induced thrombocytopenia." *Blood* 84:3691-9.

Warnock, M. L. 1968. "Intestinal phosphatase and fat absorption." *Proc. Soc. Exp. Biol. Med.* 129:768-72.

Wassef, N. M., Matyas, G. R., and Alving, C. R. 1991. "Complement-dependent phagocytosis of liposomes by macrophages: suppressive effects of stealth lipids." *Biochem. Biophys. Res. Commun.* 176:866-74.

Watanabe, K., and Salomon, D. S. 2010. "Intercellular transfer regulation of the paracrine activity of GPI-anchored Cripto-1 as a Nodal co-receptor." *Biochem. Biophys. Res. Commun.* 403:108-13.

Watanabe, K., Hamada, S., Bianco, C., Mancino, M., Nagaoka, T., Gonzales, M., Bailly, V., Strizzi, L., and Salomon, D. S. 2007. "Requirement of glycosylphosphatidylinositol anchor of Cripto-1 for trans activity as a Nodal co-receptor." *J. Biol. Chem.* 282:35772-86.

Webb, P. D., and Todd, J. 1988. "Attachment of human placental-type alkaline phosphatase *via* phosphatidylinositol to syncytiotrophoblast and tumour plasma membranes." *Eur. J. Biochem.* 172:647-52.

Weber, M. C., Groger, R. K., and Tykocinski, M. L. 1994. "A glycophosphatidylinositol-anchored cytokine can function as an artificial cellular adhesin." *Exp. Cell Res.* 210:107-12.

Wei, J. S., Zeng, H. B., Liu, S. Q., Wang, X. G., Tay, E. H., and Yang, Y. Y. 2005. "Temperate- and pH-sensitive core-shell nanoparticles self-assembled from poly/N-isopropylacrylamide-co-acryl acid-co-cholesteryl acrylate for intracellular delivery of anticancer drugs." *Front. Biosci.* 10:3058-67.

Weinstein, M. M., Goulbourne, C. N., Davies, B. S., Tu, Y., Barnes, R. H. 2nd, Watkins, S. M., Davis, R., Reue, K., Tontonoz, P., Beigneux, A. P., Fong, L. G., and Young, S. G. 2012. "Reciprocal metabolic perturbations in the adipose tissue and liver of GPIHBP1-deficient mice." *Arterioscler. Thromb. Vasc. Biol.* 32:230-5.

Weinstein, M. M., Tu, Y., Beigneux, A. P., Davies, B. S., Gin, P., Voss, C., Walzem, R. L., Reue, K., Tontonoz, P., Bensadoun, A., Fong, L. G., and Young, S. G. 2010. "Cholesterol intake modulates plasma triglyceride levels in glycosylphosphatidylinositol HDL-binding protein 1-deficient mice." *Arterioscler. Thromb. Vasc. Biol.* 30:2106-13.

Weinstein, M. M., Yin, L., Beigneux, A. P., Davies, B. S., Gin, P., Estrada, K., Melford, K., Bishop, J. R., Esko, J. D., Dallinga-Thie, G. M., Fong, L. G., Bensadoun, A., and Young, S. G. 2008. "Abnormal patterns of lipoprotein lipase release into the plasma in GPIHBP1-deficient mice." *J. Biol. Chem.* 283:34511-8.

Welsch, U., Schumacher, U., Bucheim, W., Schinko, I., Jenness, P., and Patton, S. 1990. "Histochemical and biochemical observations on milk-fat-globule membranes from several mammalian species." *Acta Histochem.* 40(suppl):59-64.

Wells, E., and Findlay, J. B. C. 1980. "The isolation of human-erythrocyte band-3 polypeptide labelled with a photosensitive hydrophobic probe." *Biochem. J.* 187:719-25.

Wieckowski, E., and Whiteside, T. L. 2006. "Human tumor-derived *vs.* dendritic cell-derived exosomes have distinct biologic roles and molecular profiles." *Immunol. Res.* 36:247-54.

Wilcox, L. A., Ezzel, J. L., Bernshaw, N. J., and Parker, C. J. 1991. "Molecular basis of the enhanced susceptibility of the erythrocytes of paroxysmal nocturnal hemoglobinuria to hemolysis in acidified serum." *Blood* 78:820-9.

Wilhelm, O. G., Wilhelm, S., Escott, G. M., Lutz, V., Magdolen, V., Schmitt, M., Rifkin, D. B., Wilson, E. L., Graeff, H., and Brunner, G. 1999. "Cellular glycosyl-phosphatidylinositol-specific phospholipase D regulates urokinase receptor shedding and cell surface expression." *J. Cell Physiol.* 180:225-35.

Wittgenstein, L 1984b. *Über Gewissheit* [*About certainty*]. Werkausgabe Band 8. Frankfurt am Main: Suhrkamp. S. 465.

Wittgenstein, L. 1970. *Philosophische Bemerkungen* [*Philosophical remarks*]. Werkausgabe Band 2. Frankfurt am Main: Suhrkamp. S. 273.

Wittgenstein, L. 1984a. *Philosophische Untersuchungen* [*Philosophical investigations*]. Werkausgabe Band 1. Frankfurt am Main: Suhrkamp. §109.

Wittgenstein, L. 2013. *Das Blaue Buch* [*The Blue Book*]. Werkausgabe Band 5. Frankfurt am Main: Suhrkamp.

Woitiski, C. B., Neufeld, R. J., Veiga, F., Carvalho, R. A., and Figueiredo, I. V. 2010. "Pharmacological effect of orally delivered insulin facilitated by multilayered stable nanoparticles." *Eur. J. Pharmaceut. Res.* 41:556-63.

Woitiski, C. B., Sarmento, B., Carvalho, R. A., Neufeld, R. J., and Veiga, F. 2011. "Facilitated nanoscale delivery of insulin across intestinal membrane models." *Int. J. Pharm.* 412:123-31.

Wolf, P. 1967. "The nature and significance of platelet products in human plasma." *Br. J. Haematol.* 13:269-88.

Wolfers, J., Lozier, A., Raposo, G., Regnault, A., Théry, C., Masurier, C., Flament, C., Pouzieux, S., Faure, F., Tursz, T., Angevin, E., Amigorena, S., and Zitvogel, L. 2001. "Tumor-derived exosomes are a source of shared tumor rejection antigens for CTL cross-priming." *Nat. Med.* 7:297-303.

Wong, H. L., Bendayan, R., Rauth, A. M., and Wu, X. Y. 2004. "Development of solid lipid nanoparticles containing ionically complexed chemotherapeutic drugs and chemosensitizers." *J. Pharm. Sci.* 93:1993-2008.

Wu, G., Guo, Z., Chatterjee, A., Huang, X., Rubin, E., Wu, F., Mambo, E., Chang, X., Osada, M., Sook, Kim, M., Moon, C., Califano, J. A., Ratovitski, E. A., Gollin, S. M., Sukumar, S., Sidransky, D., and Trink, B. 2006. "Overexpression of glycosylphosphatidylinositol (GPI) transamidase subunits phosphatidylinositol glycan class T and/or GPI anchor attachment 1 induces tumorigenesis and contributes to invasion in human breast cancer." *Cancer Res.* 66:9829-36.

Wu, Q., and Guo, Z. 2010. "Glycosylphosphatidylinositols are potential targets for the development of novel inhibitors for aerolysin-type of pore-forming bacterial toxins." *Med. Res. Rev.* 30:258-69.

Wurtman, R. 2015. "Biomarkers in the diagnosis and management of Alzheimer's disease." *Metabolism* 64(Suppl 1):S47-50.

Xie, H., Lim, B., and Lodish, H. F. 2009. "MicroRNAs induced during adipogenesis that accelerate fat cell development are downregulated in obesity." *Diabetes* 58:1050-7.

Xie, M., and Low, M. G. 1995. "Streptolysin-O induces release of glycosylphosphatidylinositol-anchored alkaline phosphatase from ROS cells by vesiculation independently of phospholipase action." *Biochem. J.* 305:529-37.

Xu, R., Greening, D. W., Zhu, H. J., Takahashi, N., and Simpson, R. J. 2016. "Extracellular vesicle isolation and characterization: toward clinical application." *J. Clin. Invest.* 126:1152-62.

Yamamoto, A., Taniguchi, T., Rikyuu, K., Tsuji, T., Fujita, T., Murakami, M., and Muranishi, S. 1994. "Effects of various protease inhibitors on the intestinal absorption and degradation of insulin in rats." *Pharm. Res.* 11:1496-500.

Yang, M., Chen, J., Su, F., Yu, B., Su, F., Lin, L., Liu, Y., Huang, J. D., and Song, E. 2011. "Microvesicles secreted by macrophages shuttle invasion-potentiating microRNAs into breast cancer cells." *Mol. Cancer* 10:117.

Yang, Q., Gou, Y., Li, L., and Hui, S. W. 1997. "Effects of lipids headgroup and packing stress on poly(ethylene glycol)-induced phospholipid vesicle aggregation and fusion." *Biophys. J.* 73:277-82.

Yang, T., Choi, M. K., Cui, F. D., Kim, J. S., Chung, S. J., Shim, C. K., and Kim, D. D. 2007. "Preparation and evaluation of paclitaxel-loaded PEGylated immunoliposome." *J. Control. Rel.* 120:169-77.

Yang, W., Peters, J. I., and Williams III, R. O. 2008. "Inhaled nanoparticles – A current review." *Int. J. Pharmaceut.* 356:239-47.

Yang, Y. Y., Wang, Y., Powell, R., and Chan, P. 2006. "Polymeric core-shell nanoparticles for therapeutics." *Clin. Exper. Pharmacol. Physiol.* 33:557-62.

Yáñez-Mó, M., Siljander, P. R., Andreu, Z., Zavec, A. B., Borràs, F. E., Buzas, E. I., Buzas, K., Casal, E., Cappello, F., Carvalho, J., Colás, E., Cordeiro-da Silva, A., Fais S., Falcon-Perez, J. M., Ghobrial, I. M., Giebel, B., Gimona, M., Graner, M., Gursel, I., Gursel, M., Heegaard, N. H., Hendrix, A., Kierulf, P., Kokubun, K., Kosanovic, M., Kralj-Iglic, V., Krämer-Albers, E. M., Laitinen, S., Lässer, C., Lener, T., Ligeti, E., Linē, A., Lipps, G., Llorente, A., Lötvall, J., Manček-Keber, M., Marcilla, A., Mittelbrunn, M., Nazarenko, I., Nolte-'t Hoen, E. N., Nyman, T. A., O'Driscoll, L., Olivan, M., Oliveira, C., Pállinger, É., Del Portillo, H. A., Reventós, J., Rigau, M., Rohde, E., Sammar, M., Sánchez-Madrid, F., Santarém, N., Schallmoser, K., Ostenfeld, M. S., Stoorvogel, W., Stukelj, R., Van der Grein, S. G., Vasconcelos, M. H., Wauben, M. H., and De Wvever, O. 2015. "Biological properties of extracellular vesicles and their physiological functions." *J. Extracell. Vesicles* 4:27066.

Yasuda, T., Ishida, T., and Rader, D. J. 2010. "Update on the role of endothelial lipase in high-density lipoprotein metabolism, reverse cholesterol transport, and atherosclerosis." *Circ. J.* 74:2263-70.

Yeung, C. H., Schröter, S., Wagenfeld, A., Kirchhoff, C., Kliesch, S., Poser, D., Weinbauer, G. F., Nieschlag, E., and Cooper, T. G. 1997. "Interaction of the human epididymal protein CD52 (HE5) with epididymal spermatozoa from men and cynomolgus monkeys." *Mol. Reprod. Dev.* 48:267-75.

Yokoyama, M. 2005. "Drug targeting with nano-sized carrier systems." *J. Artif. Organs* 8:77-84.

Young, G. P., Friedman, S., Yedlin, S. T., and Allers, D. H. 1981. "Effect of fat feeding on intestinal alkaline phosphatase activity in tissue and serum." *Am. J. Physiol.* 241:G461-8.

Young, G. P., Yedlin, S. T., and Alpers, D. H. 1981. "Distribution of soluble and membranous forms of alkaline phosphatase in the small intestine of the rat." *Biochim. Biophys. Acta* 676:257-65.

Young, S. G., Davies, B. S., Voss, C. V., Gin, P., Weinstein, M. M., Tontonoz, P., Reue, K., Bensadoun, A., Fong, L. G., and Beigneux, A. P. 2011. "GPIHBP1, an endothelial cell transporter for lipoprotein lipase." *J. Lipid Res.* 52:1869-84.

Younus, S., and Rodgers, G. 2011. "Biomarkers associated with cardiometabolic risk in obesity." *Am. Heart Hosp. J.* 9:E28-32.

Yu, Y. H., Chang, Y. C., Su, T. H., Nong, J. Y., Li, C. C., and Chuang, L. M. 2013. "Prostaglandin reductase-3 negatively modulates adipogenesis through regulation of PPARγ activity." *J. Lipid Res.* 54:2391-9.

Yuan, A., Farber, E. L., and Rapoport, A. L. 2009. "Transfer of microRNAs by embryonic stem cell microvesicles." *PLoS ONE* 4:e4722.

Yuan, J., Dunn, P., and Martinus, R. D. 2011. "Detection of Hsp60 in saliva and serum from type 2 diabetic and non-diabetic control subjects." *Cell. Stress Chaperones* 16:689-93.

Zaiou, M., El Amri, H., and Bakillah A. 2018. "The clinical potential of adipogenesis and obesity-related microRNAs." *Nutr. Metab. Cardiovasc. Dis.* 28:91-111.

Zalman, L. S., Wood, L. M., Frank, M. M., and Müller-Eberhard, H. J. 1987. "Deficiency of the homologous restriction factor in paroxysmal nocturnal hemoglobinuria." *J. Exp. Med.* 165:572-7.

Zambaux, M. F., Bonneaux, F., and Gref, R. 1998. "Influence of experimental parameters on the characteristics of poly(lactic acid) nanoparticles prepared by double emulsion method." *J. Control. Rel.* 50:31-40.

Zamze, S. E., Ferguson, M. A., Collins, R., Dwek, R. A., and Rademacher, T. W. 1988. "Characterization of the cross-reacting determinant (CRD) of the glycosyl-phosphatidylinositol membrane anchor of *Trypanosoma brucei* variant surface glycoprotein." *Eur. J. Biochem.* 176:527-34.

Zelber-Sagi, S., Ratziu, V., and Oren, R. 2011. "Nutrition and physical activity in NAFLD: an overview of the epidemiological evidence." *World J. Gastroenterol.* 17:3377-89.

Zeng, X., Goetz, J. A., Suber, L. M., Scott, W. J. Jr., Schreiner, C. M., and Robbins, D. J. 2001. "A freely diffusible form of Sonic hedgehog mediates long-range signalling." *Nature* 411:716-20.

Zhang, F., Crise, B., Su, B., Hou, Y., Rose, J. K., Bothwell, A., and Jacobson, K. 1991. "Lateral diffusion of membrane-spanning and glycosylphosphatidylinositol-linked proteins: toward establishing rules governing the lateral mobility of membrane proteins." *J. Cell Biol.* 115:75-84.

Zhang, F., Schmidt, W. G., Hou, Y., Williams, A. F., and Jacobson, K. 1992. "Spontaneous incorporation of the glycosyl-phosphatidylinositol-linked protein Thy-1 into cell membranes." *Proc. Natl. Acad. Sci. USA* 89:5231-5.

Zhang, H., and Martin-DeLeon, P. A. 2003. "Mouse epididymal Spam1 (pH-20) is released in the luminal fluid with its lipid anchor." *J. Androl.* 24:51-8.

Zhang, H., Chung, D., Yang, Y. C., Neely, L., Tsurumoto, S., Fan, J., Zhang, L., Biamonte, M., Brekken, J., Lundgren, K., and Burrows, F. 2006. "Identification of new biomarkers for clinical trials of Hsp90 inhibitors." *Mol. Cancer Ther.* 5:1256-64.

Zhang, H., Jones, R., and Martin-DeLeon, P. A. 2004. "Expression and secretion of rat SPAM1 (2B1 or PH20) in the epididymis: role of testicular lumicrine factors." *Matrix Biol.* 22:653-61.

Zhang, J., Gerhardinger, C., and Lorenzi, M. 2002. "Early complement activation and decreased levels of glycosylphosphatidylinositol-anchored complement inhibitors in human and experimental diabetic retinopathy." *Diabetes* 51:3499-504.

Zhang, Y., and Zhuo, R. X. 2005a. "Synthesis and *in vitro* drug release behavior of amphiphilic triblock copolymer nanoparticles based on poly(ethylene glycol) and polycaprolactone." *Biomaterials* 26:6736-42.

Zhang, Y., and Zhuo, R. X. 2005b. "Synthesis and drug release behavior of poly(trimethylene carbonate)-poly(ethylene glycol)-poly(trimethylene carbonate) nanoparticles." *Biomaterials* 26:2089-94.

Zhao, P., Nairn, A. V., Hester, S., Moremen, K. W., O'Regan, R. M., Oprea, G., Wells, L., Pierce, M., and Abbott, K. L. 2012. "Proteomic identification of glycosylphosphatidylinositol anchor-dependent membrane proteins elevated in breast carcinoma." *J. Biol. Chem.* 287:25230-40.

Zhou, J., Li, Y. S., and Chien, S. 2001. "Shear stress-initiated signaling and its regulation of endothelial function." *Arterioscler. Thromb. Vasc. Biol.* 34:2191-8.

Zineh, I., and Huang, S. M. 2011. "Biomarkers in drug development and regulation: a paradigm for clinical implementation of personalized medicine." *Biomark. Med.* 5:705-13.

Zitvogel, L., Regnault, A., Lozier, A., Wolfers, J., Flament, C., Tenza, D., Ricciardi-Castagnoli, P., Raposo, G., and Amigorena, S. 1998. "Eradication of established murine tumours using a novel cell-free vaccine: dendritic cell-derived exosomes." *Nat. Med.* 4:594-600.

Zolg, J. W., and Langen, H. 2004. "How industry is approaching the search for new diagnostic markers and biomarkers." *Mol. Cell. Proteomics* 3:345-54.

Zurzolo, C., and Simons, K. 2016. "Glycosylphosphatidylinositol-anchored proteins: Membrane organization and transport." *Biochim. Biophys. Acta* 1858:632-9.

# About the Author


***Günter A. Müller, PhD***

Head of Molecular Biochemistry, Lecturer
Ludwig-Maximilians University
Munich, Germany
Email: guenter.mueller@helmholtz-muenchen.de


Günter A. Müller studied biology, anthropology and human genetics as well as philosophy, politics and economy at the Ludwig-Maximilians-University Munich. He received its PhD and habilitation in genetics. After more than two decades of drug discovery research for metabolic diseases at pharmaceutical industry in Frankfurt am Main, he is currently engaged in biomarker and diabetes research at the Helmholtz Center for Health and Environmental Research and as lecturer at the Ludwig-Maximilians University in Munich. He is publishing in the areas of biochemistry, genetics, cell biology and diabetes with focus on the roles of glycosylphosphatidylinositol-anchored proteins as well as about the interface between natural (life) sciences and human (philosophy) sciences with the aim of stimulating transdisciplinary and productive interactions.

# INDEX

## A



## B



## C



## D



## E

## F



## G



## H



## I



## L



## M



## N